Lecture Notes in Computer Science 4857

Commenced Publication in 1973
Founding and Former Series Editors:
Gerhard Goos, Juris Hartmanis, and Jan van Leeuwen

J. Mark Ware   George E. Taylor (Eds.)

Web and Wireless Geographical Information Systems

7th International Symposium, W2GIS 2007
Cardiff, UK, November 28-29, 2007
Proceedings

 Springer

Volume Editors

J. Mark Ware
George E. Taylor
University of Glamorgan
Faculty of Advanced Technology
Pontypridd CF37 1DL, UK
E-mail: {jmware,getaylor}@glam.ac.uk

Library of Congress Control Number: 2007939661

CR Subject Classification (1998): H.2, H.3, H.4, H.5, C.2

LNCS Sublibrary: SL 3 – Information Systems and Application, incl. Internet/Web
and HCI

ISSN   0302-9743
ISBN-10   3-540-76923-4 Springer Berlin Heidelberg New York
ISBN-13   978-3-540-76923-1 Springer Berlin Heidelberg New York

Springer is a part of Springer Science+Business Media

springer.com

© Springer-Verlag Berlin Heidelberg 2007
Printed in Germany

Typesetting: Camera-ready by author, data conversion by Scientific Publishing Services, Chennai, India
Printed on acid-free paper   SPIN: 12194919   06/3180   5 4 3 2 1 0

Preface

These proceedings contain the papers selected for presentation at the Seventh International Symposium on Web and Wireless Geographical Information Systems (W2GIS 2007). The symposium was organized locally by the University of Glamorgan (UK) and held in Cardiff (UK) in November 2007. It was the latest in a series of annual events that started in Kyoto 2001 and now alternate locations each year between East Asia and Europe. As in previous years, the aim of the symposium was to provide an up-to-date review of advances and recent developments in theoretical and technical issues relating to Web and wireless GIS technologies.

The symposium organizers received 45 full papers in response to the call for papers. These papers went through a thorough review process, with each paper being assessed by at least three reviewers. The quality of papers was high, and 21 were selected for presentation and inclusion in these proceedings. It was particularly pleasing to note that 19 different countries are represented by the authors of selected papers. The keynote address was delivered by Jonathan Raper of City University, London, in which he talked about his research work in the areas of location-based services and mobile information needs.

We wish to thank all the authors that submitted papers to this symposium. Our special thanks go to all those who attended, and particularly those who presented. Thanks are also due to Springer LNCS for their continued support. We also thank all Program Committee members for the quality of their evaluations, and, of course, the Local Organizing Committee for the hard work they put in to making the symposium such a success. Finally, we express our gratitude to the W2GIS Steering Committee for providing help and advice.

September 2007

Mark Ware
George Taylor

Preface

Organization

Symposium Chairs

M. Ware, University of Glamorgan, UK
G. Taylor, University of Glamorgan, UK

Steering Committee

M. Bertolotto, University College Dublin, Ireland
J.D. Carswell, Dublin Institute of Technology, Ireland
C. Claramunt, Naval Academy, France
M. Egenhofer, NCGIA, USA
T. Tezuka, Kyoto University, Japan
C. Vangenot, EPFL, Switzerland

Local Organizing Committee

S. Cutts, University of Glamorgan, UK
M. Langford, University of Glamorgan, UK
G. Savage, University of Glamorgan, UK
N. Thomas, University of Glamorgan, UK
K. Verheyden, University of Glamorgan, UK

Program Committee

P. Agouris, University of Maine, USA
S. Anand, University of Nottingham, UK
M. Arikawa, University of Tokyo, Japan
A. Bouju, University of La Rochelle, France
I. Cruz, University of Illinois, USA
M.L. Damiani, DICO - University of Milan, Italy
R.A. de By, ITC, The Netherlands
M. Duckham, University of Melbourne, Australia
M. Gahegan, Penn State, USA
G. Higgs, University of Glamorgan, UK
K. Hornsby, University of Maine, USA
B. Huang, Chinese University of Hong Kong
C.S. Jensen, Aalborg University, Denmark
D. Kidner, University of Glamorgan, UK
B. Kobben, ITC, The Netherlands

Y. J. Kwon, Hankuk Aviation University, Korea
J. Li, University of Leicester, UK
S. Li, Ryerson University, Canada
H. Martin, University of Grenoble, France
P. Muro-Medrano, Universidad de Zaragoza, Spain
S. Mustiere, IGN, France
S. Nittel, University of Maine, USA
D. Papadias, HKUST, Hong Kong
G. Percivall, Open Geospatial Consortium, USA
D. Pfoser, Computer Technology Institute, Greece
M. Raubal, University of California SB, USA
C. Ray, Naval Academy Research Institute, France
M. Schneider, University of Florida, USA
S. Spaccapietra, EPFL, Switzerland
Y. Tao, City University of Hong Kong
Y. Theodoridis, University of Piraeus, Greece
S. Timpf, University of Wuerzburg, Germany
A. Voisard, Fraunhofer ISST and FU Berlin, Germany
R. Weibel, University of Zurich, Switzerland
S. Winter, The University of Melbourne, Australia
O. Wolfson, University of Illinois at Chicago, USA
S. Yi, Huazhong University of Science and Technology, PR China
A. Zipf, Mainz University of Applied Sciences, Germany

Sponsors

Table of Contents

Continuous Perspective Query Processing for 3-D Objects on Road Networks

Joon-Seok Kim, Kyoung-Sook Kim, and Ki-Joune Li

Department of Computer Science and Engineering
Pusan National University, Pusan 609-735, South Korea
{joonseok,ksookim,lik}@pnu.edu

Abstract. In order to provide a streaming service of 3-D spatial objects to the mobile clients on a street, we propose a new query type, called continuous perspective query. The perspective query differs from conventional spatial queries in that the levels of details (LOD) of query results depend on the distance between the query point and spatial objects. The objects in the vicinity are to be provided with a higher LOD than those far from the query point. We are dealing with continuous queries, and the LODs of the results are also changing according to the distance from the mobile query point. The LOD of the result object changes from a low LOD to a high LOD as the mobile query point approaches to the object. In this paper, we also propose a processing method for continuous perspective query to reduce the processing cost at a server and communication cost between the server and the mobile clients.

Keywords: continuous perspective query, road network, LOD, 3-D objects.

1 Introduction

Most location-based services provide mobile users with the information of spatial objects such as buildings and roads in 2-D or 2.5-D. However we may achieve a higher quality of services when the 3-D information is offered such as 3-D geometry, and colors and texture on the facets of the objects. This information enables more realistic visualization of objects than simple 2-D data.

One of the main problems in providing 3-D information to mobile clients comes from the large size of data. It is almost impossible to store a large volume of 3-D data in a mobile device such as a PDA and a mobile phone with a relatively small size of memory. Furthermore, the transmission of 3-D data is limited due to the narrow bandwidth of wireless communication.

The basic idea of this paper is based on the observation that we do not need a detail information on an object which is far from the query point but only its simplified sketch in rendering aspects. However, more information is progressively required as the query point becomes closer to the object to describe it in detail. In this paper, we propose a new type of query, called continuous

J.M. Ware and G.E. Taylor (Eds.): W2GIS 2007, LNCS 4857, pp. 1–15, 2007.

perspective query for this purpose. This query is a kind of continuous query, where the levels of details (LOD) of query results depend on the distance between the query point and spatial objects. The objects closely located to the query point are to be provided with a higher LOD than those far from the query point.

Since we are dealing with continuous queries, the LODs of the results are also changing according to the distance from the mobile query point. The LOD of the results should be upgraded as the mobile query point approaches to the object. Also an efficient protocol of upgrades reduces the communication cost and processing cost at the server. For this reason, a method is proposed to process the continuous perspective query and upgrade LOD in this paper.

The remainder of paper is organized as follows. In section 2, we review related work in perspective query and continuous query. We describe the basic concept and define continuous perspective query in the next section. In section 4, we propose a processing method for the query for the mobile clients on streets. An analysis of performance is presented in section 5 and we conclude the paper in section 6.

2   Related Work and Motivation

A perspective query is to request data which have different levels of detail according to the importance of data. In GIS and LBS, this type of query is used to optimize the processing of geo-visualization. We may simplify a certain level of geometries of objects that are relatively far from the viewpoint to reduce the size of data. For this type of query, several researches have been done since last few years. For example, a processing method was proposed for perspective query by selecting proper LOD(Level of Detail) based on the range from the viewpoint in [4]. By this method, the entire space is divided into three ranges according to the distance from the viewpoint and a proper LOD is assigned to each range from high LOD to low LOD.

LOD-R-trees was proposed to index spatial objects with LOD in [2], which is a new data structure combined with R-trees and LODs. It stores the graphic data of spatial objects at each node and returns properly simplified LOD according to the distance. Similarly, V-Reactive tree was proposed in [3], which is another variant of R-trees for multiple LODs. It can be used for implementing the generalization of multi-LODs technique.

However the mobility of viewpoint has not been considered by these work. In most applications of perspective query such as LBS and navigation services, the spatial query condition continuously changes for a given period of time. During this time period, the queries are to be repeatedly evaluated for providing the correct information as the spatial query condition changes. This query is called continuous query [1]. Recently, a lot of attention has been paid to continuous range query and continuous k-nearest neighbor query as shown in table 1.

Among these work, we refer the work related to the continuous range query processing because a perspective query can be regarded as a range query. A

Table 1. Classification of movement patterns and types of continuous queries

mobility of (query, object) — types	Range	K-NN
(static query, dynamic objects)	[14,6,8,10]	[12,15]
(dynamic query, static objects)	[5,14,10]	[12,11,13,7]
(dynamic query, dynamic objects)	[5,14,10]	[12]

dynamic query was defined as a temporally series of snapshot queries in [5], where the authors proposed an index data structure for the trajectories of moving objects and described how to evaluate dynamic queries efficiently which represent predictable or non-predictable movement of an observer. In [6], a velocity constrained indexing and query indexing (Q-index) have been proposed by S. Prabhakar et. al. to evaluate continuous range queries. By indexing queries instead of objects, the index structures can avoid frequent updates and expensive maintenance. By assuming that there is a sufficient amount of memory and computation power, Ying Cai et al. claimed in [8] that range queries relevant to an object can be sent to it to quickly determine whether an update be needed. A scheme called Motion Adaptive Indexing (MAI) was proposed by B. Gedick in [9]. This enables optimization of continuous query evaluation according to the dynamic motion behavior of the objects. In that paper, the authors introduced the concept of motion sensitive bounding boxes (MSB) to model and index both moving objects and moving queries.

Mokbel et al. proposed a method based on shared execution and incremental evaluation of continuous queries, called SINA in [10]. Shared execution is performed for query evaluation as a spatial join between the mobile objects and the queries. The incremental evaluation means that the query processing system produce s only positive or negative updates from the previous result. D. Stojanovic et al. proposed a method for continuous range query processing in [14], characterized by the mobility of objects and queries that follow paths in an underlying spatial network. They introduced an additional pre-refinement step which generates main memory data structures to support periodical and incremental refinement steps. However most of work has focused on processing continuous range queries not perspective queries with LODs.

Therefore, we integrate continuous query and perspective query into a new query type for 3-D data with multi-level of details while the previous work have separately considered these two query types. In our work, we consider a query processing method to reduce the communication and processing cost. Furthermore, the above methods for continuous queries are based on the movement of mobile objects in a two dimensional Euclidean space except the work in [14]. In most cases, the streaming services of mobile clients are required on streets for pedestrians or car drivers. For this reason, we also focus on the query of mobile clients who move on road networks instead of Euclidian spaces.

3   Basic Concepts of Continuous Perspective Query

In this section, we define continuous perspective query through a combination
of perspective query and continuous query. Figure 1 illustrates an example of
continuous perspective queries. The results of a perspective query change as the
viewpoint moves from $t_1, t_2, t_3, \ldots$, to t_n. For example the LOD of object b
upgrades from low LOD at t_1, medium LOD at t_2, to high LOD at t_3. Hence,
continuous perspective query is considered as an extended perspective query
type to reflect the mobility of query point.

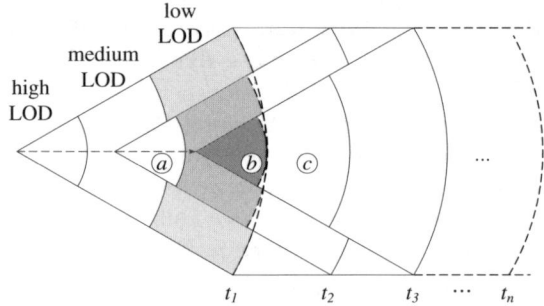

Fig. 1. Example of continuous perspective query

3.1   Perspective Query

A perspective query is intended first to search spatial objects in view range, and
second to determine their LOD. The query is given with three parameters, which
are a viewpoint, a view direction, and a view angle as follows,

- viewpoint: the query point in 3D space is given as $v_p = (x, y, z)$ in figure 2(a).
- view direction: the direction of the viewpoint in 3D space is given as $v_{dir} = (\varphi, \theta)$ in figure 2(a), where φ is the horizontal angle of view in the plane XY and θ is the vertical angle from the XY plane.
- view angle: the view angle is given as $v_{ang} = (\alpha, \beta)$, where α and β are the vertical and horizontal extents of view angle as figure 2(b).

A query region, which is surrounded by four triangular bounding planes as
figure 2(c), is defined as $q_{reg} = (v_p, v_{dir}, v_{ang})$. Then we define perspective query
as follows;

Definition 1. *Perspective Query*

$$Q_p(q_{reg}) = \{(s, l)| \; s \in S, l \in L\}$$

S in this definition represents the set of spatial objects and $L = \{LOD_1, LOD_2,
..., LOD_n\}$ is the set of LODs.

Figure 3 shows an example of perspective query and its results. Given a set of
six spatial objects a, b, c, d, e, f and a query region $q_{reg} = (v_p, v_{dir}, v_{ang})$, then the

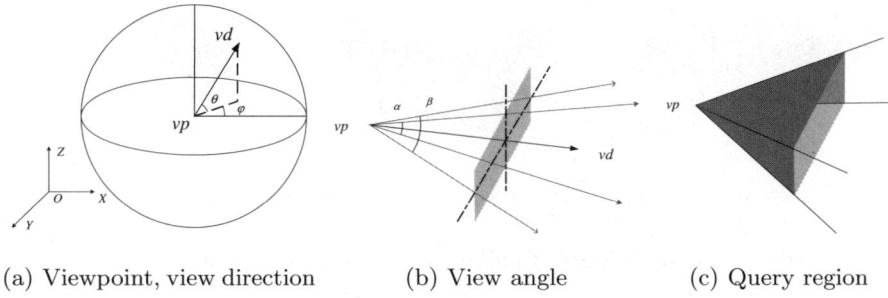

(a) Viewpoint, view direction   (b) View angle   (c) Query region

Fig. 2. Parameters of perspective query and query region

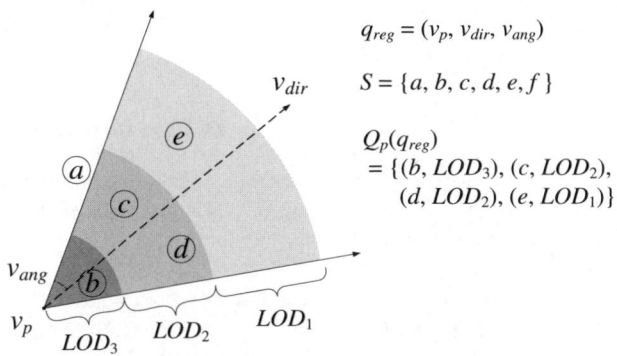

$q_{reg} = (v_p, v_{dir}, v_{ang})$

$S = \{a, b, c, d, e, f\}$

$Q_p(q_{reg})$
$= \{(b, LOD_3), (c, LOD_2),$
$(d, LOD_2), (e, LOD_1)\}$

Fig. 3. Example of perspective query

result of the query is $\{(b, LOD_3), (c, LOD_2), (d, LOD_2), (e, LOD_3)\}$. The object a is not in the result because it is out of query region and object f does not belong to the result since the distance from the viewpoint exceeds the maximum threshold. Objects b, c and e are in the range of LOD_3, LOD_2 and LOD_1, respectively.

3.2 Continuous Perspective Query

Continuous perspective query is to continuously update results of perspective query as time goes by. As range query or k-NN query become continuous query by the movement of query point, a perspective query becomes a continuous perspective query according to the change of the parameters (v_p, v_{dir}, v_{ang}). This is presented in figure 4.

Let $q_{reg}(t)$ be the query region at time t. Then we can represent perspective query as a function of time, $Q_p(q_{reg}(t))$ as follows, where T is a set of sampled time.

Definition 2. *Continuous Perspective Query*

$$Q_{cp}(T) = \{(t, Q_p(q_{reg}(t))) \mid t \in T\}.$$

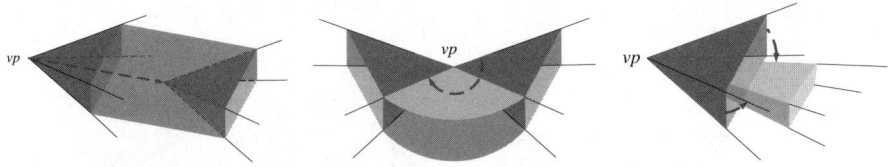

(a) Change of viewpoint   (b) Change of view direction   (c) Change of view angle

Fig. 4. Changes of parameters in perspective query

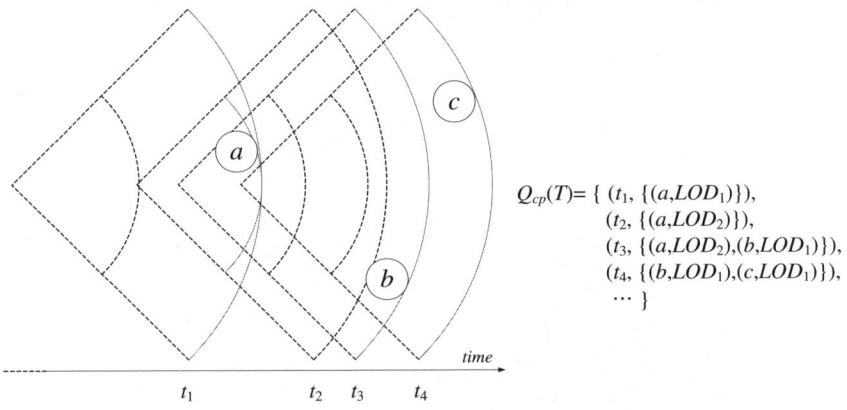

$$Q_{cp}(T)= \{ \ (t_1, \{(a, LOD_1)\}), \\ (t_2, \{(a, LOD_2)\}), \\ (t_3, \{(a, LOD_2), (b, LOD_1)\}), \\ (t_4, \{(b, LOD_1), (c, LOD_1)\}), \\ \cdots \ \}$$

Fig. 5. Continuous perspective query

Figure 5 illustrates a continuous perspective query for the sampled times t_1, t_2, t_3, and t_4, where there are only two LODs, which are LOD_1 and LOD_2. While the query result at t_1 is $\{(a, LOD_1)\}$, the LOD is upgraded to LOD_2. It becomes $\{(a, LOD_2), (b, LOD_1)\}$ at t_2, since a new object b is included in the query region.

3.3   Continuous Perspective Query on Road Networks

In this paper, we restrict the continuous perspective query to road network space for two reasons. First, most real applications of perspective query are related to the navigation of vehicles and the information about 3-D spatial objects along a road is very useful to drivers.

Second, we can easily obtain the view direction and angles of a vehicle on a road. Once a vehicle is on a given road, we can compute the view direction of the driver from the direction of the road. We can also define the view angle of the driver in spite of slight difference between drivers and vehicle types. This means that the third parameter of continuous perspective query, v_{ang} is fixed, and the second parameter v_{dir} depends on the road where the vehicle is located. For these reasons, we focus on road network space rather than Euclidian space.

Then we redefine the query region of continuous perspective query for road network space based on these assumptions and observations. First the position on road networks is given as (seg_{ID}, d), where seg_{ID} is the segment ID of a road segment and d is the displacement from the starting point of the road segment. As a consequence, the query region is defined, differently from the definition of the previous section, as $q_{reg} = (seg_{ID}, d, dir_{RS}, v_{ang})$, where dir_{RS} is the direction of the road segment, and v_{ang} is given a priori.

For the reason of convenience, we define rsi, which means road segment interval, as $rsi = (seg_{ID}, d_s, d_e)$, where d_s and d_e are the starting and ending displacements respectively within a road segment seg_{ID}. And we also define a sequence of rsi, $SEQ_{rsi} = (rsi_1, rsi_2, rsi_3, \ldots, rsi_n)$. In fact, a SEQ_{rsi} implies the route of a moving object on road networks as shown in figure 6(b). These definitions are illustrated in figure 6(a) and 6(b).

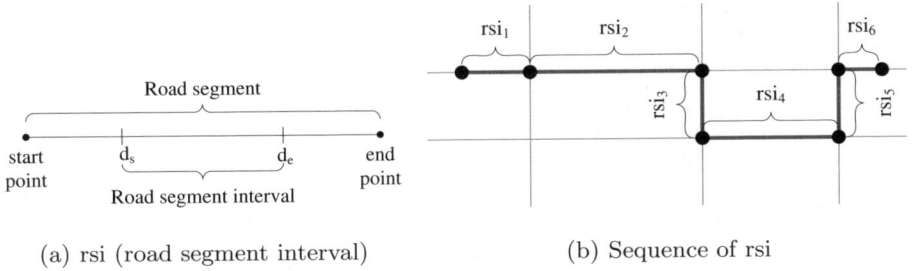

(a) rsi (road segment interval)   (b) Sequence of rsi

Fig. 6. Road segment interval and sequence of it

4   Query Processing for Continuous Perspective Query

In this section, we propose a query processing method for continuous perspective query on road networks. The objectives of proposed method are to reduce the processing and communication costs.

4.1   Requirements for Processing Continuous Perspective Query

Before explaining the query processing method, we introduce a notion of *change point*, which will be required for the query processing. A set of change points $CP(rsi)$ on a road segment interval rsi is defined as a set of points where the result of a continuous perspective query should be changed as the view position is moving along the road segment interval rsi. In fact, the change of the query result takes place when the view position is passing though specific points on the road as shown in figure 7.

Then we define the continuous perspective query on road segment and road networks differently from the definition 2.

Definition 3. *Continuous Perspective Query on Road Segment*

$$Q_{cprs}(rsi) = \{(p, Q_p(p))|p \in CP(rsi)\}.$$

Definition 4. *Continuous Perspective Query on Road Networks*

$$Q_{cprn}(SEQ_{rsi}) = \bigcup_{ris \in SEQ_{rsi}} Q_{cprs}(rsi).$$

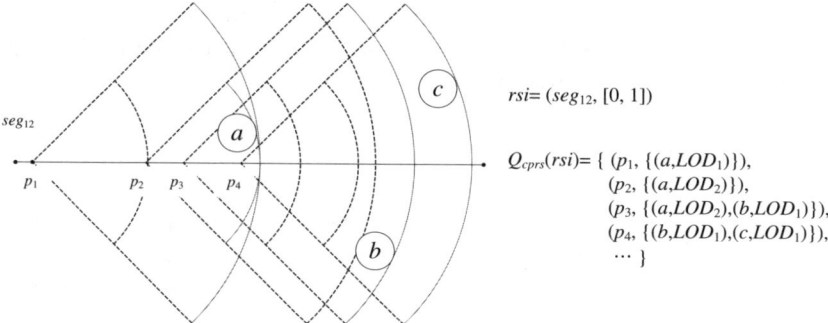

Fig. 7. Change positions and continuous perspective query on road segment

Figure 7 shows examples of change positions and continuous perspective query on a road segment. The sampling time in figure 5 is replaced with the change positions in figure 7.

The conceptual processing steps for continuous perspective query is summarized as follows. First, the view point from a mobile client is periodically reported to the server where the 3-D databases are stored. Second, the server sends the query result to the client. Since the query result is a stream, the server sends only the difference from the previous result to the mobile client. It means that the server does not need to send the query result if no difference of the result is found. And the server tells whether there is any difference by means of the change point. If the viewpoint is passing though a change point, it means that a difference is taking place and the difference is to be sent to the mobile client.

The information about the difference is described as an *update* $u = (p, s, LOD, op)$, where p is a change position, s and LOD are a 3-D object and its LOD respectively, and, op is an operation, which is one of add, remove, or update. This conceptual query processing steps are illustrated by figure 8 and summarized as follows,

- step 1: The viewpoint is sent to the server.
- step 2: The server checks if the viewpoint is passing through a change point. Then the server sends the update information u to the client.
- step 3: Repeat step 1 and step 2.

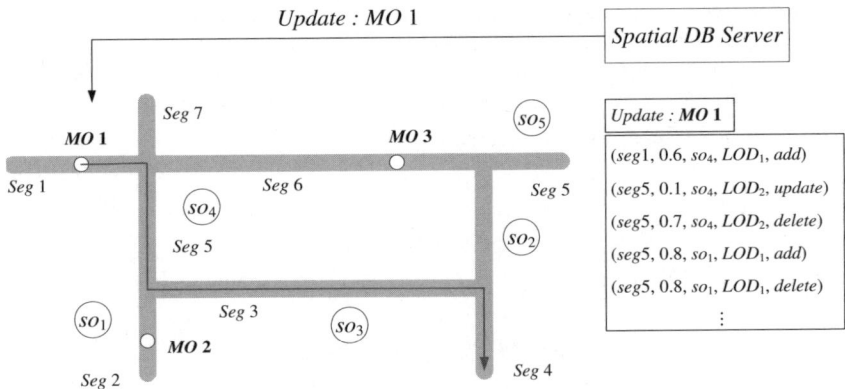

Fig. 8. Conceptual query processing steps

4.2   Query Processing by Pre-fetching

In order to reduce the communication between the server and clients, and to improve the response time at clients, we can pre-fetch possible updates when a viewpoint is entering to a road segment interval. Then the continuous perspective query processing on a road segment consists of the following steps;

- step 1: the client sends the road segment interval (rsi) to the server when entering the rsi.
- step 2: the server computes the query region from the rsi.
- step 3: the server searches the change points contained in the query region.
- step 4: the server computes the update at each change point.
- step 5: the server sends the sorted set of updates on the rsi to the client.
- step 6: repeat from step 1 to step 5.

Figure 9 explains the query processing, where only one LOD is defined for the reason of simplicity. First figure 9(a) shows the step 2 and 3 that the server computes the query region and searches the spatial objects contained by the query region. The object a and h are excluded from the candidate objects because they are not contained by the query region. Figure 9(b) represents the processing step 4 and 5. The objects b and i are removed from the query result since they are out of the query region. And the change positions are computed from the searched objects (c, d, e, f, g). Finally, the sorted list of updates according to change positions is sent to the client.

After the client receives the the sorted list, it triggers each update in the sorted list, when the viewpoint passes through a change point in the list. In order to compute and send the sorted list of updates, we need a preprocessing at the server side, which will be explained in the next subsequent section.

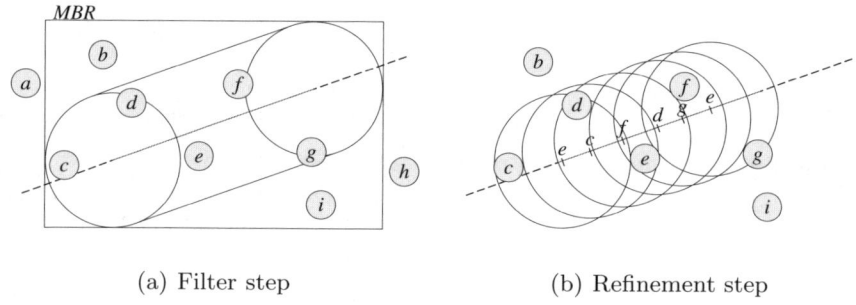

(a) Filter step   (b) Refinement step

Fig. 9. Query processing by rsi

4.3   Data Structures for the Preprocessed Results

Since the computation of change positions and the corresponding updates are fixed once a road segment and 3-D objects are given, we do not need to repeat unnecessary computations. Instead we can simply get the pre-computed list of a given road segment and reduce the overhead on the server.

MappingTable	
RSI ID	*Page No.*
...	...
1440	231
1441	242
...	...

231				
RSI ID	1440			
displacement	*SOID*	*LOD*	*operation*	*size(byte)*
...	...	...	...	...
0.11	8412	1	add	132
0.15	3231	2	delete	1523
0.21	63263	3	delete	17123
0.3	15234	1	add	104
0.31	44141	3	update	23412
...	...	...	...	...

Fig. 10. Structure of sorted offset list

We preprocess the sorted list of each road segment to efficiently retrieve the results and provide the stream of the results to mobile clients. Figure 10 shows the data structures of the sorted list of updates. First a mapping table contains the identifier of road segment interval and the disk page number where the pre-computed sorted list is stored. Second the pre-computed sorted list is stored at the corresponding disk page as shown by figure 10. The size of each object to send to client is also included as an attribute, since this information is useful for the transfer between the server and client.

5   Evaluation of Performance

In this section, we investigate the performance of the proposed query processing method. The processing cost of the query includes the cost of the server, the communication cost, and the cost at the client side. The performance bottleneck of the entire system comes from the costs of the server and communication, which we will discuss in this paper.

5.1   Environment of Experiments

For our experiments, we prepared a set of 3-D spatial objects extracted from the real building data in Daejon city area. The road network consists of 5247 road segments and the total length is $457km$. Since no data from a massive number of real vehicles are currently available, we generated 10000 moving objects as mobile clients using a synthetic generator based on a road network. Table 2 shows description of data set for our test.

Table 2. Description of data set

parameters	values
number of segments	5247
average length of segments	$80m$
sum of length of segments	$457km$
average number of connections	1.74
number of buildings	280000
density of building	72.8%
size of 3-D data	$3Gb$
number of mobile clients	10000

This data set consists of three LODs, where LOD 1 describes only the simplified geometry of building objects, LOD 2 includes the detail geometry and texture of buildings, and the real facet image of buildings is contained in LOD 3. The size of a building data is depending on its LOD. Table 3 shows approximate average sizes of a 3-D object at different LODs.

Table 3. Approximate sizes of 3D object for different LODs

	size
LOD 1	200 Bytes
LOD 2	1 K Bytes
LOD 3	100 K Bytes

In addition, we defined the several visible ranges of each LOD as table 4 to analyze the cost with query ranges. For the LOD type 1, the lengths of query range are defined so that the LOD 1 of a 3-D object with 10 m height be displayed as a 5 mm object on a screen of 200 mm x 100 mm. For the LOD 2, this object should be displayed as an 1 cm object on the screen, and LOD 3 be as 5 cm object. Then for LOD type 1, the lengths of the query ranges for LOD 1, LOD 2, and LOD 3 become 117 m, 58 m, and 11 m, respectively. Similarly, the lengths of query range for LOD type k are determined as k times of the lengths in LOD type 1 as shown by table 4.

Table 4. Lengths of query ranges for several LOD types

	LOD1	LOD2	LOD3
LOD type1	117m	58m	11m
LOD type2	234m	116m	22m
LOD type3	351m	174m	33m
LOD type4	468m	232m	44m
LOD type5	585m	290m	55m

5.2   Cost at Server-Side

First of all, we analyze the cost at the server side that is mainly determined by the costs for searching and pre-fetching the sorted list, and reading the data of 3D objects, when the viewpoint is entering a new road segment interval. Let us assume that all data are stored on disk except the mapping table. Then the processing cost for new road segment interval C_{RSI} is given as

$$C_{RSI} = C_{MT} + C_{SDL} + C_{DATA}$$

where C_{MT} is the cost for accessing the mapping table, C_{SDL} is for reading the disk pages of the sorted list for a given road segment interval, and C_{DATA} for retrieving the data of 3-D objects on the road segment interval. However, since the size of the mapping table is small and can be stored in main memory, we ignore the cost for the mapping table C_{MT}. Thus we concentrate the number of disk page accesses for C_{SDL} and C_{DATA}.

The cost for accessing the sorted list disk pages depends on the number of tuples on a road segment interval. If the length of the road segment interval is sufficiently short, the all change positions can be stored at a single disk page. But as grows the length, the number of disk pages increases. Figure 11 shows this relationship by experimental results with respect to several LOD types. Since the size of a tuple of the sorted list is 16 bytes, 256 tuples are stored at a disk page of 4 K bytes. It means that we need 4 disk pages to store a sorted list of LOD type 1 when the length of road segment interval is 200 meters. We observe that the number of tuples to transfer is proportional to the length of road segment interval, which is obvious when the density of 3-D objects is uniform.

The cost for retrieving the data of 3-D objects is determined by the size of 3D data and the placement on the disk. As shown in table 3, we see that the

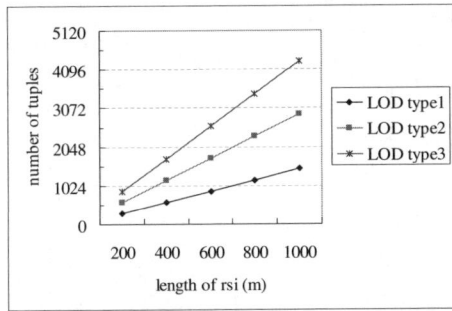

Fig. 11. Number of tuples at a road segment interval

size of 3-D object for LOD 3 is quite large and it is practically the bottleneck of the server performance. The processing cost at the server is determined by C_{DATA} than C_{SDL} due to the large size of 3-D object of LOD 3. This fact is clearly illustrated in table 5, which shows the total cost C_{RSI} of the server.

Table 5. Cost for retrieving data and the total cost at the server (disk page: 4K bytes)

length of rsi	200	400	600	800	1000
C_{Data}	468.6	937.2	1405.8	1874.5	2343.1
C_{SDL}	1	2	3	4	5
C_{Total}	469.6	939.2	1408.9	1878.5	2348.1

5.3 Cost for Communication

For the communication cost, we consider two measures; the number of communications and the size of data to transfer. There is a trade-off relationship between these two measures. Basically, we can reduce the communication cost by reducing the number of communications from the server to mobile clients. It depends on the number of entrances to a new road segment interval. Thus we may enlarge the length of road segment intervals to reduce the number of communications. On the other side, a long road segment interval results in a large amount of data of a transfer.

The size of data to transfer depends on the number of updates. Figure 12 show the relation between the amount of data to transfer and the length of a road segment interval. For example, the server should transfer about 10 K bytes per 1 meters as shown in figure 12. Suppose that a vehicle is moving in 36 km/h. Then the server should transfer 100 K bytes per second. It means that we should either provide a high speed wireless communication media or reduce the size of LOD 3 data to support the continuous perspective query.

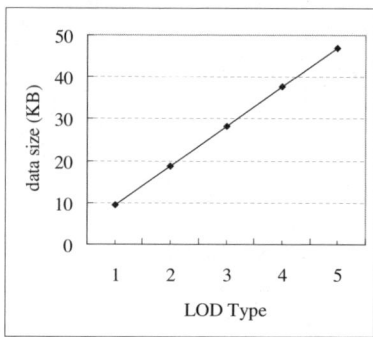

Fig. 12. Amount of data to transfer per 1 m

6   Conclusions

In this paper, we proposed a new type of query, called continuous perspective query to support stream services of 3D data to navigation users on the street. In comparison with the conventional spatial query such as range query, the results of perspective query depends on the distance between the query point and spatial objects and multiple LODs are applied to the results according to the distance. The contributions of our work are summarized as follows;

- we have introduced the continuous perspective query type. The query may be used for providing streaming service of 3-D data to navigation users along streets.
- an efficient query processing method has been proposed, which is based on pre-fetching and preprocessing techniques.

In this paper, we excluded the data modelling and progressive transfer issues from the scope of the study. We assumed that a complete data of a 3D object should be transferred when upgrading its LODs, for example from LOD 2 to LOD3. However, we could reduce the amount of data to transfer by a progressive transfer of 3D geometry data. And a careful data model is required to realize the progressive transfer. For this reason, our future work will include the data modelling and progressive transfer for continuous perspective query processing.

Acknowledgment. This research was partially supported by the STAR Project (Grant 20050682).

References

1. Terry, D.B., Goldberg, D., Nichols, D., Oki, B.M.: Continuous Queries over Append-Only Databases. In: Proceedings of ACM SIGMOD Conference, pp. 321–330 (1992)

2. Kofler, M., Gervautz, M., Gruber, M.: R-trees for Organizing and Visualizing 3D GIS Databases. Journal of Visualization and Computer Animation 11(3), 129–143 (2000)
3. Li, J., Jing, N., Sun, M.: Spatial Database Techniques Oriented to Visualization in 3D GIS. In: Proceedings of the 2nd International Symposium on Digital Earth 2001 (2001)
4. Moller, T., Haines, E., Akenine-Moller, T.: Real-Time Rendering, 2nd edn. A.K. Peters (2002)
5. Lazaridis, I., Porkaew, K., Mehrotra, S.: Dynamic Queries over Mobile Objects. In: Jensen, C.S., Jeffery, K.G., Pokorný, J., Šaltenis, S., Bertino, E., Böhm, K., Jarke, M. (eds.) EDBT 2002. LNCS, vol. 2287, pp. 269–286. Springer, Heidelberg (2002)
6. Prabhakar, S., Xia, Y., Kalashnikov, D., Aref, W., Hambrusch, S.E.: Query Indexing and Velocity Constrained Indexing: Scalable Techniques for Continuous Queries on Moving Objects. IEEE Transactions on Computers 51(10), 1124–1140 (2002)
7. Tao, Y., Papadias, D., Shen, Q.: Continuous Nearest Neighbor Search. In: Bressan, S., Chaudhri, A.B., Lee, M.L., Yu, J.X., Lacroix, Z. (eds.) CAiSE 2002 and VLDB 2002. LNCS, vol. 2590, pp. 287–298. Springer, Heidelberg (2003)
8. Cai, Y., Hua, K.A., Cao, G.: Processing Range-Monitoring Queries on Heterogeneous Mobile Objects. In: Proceedings of Mobile Data Management Conference 2004, pp. 27–38 (2004)
9. Gedik, B., Wu, K.-L., Yu, P., Liu, L.: Motion Adaptive Indexing for Moving Continual Queries over Moving Objects. In: Proceedings of the CIKM Conference 2004, pp. 427–436 (2004)
10. Mokbel, M.F., Xiong, X., Aref, W.G.: SINA: Scalable Incremental Processing of Continuous Queries in Spatio-temporal Databases. In: Proceedings of SIGMOD Conference 2004, pp. 623–634 (2004)
11. Kolahdouzan, M.R., Shahabi, C.: Continuous K-nearest neighbor queries in spatial network databases. In: Proceedings of STDBM 2004, pp. 33–40 (2004)
12. Xiong, X., Mokbel, M.F., Aref, W.G.: SEA-CNN: Scalable Processing of Continuous K-Nearest Neighbor Queries in Spatio-temporal Databases. In: Proceedings of ICDE 2005, pp. 643–654 (2005)
13. Cho, H.-J., Chung, C.-W.: An Efficient and Scalable Approach to CNN Queries in a Road Network. In: Proceedings of VLDB 2005, pp. 865–876 (2005)
14. Stojanovic, D., Djordjevic-Kajan, S., Papadopoulos, A.N., Nanopoulos, A.: Continuous Range Query Processing for Network Constrained Mobile Objects. In: Proceedings of the 8th ICEIS 2006, pp. 63–70 (2006)
15. Mouratidis, K., Yiu, M.L., Papadias, D., Mamoulis, N.: Continuous Nearest Neighbor Monitoring in Road Networks. In: Proceedings of VLDB 2006, pp. 43–54 (2006)

A GIS Based Approach to Predicting Road Surface Temperatures

Richard Fry[1], Lionel Slade[2], George Taylor[1], and Ian Davy[3]

[1] GIS Research Unit, Faculty of Advanced Technology, University of Glamorgan,
Pontypridd, South Wales. CF37 1DL. +44 (0)1443 480 480
rfry@glam.ac.uk, getaylor@glam.ac.uk
[2] Geobureau Ltd, 47 Cowbridge Road, Pontyclun, Rhondda Cynon Taf,
South Wales CF72 9WS. +44 (0) 1443 229999
lionel.slade@geobureau.co.uk
[3] Aerospace & Marine International (UK), Suite G3, Banchory Business Centre
Banchory, Aberdeenshire, AB31 5ZU
idavy@weather3000.com

Abstract. Treatment of transport networks of the winter period is a key responsibility of maintenance teams whether in the private or public sector. With tightening budgets and increasing statutory requirements, maintenance teams are looking to improve the treatment of networks by improving ice prediction techniques. Improved forecasting techniques leads to financial savings, reduction in environmental damage and degradation of road surfaces and reduces potential litigation cases. This paper outlines a GIS desktop based approach to surveying a route and the results of integration into Aerospace and Marine International's GRIP (Geographical Road Ice Predictor) model in recent trials held across the county of Hampshire as to the effectiveness of the methods used. It also discusses development work currently being undertaken to display the forecast data via an internet mapping portal using ESRI ArcGIS Server.

Keywords: GIS, Ice Prediction, Roads, ArcGIS Server, ASP.NET.

1 Introduction

Ice prediction models have been developed in many countries where a temperate climate results in marginal temperatures - when temperatures are at or below freezing and a gritting run would be advised- at night. A variety of techniques have been developed with the aim of better predicting ice on road networks across the world.

Network forecasting models can be broken down into two key concept areas:

- Temporal component that consists of a standard road weather prediction model. This uses forecast meteorological data to produce a *Road Surface Temperature* (RST) forecast curve.
- Spatial component that uses geographical attribute data to modify the forecast curve on a site-specific basis.

J.M. Ware and G.E. Taylor (Eds.): W2GIS 2007, LNCS 4857, pp. 16–29, 2007.

1.1 Temporal Component

Thornes' [11] model uses a zero-dimensional energy balance approach. The model uses standard 3-hourly forecast meteorological data to produce the 24-h RST forecast curve. This forecast is provided to the winter maintenance engineer at midday so that early decisions can be made regarding the salting of the road network. In the model a forecast for a particular night may be marginal with regard to temperatures – i.e. just below freezing. Should the surface receive an input of moisture, the forecast site will need salting, however, large sections of the road network will not. A sensitivity test of the model is described in Thornes and Shao [12] and the temporal forecasting ability of the model is covered in Parmenter and Thornes [13]. Both studies indicated that Thornes' 1984 model has significant forecasting ability and compares favourably with other road weather models. The model used in this research is Davy's 2005 GRIP model which produces a forecast at higher frequencies – three instances per hour – and at finer resolutions - 100m spacing along a network.

1.2 Spatial Component

Davy [6] first introduced a spatial component into the weather model by replacing geographical constants with variables. This work was further explored by Chapman et al. [4] who also added a spatial component to the Thornes [11] model. In the original models, latitude, land use, and road construction were all constant. In Davy's GRIP model and the revised model described in Chapman and Thornes [9], these variables along with altitude and traffic are parameterised.

The use of Network Forecasting has become a principal component in helping a highway engineer to make an informed decision on whether to grit a route on a given night. Traditionally this network forecast has relied on a thermal survey of the road network as a main input to a forecast model. Although recording actual surface temperatures thermal mapping is expensive and requires repeating to achieve a representative set of climate scenarios. Recent work in the field has seen a move away from thermal mapping to better categorise and model local climate conditions.

2 The Hampshire Trial

The county of Hampshire covers an area of approximately 3,738 km2 in the south of England and at its widest points is approximately 90 km east-west and 65 km north-south. The county borders (clockwise from West), Dorset, Wiltshire, Berkshire, Surrey and West Sussex. The 2001 census gave the population of the administrative county as 1.24 million. Fig. 1 illustrates the trial area.

Hampshire has a relatively mild climate consistent with the rest of the southern areas of the British Isles. For the period from 1971 – 2000 Hampshire, and the rest of the south of England has an average annual temperature of 13.6 °C, average annual rainfall of 781.5 mm, and average sunshine of 1515 hours per year.[1] The relatively

---

[1] Met Office, 2000 – Annual and monthly Averages 1971 -2000
(http://www.metoffice.gov.uk/climate/uk/averages/19712000/areal/england_s.html)

humid climate experienced by Hampshire makes it particularly susceptible to Hoar Frosts. Hoar Frosts form when the relative humidity in supersaturated air is greater then 100%. The formation of hoar frost is similar to the formation of dew with the difference that the temperature of the object on which the hoar frost forms is well below 0°C, whereas this is not the case with dew.

Fig. 1. Hampshire Location Map

The Hampshire County Council trials were carried out in conjunction with GeoBureau (GB), Aerospace & Marine International (AMI), Findlay Irvine (FI) and Hampshire County Council (HCC). The trial involves providing four forecasts per hour for HCC road network to aid in the gritting decision making process. GeoBureau's role in this trial was to provide AMI – the forecaster – with a predefined set of geographical variables. These variables would, in turn, provide key inputs into the forecasting model with the resulting forecast being uploaded to the FI Network forecast service and ultimately viewed by the HCC highway engineers in their decision making process.

HCC gritting policy consists of forty-four routes totalling 3,368 km in length (Fig. 1). The routes covered a diverse landscape ranging from densely populated urban areas, to open plains and dense woodlands. At strategic positions along the routes 8 Findlay Irvine weather stations have been installed to monitor local meteorology conditions. These stations are used to calibrate the weather model and monitor its accuracy.

With the advancements in computing power, data quality and data availability it is now becoming more feasible to model aspects of the real world on larger scales. These advancements have meant that formerly impractical and un-feasible tasks are now being reconsidered.

In order to extract the necessary information for the weather model a high resolution 3D model of Hampshire was built. This involved extracting approximately 2.5 million building footprints, digitizing 3,368km of road network within ±1m, a DEM to provide aspect, gradient and height above sea level information, vegetation height and position, and calculating a population index to act as a proxy for the Urban Heat Island effect. From this model an analysis of the forty-four gritting routes was performed for five thousand forecast points. For each point in the model over 20 parameters were calculated. These main parameters are summarised in table 1:

Table 1. Parameter Groupings derived from GIS model

PARAMETER GROUP	DATA SOURCE DERIVED FROM
Skyview and Shading Calculations	Building heights and Vegetation Heights.
Urban Heat Island Effect	Calculated Population index.
Road Surface Type	Road Class
Sea warming effects	Distance and bearing to coast
Wind Effects	Topographical elevation model
Altitude	Elevation model

3 Accuracy of Model

Recent research into factors influencing road surface temperature has focused on the measurement of skyview factor (e.g. Chapman et al. [3]; [4]; [5];Bradley et al. [1]; Chapman & Thornes [2]; Gustavsson [7];[8]), which has been found to be the dominant control on road surface temperatures together with land use, at high levels of atmospheric stability (Chapman et al. [3]). The skyview factor is a Fig. between 0 and 1 which, in this case, represents what proportion of the sky is visible from a given point on the road. As Thornes et al [9] state after Thornes [10], skyview factors play an important role in the radiation budget, and influences road surface temperature by obstructing the incoming solar radiation to the road surface and controlling the loss of long-wave radiation over the night time period. One of the key aspects of this model was to build an accurate road canopy – a representation of what is directly above a point on a road network- to calculate the skyview factor. Therefore to assess the model and its effectiveness it was crucial to see how accurate the building, vegetation heights and vegetation positions were and to make an assessment of the calculations and assumptions that were made when building the model.

A survey of representative building types was made using the Leica DISTO™ and ArcPad as the data collection tool. The DISTO™ has reported accuracies of ±0.5 - 100mm over larger distances[2] (200m); given that the measurements were made at a

---

[2] Leica DISTO FAQ. (http://www.leica-geosystems.com/cpd/en/support/lgs_4892.htm)

maximum of 25m away from the structure it is assumed – with some confidence - that the field measurements are representative of an actual building height. A statistical analysis of the results showed that the mean and the standard deviation for the calculated results were 5.02m and 0.94m respectively. The same results for the observed field results were 5.54m for the mean and 1.64m for the standard deviation. An analysis of the differences showed that the mean average error was 1.02m with a standard deviation of 0.65m. Fig. 2 is a graphical representation of the estimated heights and observed heights.

Table 2. Statistical Summary of Fieldwork Measurements compared to Estimated Measurements

STATISTIC	CALCULATED	OBSERVED	DIFFERENCES
Mean (m)	5.024824	5.536705882	1.021058824
Standard Deviation (m)	0.943075	1.638951747	0.647860992

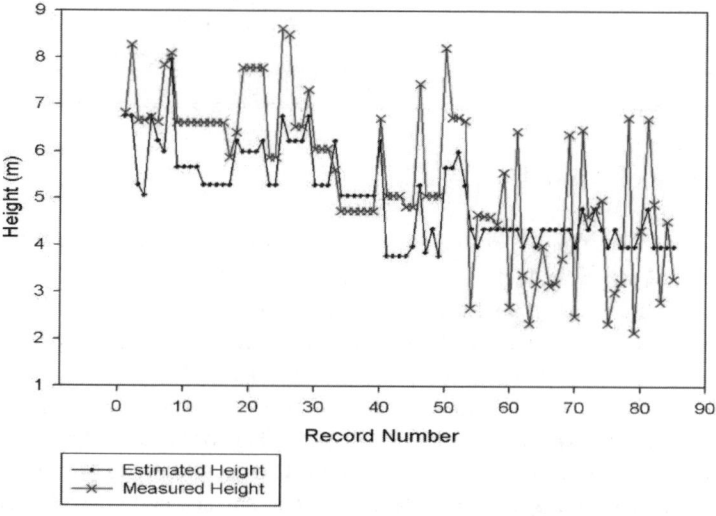

Fig. 2. Estimated Heights plotted against Observed Heights

Table 2 shows that mean error between the estimated and the actual field results 1.02m which was within the acceptable tolerance levels for the calculations. The distribution curves of the observed and expected datasets show that there are a greater proportion of results in the 6m region; however the distribution curve for the observed datasets does have a negative skew and therefore results in a substantial overlapping of the expected normal distribution curve (Fig. 3). The trends shown in Fig. 3 also show that the height differences from building to building are consistent in both data sets. This suggests that the method used to calculate is relatively accurate both in terms of the accuracy of the building heights and of the relative size differences to the building. These results showed that model, the GIS techniques used and the data was accurate enough to provide the basis for ice prediction weather forecasts.

It is expected that further fieldwork to be carried out over the next few months investigating the vegetation aspects of the model will further back up the validity of using a desktop model to measure skyview factors.

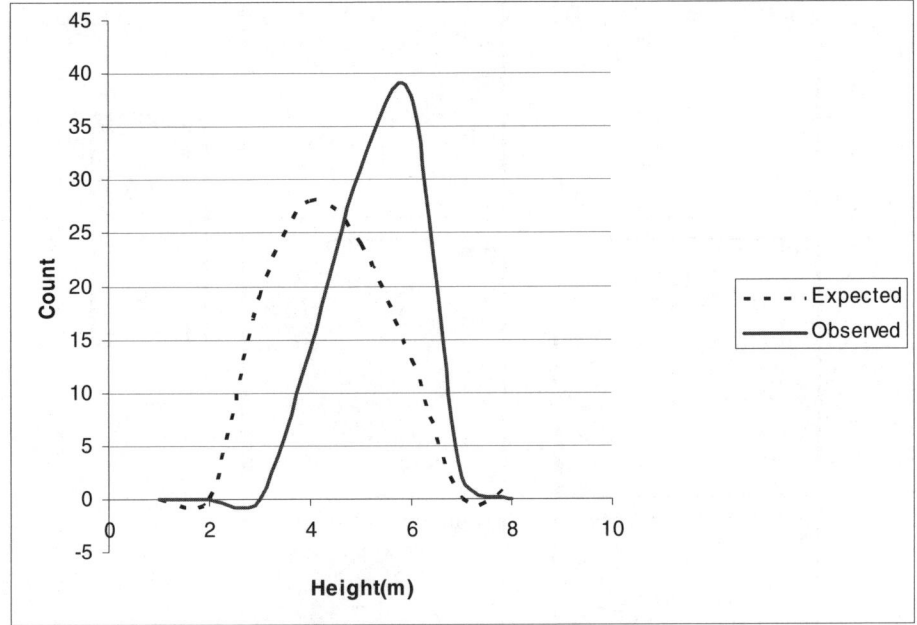

Fig. 3. Normal Distribution Curves of observed and expected results

4 Ice Prediction Forecasts

The aim of this research is to replace more traditional ice prediction techniques – and in particular the surveying aspect used in these techniques - with a desktop model. Working with HCC provided an ideal opportunity to develop, test and refine the modelling technique and monitor the results in terms of forecast accuracy in a live trial. Using the inputs from the GIS model the forecaster was able to provide 20 minute updates via an internet mapping service to HCC. These provided predicted road surface temperatures to the highway managers – upon which they were able

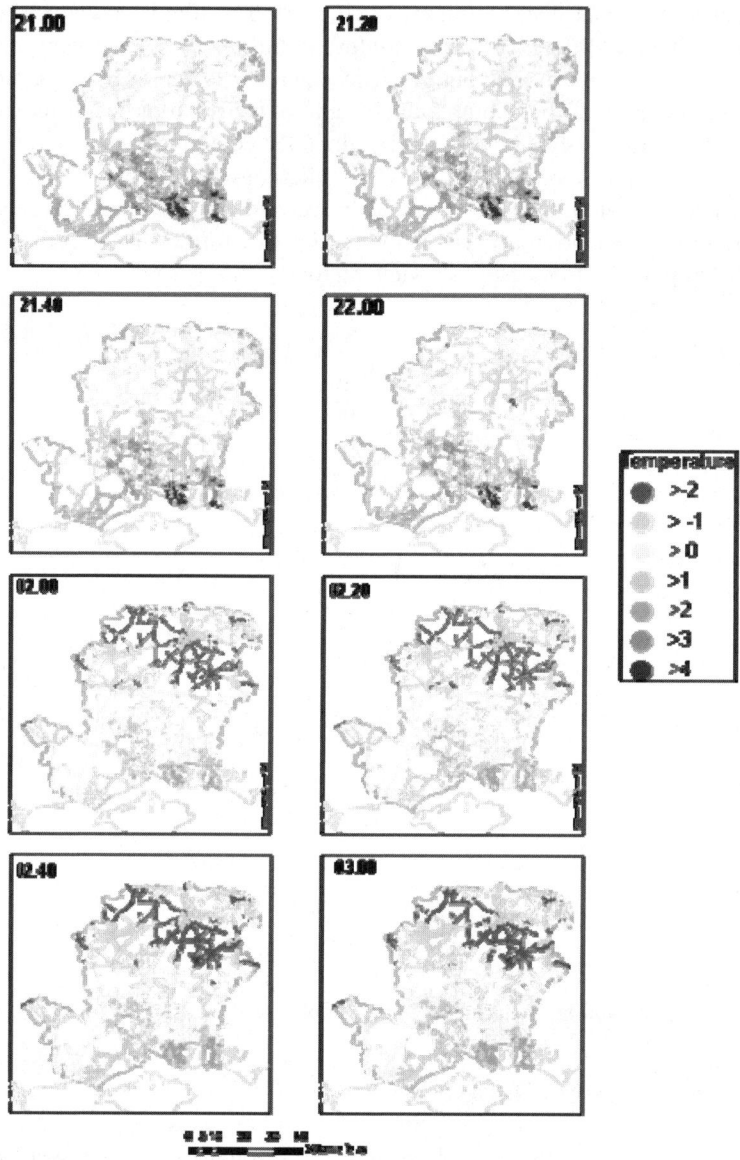

Fig. 4. Network Forecasts for the 9th/10th December 2006

to make informed decision on whether to grit a route. An example of a forecast map from the night of the 9th and morning of the 10th of December is shown in Fig. 4.

Fig. 5 shows the results from the predicted road temperatures, shown in Fig. 4, plotted against actual recordings taken from the Wootton weather station illustrated in Fig. 1.

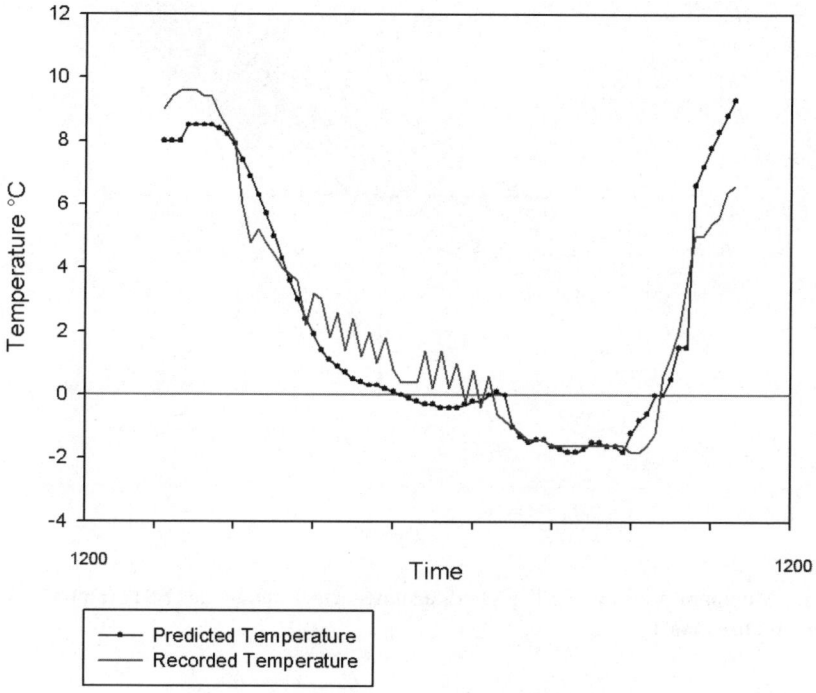

Fig. 5. Temperature Comparison Graph

Fig. 5 shows that the predicted temperature and the actual recorded temperature are very similar across an evening and that the temperature changes follow the same pattern through the night. The accuracy of the temperatures recorded can also be analysed over a longer period when you look at the minimum temperatures recorded over a period. Fig. 6 shows minimum forecast RST plotted against actual minimum RST for season 2006/2007. The error zones are created with two types of error; Frost Forecast – No frost occurred (FN), and no frost forecast and frost occurred (NF). The more severe error are the NF errors, where the road temperature actually fell below zero when it was forecast to stay above zero and FN errors are where road temperature stayed above zero when forecast to fall below zero.

Overall accuracy of all temperatures predicted by the model currently stands at 97.1 % for the 2006-2007 season.

Fieldwork being conducted at the moment to gain a wider range of temperature readings across the network will add to the data collated and provide a larger sample to analyse. This will be reported on in due course.

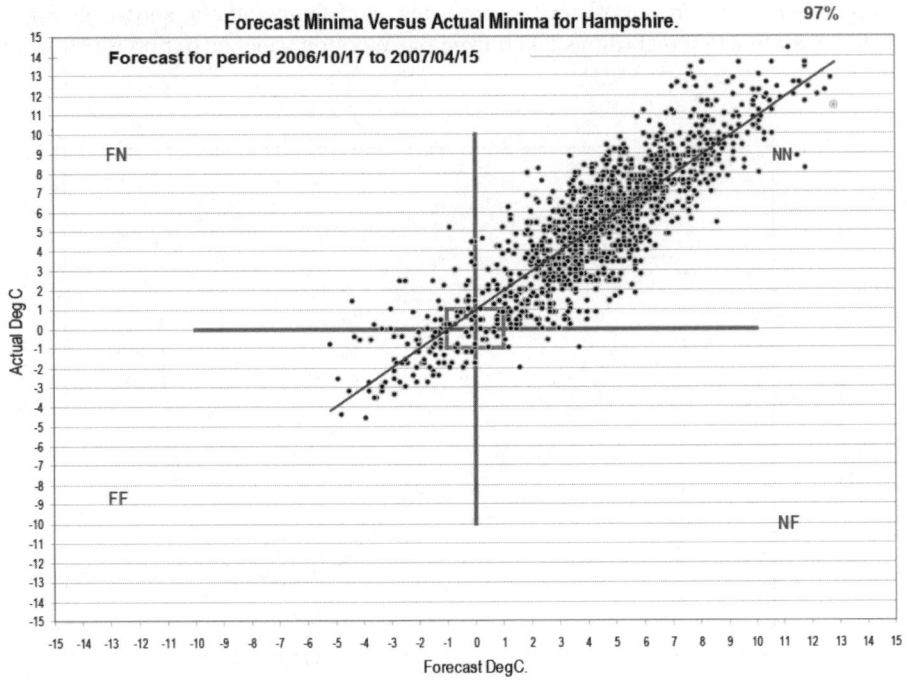

Fig. 6. Minimum forecast RST plotted against actual minimum RST (reproduced with permission from AMI)

5 Network Forecasting Display Interface

The ability for weather forecaster to be able convey his/her forecast to the road maintenance engineers using a clear and unambiguous system is as important as the producing the forecast. This system has two key end user prerequisites, namely:

- To be readily accessible from any PC
- Be available at any time on any given day during the winter period.

An ArcGIS Server implementation through an ASP.NET website provided the necessary GIS functionality and met the end user prerequisites. The development of Interface is currently an ongoing process, due to this section reports on the current status of the development.

The first steps were to design and develop an SQL Server database to handle the forecast data and GIS derived forecast point data. The database was first modelled as Entity Relationship Diagram in VISIO to clarify the data structures and flow. This is illustrated in Fig. 7.The next step was to write a program to read an UNIX formatted comma separated values text file and populate the appropriate tables with data. This was programmed using VB.NET with the ability to be set as a scheduled task to run every time a new forecast file is uploaded to the server. Once the database and associated tools and been modelled and programmed the interface could be designed and developed.

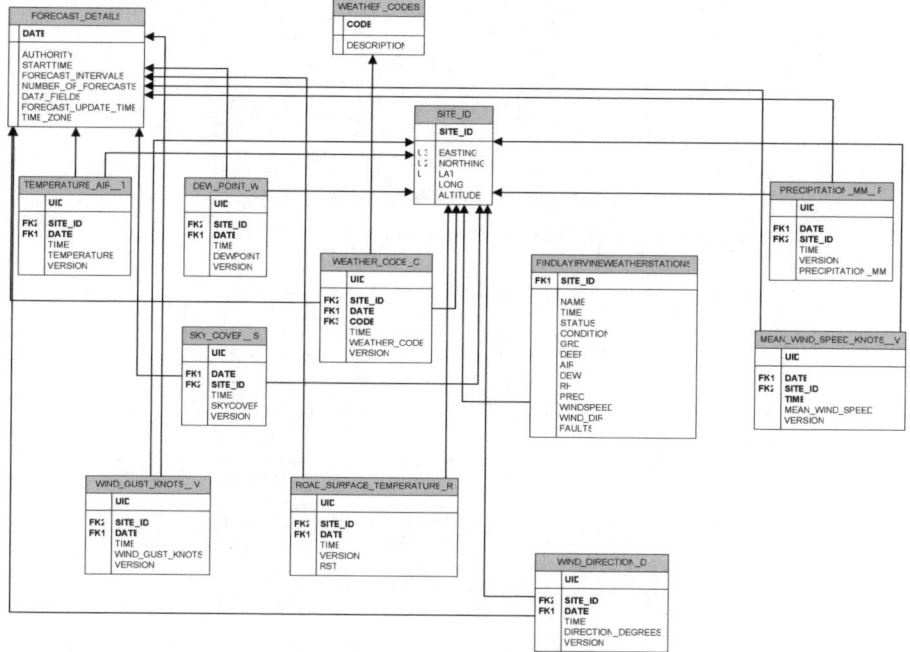

Fig. 7. Entity Relationship Model for Forecast interface Database

The website core design was based on the ESRI mapping template and conforms to the ESRI web mapping standards. Further to this most of the fundamental interface design principles have been adhered to. For example in this web mapping application the main focus is to provide clear unambiguous information to the end user and therefore a large mapping area dominates the browser. This is coupled with a clear legend which enables the end-user to clearly identify what they are looking at. Also an animated loading icon appears whenever there is a change in the mapping data drawing the user's eye to the change of state in the information being displayed. There are a lot tools available in the application such as magnifiers, information buttons, various navigation buttons and route selection buttons. However these functions are often found as part of a toolbar or toolset that is accessed by expanding the appropriate toolbar area. This results in the end user having a large toolset but not being overwhelmed upon application load. These tools are further enhanced with tooltips which give the end-user a brief description as to what the tool does. There is also a detailed help section should the user require further assistance.

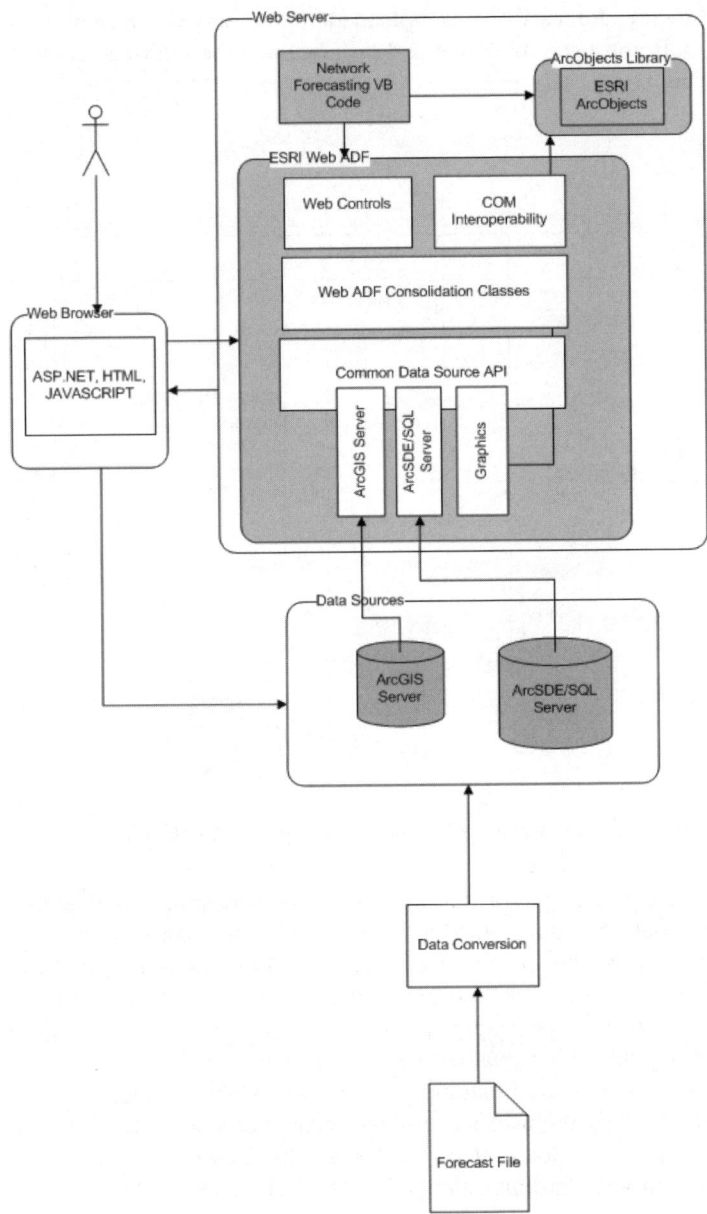

Fig. 8. Enterprise Application Diagram of web mapping interface

The next step was to develop the main functionality for the interface, specifically the display and render of the forecast points and each points associated road surface temperature for the time interval being viewed. There are different methods available when developing in ArcGIS Server, in this case a combination of fine grained

ArcObjects programming using COM integration, AJAX and the ESRI .NET ADF were used to develop the main GIS functionality. This enabled code developed in ArcMAP to be reused by calling the relevant COM objects through the server object container. This method requires careful management and disposal of the COM objects so that performance isn't inhibited, but does provide the developer with the same ArcObjects as would be available when developing in a windows environment. This method is only available if the server hosting the website and ArcGIS server are running of the same machine as it uses local objects and connections. This is illustrated in Fig. 8.

The results of this functionality are the ability to view and animate a specific route over a 24 Hour period. The animation cycles through the forecast intervals displaying each or all routes and their relevant road surface temperatures for the given forecast interval. The animation can be viewed at three speeds, paused and rewound. This enables the end user to view and review the forecast in fine detail and review the information that is available. Other basic functions that have been developed are the integration of callback functions with the ESRI Web ADF to allow partial page updates of the map interface zoom and pan functionalities, 'identify' functionalities where the user can click on a point and all the relevant information about that point is displayed and other basic navigation functions such as an overview map and a navigation compass.

Other functionality is currently been added to the mapping interface to provide the road maintenance engineers with as much information as possible so that an informed decision can be made. This includes graphical representations of the thermal fingerprint over given night for a given route providing the highway engineers alternatives to the mapping interface, historic tabular data and printable 24 hour reports highlighting routes that fall below the critical temperature of $1°C$, integration with outstations to provide real time temperature updates from the site and historic mapping interface based around calendar functionality to provide the end user with a record of forecasts. The historical elements to the website could be particularly important in litigation case against a local authority as it will provide an accurate record of the forecast that night and thus provide the case with scientifically based evidence. A screen shot of the current development can be seen in Fig. 8.

Once the development of the site is complete it will be tested thoroughly for robustness and then assessed for accessibility, layout and functionality using an internet survey. The results of this, along with a full review of the GIS Interface will be reported upon in due course.

Upon completion of a successful trial work will begin on creating dynamic gritting routes from a forecast. This will be based on the ITN road network and the forecast data which will be analyzed to identify cold sections of the road network. From this an algorithm will dynamically produce a set of routes which will be then uploaded to the gritting vehicles onboard GPS system using a mobile data network. The vehicle driver will be presented with a gritting route on his navigation system, the technology already exists whereby a route is gritted – density and type of gritting solution - based on location and it is anticipated system will integrate with this existing technology. If successful this will result in a more efficient gritting regime which will save time, money and resources as well as reduce the impact of gritting on the environment.

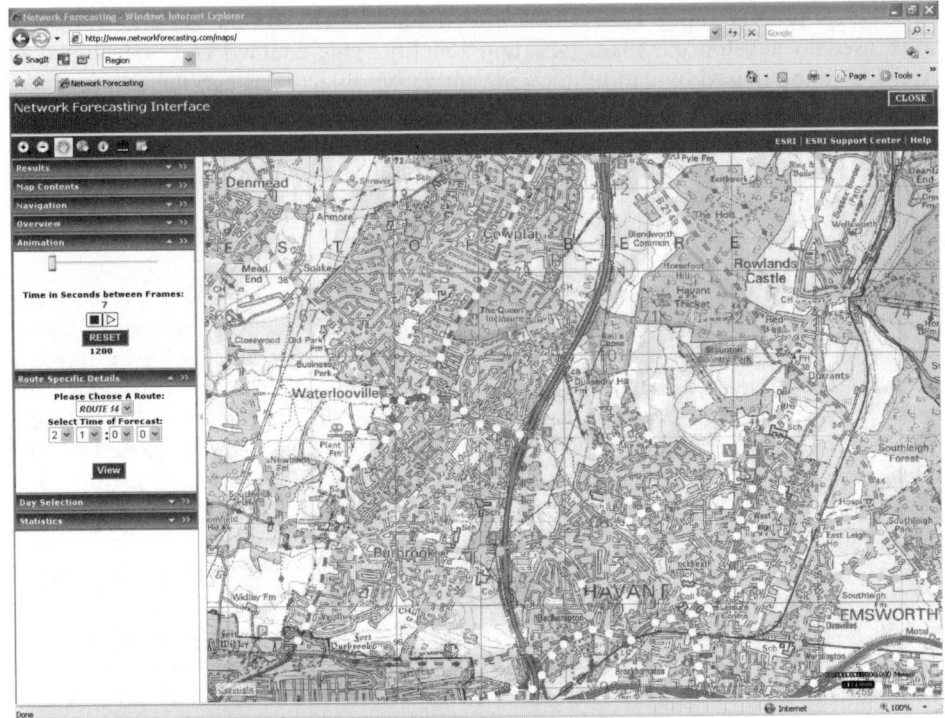

Fig. 9. Screen Capture of Network Forecasting Interface

6   Conclusions and Further Work

The initial results from the Hampshire trials indicate that the GIS techniques used to create the 3D model can enhance, if not replace, the surveying techniques used in previous ice prediction models. The model has saved several unnecessary gritting runs already this season which will, given a continued reduction in unnecessary gritting runs, reduce the environmental impact of gritting, reduce road maintenance needs, reduce accidents caused by over gritting and ultimately save money.

However the model is not perfect and with the increase in the quality and accessibility of spatial data, coupled with the improvement in GIS techniques and computing power, the accuracy of the model can only increase.

Further work is being carried out on the technique outlined in this paper. These include trials in different counties, further validation of results and refinement in techniques. Also an analysis and trials of the techniques in other types of GIS software (e.g. MapInfo, AutoCad Map, Manifold) will be carried out to assess the robustness of the techniques developed. Further development and assessment work is also being carried out on the display interface with a view to creating a dynamic routing system and integrating it wirelessly with onboard GPS systems for gritting vehicles. These developments and results will be reported upon in the future.

Acknowledgements

This work is being carried at as part of an EPSRC funded PhD in conjunction with GeoBureau Ltd, Aerospace and Marine International and Findlay Irvine. The trial is being conducted through Findlay Irvine in conjunction with Hampshire County Council. The author would like to thank all parties for their cooperation and time whilst conducting this research.

Biography

Richard Fry is currently an EPSRC funded PhD Research student at the University of Glamorgan. His current qualifications include BSc (HONS) Geography and MSc in GIS. Prior to the commencement of his PhD he worked in a local government environment as a GIS analyst.

References

1. Bradley, A.V., Thornes, J.E., et al.: Modelling spatial and temporal road thermal climatology in rural and urban areas using a GIS. Climate Research 22(1), 41–55 (2002)
2. Chapman, L., Thornes, J.E.: The use of geographical information systems in climatology and meteorology. Prog. in Phys. Geog. 27, 313–330 (2002)
3. Chapman, L., Thornes, J.E., et al.: Modelling of road surface temperature from a geographical parameter database. Part I: Statistical. Meteorological Applications 8(4), 409–419 (2001a)
4. Chapman, L., Thornes, J.E., et al.: Modelling of road surface temperature from a geographical parameter database. Part 2: Numerical. Meteorological Applications 8(4), 421–436 (2001b)
5. Chapman, L., Thornes, J.E.: A geomatics based road surface temperature prediction model. Science of the Total Environment 360, 68–80 (2006)
6. Davy, I.: The 9th Winter Maintenance Conference & Exhibition, Speakers Papers, Surveyor Magazine (October 2000)
7. Gustavsson, T.: A study of air and road surface temperature variations during clear windy nights. International Journal of Climatology 15, 919–932 (1995)
8. Gustavsson, T.: Thermal mapping - A technique for road climatological studies. Meteorological Applications 6(4), 385–394 (1999)
9. Thornes, J.E., Cavan, G., et al.: XRWIS: The use of geomatics to predict winter road surface temperatures in Poland. Meteorological Applications 12(1), 83–90 (2005)
10. Thornes, J.E.: Thermal mapping and road weather information systems for highway engineers. In: Perry, A.H., Symons, L.J. (eds.) Highway Meteorology, pp. 39–67. E&FN Spon, London (1991)
11. Thornes, J.E.: The Prediction of Ice Formation on Motorways PhD Thesis, unpublished, University of London (1984)
12. Thornes, J.E., Shao, J.: Spectral analysis and sensitivity tests for a numerical road surface temperature prediction model. Meteorol. Mag. 120, 117–124 (1991)
13. Parmenter, B.S., Thornes, J.E.: The use of a computer model to predict the formation of ice on road surfaces. Transport and Road Research Laboratory Research Report 1, 1–19 (1986)

Thematic Clustering of Geographic Resource Metadata Collections

J. Lacasta, J. Nogueras-Iso, P.R. Muro Medrano, and F.J. Zarazaga-Soria

University of Zaragoza, Zaragoza, Spain
{jlacasta,jnog,prmuro,javy}@unizar.es

Abstract. Spatial Data Infrastructures at national or higher levels usually comprise the access to multiple geographic data catalogs. However, with classical search systems, it is difficult to have a clear idea of the information contained in the metadata holdings of these catalogs. This paper describes a set of clustering techniques to create a thematic classification of geographic resource collections based on their associated metadata.

1 Introduction

Spatial Data Infrastructures (SDI) provide the framework for the optimization of the creation, maintenance and distribution of geographical information at different organization levels (e.g., regional, national, or global level) and involving both public and private institutions [1]. Geographic data catalogs are one of the main components of an SDI as they provide services for searching geographic information by means of their metadata and according to particular search criteria. But from the perspective of a European or Global SDI, the problem is that usually there is not a unique geographic data catalog. Quite the opposite, hundreds of geographic data catalogs contribute from the different nodes of an SDI. In such situations, where searches are distributed to different metadata catalogs, it is difficult to have a clear idea of the information described in each metadata collection. Therefore, it would be interesting to provide the mechanisms that give a general overview of the contents of each geographical catalog. Furthermore, this is even useful for individual geographic data catalogs in order to facilitate a summary of the contents for those users accessing for the first time. As mentioned in [2, p. 20], the user task in a discovery scenario might be one of browsing instead of retrieval (i.e., browsing instead of performing blind searches).

Focused on the context of geographic metadata, this paper studies the thematic classification and structuring of resources by means of pattern analysis techniques, known as clustering. These techniques are useful to find groups of metadata records (clusters) that share similar values in one (or more) metadata elements according to a mathematical correlation measure (e.g. the Euclidean distance, Bernoulli, Gaussian or polynomial functions among others). It can be said that elements in a cluster share common features according to a similarity

J.M. Ware and G.E. Taylor (Eds.): W2GIS 2007, LNCS 4857, pp. 30–43, 2007.

criteria [3]. The clusters obtained as output of these techniques may help in two important issues of information retrieval systems operating on a network of distributed catalogs. On the one hand, the information of the clusters (e.g., names) may serve as metadata for describing the whole collection of metadata records. On the other hand, if the user is interested in browsing (instead of searching), the output clusters provide the means for structure guided browsing.

The novelty of the paper is to improve clustering techniques with the hierarchical relations that may be derived from keywords found in metadata records, whenever these keywords have been selected from well-established lexical ontologies. As mentioned in [4], an ontology is defined as an explicit formal specification of a shared conceptualization; i.e. the objects, concepts, and other entities that are assumed to exist in an area of interest and the relationships that hold among them. And lexical ontologies could be defined as ontologies with weak semantics where the main aim is to establish the terminology used in a domain together with a reduced set of general relations (hypernymy, hyponymy, synonymy, meaning associativity). In the geospatial community it is widely accepted to use thesauri, which can be considered as lexical ontologies, to facilitate the creation of metadata describing geospatial resources. The techniques presented in this paper make profit of this frequent use of thesauri.

The rest of this paper is organized as follows. Section 2 revises the state of the art in classification techniques. Section 3 proposes several methods of clustering taking into account the use of lexical ontologies. Section 4 shows some results obtained from the application of these techniques. Finally, the last section draws some conclusions and further work is discussed.

2   State of the Art in Classification Techniques

This section reviews the main scientific contributions about classification of metadata collections making a special emphasis in those focused in the graphical visualization of the obtained groups.

For instance, [5,6] show a system to access metadata collections through a network of intelligent thumbnails, being those thumbnails either concepts or locations. For instance, in the case of locations, the network of thumbnails corresponds to the hierarchical structure of administrative toponyms.

The use of topic maps [7] is another way to structure the classification of a resource collection. Topic maps are a representation of knowledge, with an emphasis on the find-ability of information. They can be directly used for exploratory search of a resource collection, providing an overview of the collection content. For instance, [8] presents a system that creates a graphic topic map to enhance the access to a medical database using a graphical representation to locate the different topics over a graph of the human body. Within the context of the geospatial community, [9] describes a method to generate a topic map from a metadata collection based on the keywords section of metadata. It makes profit of the Knowledge Organization System (KOS) (e.g., classification schemes, taxonomies, thesaurus or ontology) that has been selected to pick up those terms in the keywords section. The assumption of using a selected vocabulary or KOS enables the creation of hierarchical topic maps thanks to the semantic relations existent in the selected KOS. Moreover, [6] proposes the use of XTM [10] as the topic map exchange format, which can be easily visualized by a wide range of tools compliant with this format. However, in distributed systems like an SDI, where the different catalogs store thousands of metadata records, the topic map summarizing the thematic contents of a collection may still be fairly complex. In this sense, [9] already identifies the necessity of obtaining a reduced set of representative terms from the generated topic map.

Clustering techniques [11] have proved to produce good results in the classification of big collections of resources for different purposes (data mining, signal analysis, image processing...). Two main different types of classification can be highlighted:

Hard clustering: It associates each element of the collection to a unique cluster. An example of this category is *K-means* and its variants [12][13].

Fuzzy clustering: It associates each element to each cluster with a different probability. It includes fuzzy and probabilistic techniques between others. Some examples are fuzzy *C-means* [14] and finite mixture models (as Bernoulli or polynomial) [15].

The MetaCombine project [16] is a good example of building services over heterogeneous metadata. In particular, it focuses on providing a browsing service, i.e. the exploration or retrieval of resources (OAI and Web resources) through a navigable ontology. In addition, it shows the effectiveness of different clustering techniques for heterogeneous collections of metadata records according to different factors, as the time used to locate a resource, the number of clicks needed to reach it or the number of failures in locating the resource.

Other works to be highlighted in this area are the following. [3] provides a visual data mining approach to explore in a easier way the complex structure/syntax of geographical metadata records. [17] also describes a system to cluster a collection of metadata into clusters of similar metadata elements using the cluster algorithm of [18]. [19] describes a system to show graphically traditional representations of statistical information about a collection to facilitate the identification of patterns.

3   Techniques of Classification

The different classification techniques shown in the state of the art section group the resources by the similarity of some properties but do not have into account the relations that can exist between the analyzed properties. This section describes how to adapt some of those techniques to make profit of the hierarchical structure of the thesaurus concepts used to fill the keyword section of metadata records. The two following techniques make use of this information to improve the results obtained by traditional hard and fuzzy clustering approaches:

Clustering keywords selected from thesauri: The keywords of the metadata records are transformed into numerical codes that maintain the hierarchical relations between thesaurus terms. These numerical codes are then
   grouped by means of clustering techniques.

Clustering keywords as free text: Hard and fuzzy clustering techniques are
   directly applied over the text found in the keywords section. But previous
   to the clustering, some keyword expansion techniques are used to improve
   the results.

3.1 Selection of Algorithms for Hard and Fuzzy Clustering

As mentioned in [11], there are several techniques of hard clustering. From them,
the *K-means* family of algorithms [12,13] has been selected for the test with hard
clustering algorithms. It has been selected by its simplicity and by its general
use in other areas of knowledge to find patterns in collections of data and by the
availability of numerous tools to perform it. The *K-means* algorithm is based
on the partitions of N records into K disjoint subsets S_j (being $j \in 1..K$) that
minimize the sum of the distances of each record to the center of its cluster.
The mathematical function used to measure this distance varies depending on
the *K-means* implementation (Correlation functions, Spearman Rank, Kendall's
Tau. . .). The function that has been used here for the experiments is the Euclidean distance, which is one of the most frequently applied.

The implementation of *K-means* used in the experiment section is the proposed in equation 1. The objective of this equation is to minimize the value of J,
where x_n is a vector representing the *n-th* data record and v_j is the S_j centroid,
being v_j calculated with formula 2, where N_j is the number of elements contained
in S_j. The algorithm starts assigning randomly the records to the K clusters.
Then two steps are alternated until a stop criterion is met (number of iterations
or stability of J). Firstly, the centroid is computed for each record. Secondly,
every record is assigned to the cluster with the closest centroid according to the
Euclidean distance.

$$J = \sum_{j=1}^{K} \sum_{n \in S_j} |x_n - v_j|^2 \tag{1}$$

$$v_j = \frac{\sum\limits_{n \in S_j} x_n}{N_j} . \tag{2}$$

The main problem of this implementation is the need to select the number of
clusters to create, because it is not able to automatically adjust the number of
cluster returned. Other more advanced implementations use adaptive techniques.
For instance, ISODATA algorithm [20] creates or joins clusters when needed,
returning a number of clusters adjusted to the collection distribution.

With respect to the fuzzy clustering family, it includes techniques as fuzzy *C-
means* (a fuzzy variant of *K-means*) or finite mixture models (Bernoulli, Gaussian or Polynomial). They assign a metadata record to each cluster with a probability value. Usually, the cluster that has the higher probability is considered

as the cluster to which the record belongs, but in doubtful situations more than a cluster can be selected. Fuzzy clustering allows to measure to what extent a metadata record belongs to a cluster, distinguishing between the records that are clearly contained in a subgroup from those that may belong to several clusters. This distinction makes possible to sort the records of each subgroup by their degree of membership and to include a record in more than a cluster with different levels of relevance.

The fuzzy algorithm selected for the experiments has been the *C-means* algorithm proposed in [14]. It minimizes the distance of each record to every cluster centroid, adding the probability of the record to belong to each cluster. It minimizes the function in equation 3, where $A_j(x_n)$ stands for the probability of record n to be in the cluster j. Additionally, this algorithm takes into account the constraints in equation 4. The exponent m is a weighting parameter used to adjust the fuzziness of the clustering algorithm. As it can be seen, the function to minimize is similar to the one used in the *K-means* algorithm. The difference in this case is that the Euclidean distance of each record to a cluster center is multiplied by the probability of being such element in that cluster.

$$ J = \sum_{j=1}^{K} \sum_{n \in S_j} (A_j(x_n))^m \, |x_n - v_j|^2. \tag{3} $$

$$ (A_j(x_i)) \in [0, 1] \ , \ \sum_{j=1}^{K} A_j(x_i) = 1 \ for \ all \ j. \tag{4} $$

3.2   Thematic Clustering Using the Thesaurus Structure

To detect the main themes of the collection, this first approach encodes the keywords of the metadata records into numerical values that preserve the hierarchical relations among the thesaurus concepts. Then, these encodings are clustered. This process let us group metadata records that share similar thematic characteristics without losing the benefits provided by the hierarchical structure of the thesaurus used to fill the metadata records.

The goal here is to generate a numerical identifier for each concept of the thesaurus that describes the hierarchical relations of the concepts, that is, its position in the thesaurus). This identifier is similar to the Dewey Decimal Classification System [21]. It consists of a set of numbers where each number indicates the position of the term in the thesaurus branch to which it belongs (a branch is a tree whose root is a top concept with no broader concepts and contains all the descendants of this concept in the "broader/narrower" hierarchy). For example, as shown in figure 1, the identifier *"02 03 00"* indicates that the term *Geotechnology* is the third child of the second top term (*Earth Science*) of the thesaurus.

The result of clustering using these identifiers is the detection of groups of records containing concepts that are close in the thesaurus structure. These

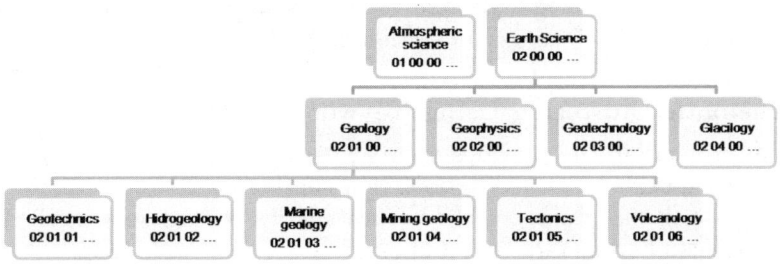

Fig. 1. Numerical encoding of concepts

clusters are then processed to obtain a reduced set of branches that concentrate most of the keywords used in the metadata records. Since these selected branches group most of the keywords in the metadata collection, they can be considered as its main themes. The approach consists of four sequential steps:

- As an initial step of this approach, the metadata collection has been processed to extract a list of pairs $< keywordInRecord\_URI, keyword\_id >$. In these pairs, $keywordInRecord\_URI$ represents a string that consists of a record identifier and a keyword term found in a record with this identifier. The second element in the pair ($keyword\_id$) is the numerical identifier of the term found in the record. This list of pairs has been used as input for the hard and fuzzy clustering.
- The second step of the approach has been the application of the hard and fuzzy clustering algorithms. The result of hard clustering is a list of clusters indicating the set of $keywordInRecord\_URI$s contained in each cluster. Fuzzy clustering returns a matrix where the rows represent the different $keywordInRecord\_URI$s and the columns represent the identified clusters. Thus, each row in the matrix contains the membership degree of $keywordInRecord\_URI$ to each cluster. The cluster with the highest probability is selected as the right cluster for each $keywordInRecord\_URI$.
- Obviously, the results obtained in the second step are not the final output, because each record may have been assigned to several clusters, as many clusters as the number of times it appears in a $keywordInRecord\_URI$. Therefore, in order to detect the most proper cluster for each record, a third step has consisted in applying the following heuristic: the records have been associated to the clusters where they appear more times.
- Finally, the fourth step determines the names of the clusters. The numerical codes representing the keywords in each cluster have been processed back to obtain the terms behind them. And the common ancestors of the keywords have been analyzed to select the cluster names. The name given to each cluster is the most specific common ancestor of the keywords in the records grouped in the same cluster. To make the name more representative, there is an exception in this rule when all the records in a cluster can be grouped under a maximum number of three sub-branches. In that case the name of

the cluster is the concatenation of the sub-branches names (e.g., "Product, Materials, Resource" cluster in section 4.2).

3.3   Thematic Clustering Using Keywords as Free Text

The main problem in the previous approach is that it is very much dependent on taking as input metadata collections that contain terms from selected vocabularies, and that the choice for those vocabularies is relatively small. But, in practice, one can find metadata containing terms from a wide range of thesauri, or even keywords typed randomly by the users. Therefore, it seems relevant to explore other alternatives with not so restrictive prerequisites that facilitate the thematic characterization of collections described with heterogeneous metadata. This section studies the application of clustering techniques considering the keywords section as a free set of terms that do not belong necessarily to a selected vocabulary or thesaurus structure.

The keywords of the metadata records are clustered using the vector space model. In this model, documents are encoded as N-dimensional vectors where N is the number of terms in the dictionary, and each vector component reflects the relevance of the corresponding term with respect to the semantics of each document in the collection [22]. This relevance is directly proportional to the number of occurrences of a keyword in a metadata record. Additionally, not only the terms that exist in the keywords section of the metadata records are included but also all the ancestors (*broader* terms) of those terms extracted from a thesaurus. That is to say, making profit of thesaurus structure, we expand the text found in keywords sections when the terms belong to selected thesauri. Additionally, the use of ancestors help to increase the relevance of some keywords.

Since metadata collections can be very heterogeneous, the output clusters can be seen as a set of different homogeneous sub-collections centered in very different themes. Thus, the naming of clusters may be quite problematic. The names of clusters should represent the theme of each cluster.

The generation of the main theme of each sub-collection (its name) is created differently when hard or fuzzy clustering is applied. When hard clustering is used, the most frequent keywords in the output cluster are the ones selected to form part of the name. With fuzzy clustering the selected name contains the terms with the highest degree of cluster membership.

4   Experiments

The techniques described in the previous section have been validated using hard *K-means* and fuzzy *C-means* clustering techniques. The obtained results have been analyzed to find the situations in which the use of each technique is most adequate.

4.1   Description of the Metadata Corpus

In order to validate the presented techniques, the content of the Geoscience Data Catalog at the U.S. Geological Survey (USGS) has been used. The USGS is the science agency of the U.S. Department of the Interior that provides information about Earth, natural and living resources, natural hazards, and the environment. Despite being a national agency, it is also sought out by thousands of partners and customers around the world for its natural science expertise and its vast earth and biological data holdings. The USGS metadata collection has been processed as indicated in [23] until a collection of 753 metadata compliant with CSDGM [24] standard were obtained. From them, the 623 keywords that have values from the GEMET thesaurus [25] have been selected for the experiments.

4.2   Thematic Clustering Using the Thesaurus Structure

The collection selected for the experiment contains keywords with terms picked up from the GEMET thesaurus. Therefore, this thesaurus has been processed to generate identifiers in the form already described in section 3.2. Because none of the branches of this thesaurus has more than 99 terms as *narrower* of a concept, two digits were enough to encode each level of the thesaurus.

In the case of hard *K-means* algorithm, five clusters (K=5) were asked, with the objective of obtaining the five most used branches of GEMET in the collection. However, the results obtained were disappointing because the algorithm did not converge, producing clusters with heterogeneous keywords.

The results obtained with fuzzy *C-means* algorithm were quite better. The clusters produced contained groups of records with similar keywords. The clusters obtained as output are the following:

Hydrosphere, Land: 19 records about the following topics (including hierarchical path):
  - Natural Environment, Anthropic Environment. Hydrosphere
  - Natural Environment, Anthropic Environment. Land

Biosphere: 10 records about the following topic (including hierarchical path):
  - Natural Environment, Atrophic Environment. Biosphere

Product, Materials, Resource: 211 records about the following topics (including hierarchical path):
  - Human activities and products, effects on the environment. Products, Materials. Materials
  - Human activities and products, effects on the environment. Products, Materials. Product
  - Human activities and products, effects on the environment. Resource

Chemistry, Substances and processes: 39 records about the following topics (including hierarchical path):
  - Human activities and products, effects on the environment. Chemistry, Substances and processes

Research Science: 291 records about the following topic (including hierarchical path):
  - Social Aspects, Environmental, politics measures. Research Science

These results exhibit how the metadata records of the collection are related, showing a broad thematic view of the collection. The use of upper level branches (the most generic) indicates disperse keywords in the metadata records, the use of lower level branches (more specific) expose a high relation with a specific theme and the lack of a branch of the thesaurus indicates that the metadata in the collection are not related to that theme.

4.3   Thematic Clustering Using Keywords as Free Text

In this second approach, the use of hard and fuzzy clustering algorithms have produced quite different results. The details of each experiment are detailed next.

Keywords as free text and hard clustering. The keywords of the USGS metadata records were transformed into a $C(NxM)$ matrix where N is the number of metadata records to cluster, and M the set of keywords contained in the collection plus the terms added from the GEMET thesaurus. An element $C(i,j)$ of the matrix takes value 1 when the term j is contained in the metadata record i or j is an ancestor in GEMET hierarchy of a term in record i. Otherwise, $C(i,j)$ takes value 0. The *K-means* algorithm were applied to the matrix asking for five clusters ($K = 5$). Then, the clusters obtained were processed to generate a representative name (combination of the most frequent keywords, or the name of the thesaurus branch that contains them), producing the following results:

Natural gas, Geology, Earth Science: This cluster consists of 151 metadata records all of them containing the keywords *Natural gas*, *Geology* and *Earth Science*.

Coal: 116 of its 117 metadata records contain the *Coal* keyword, and the other has *Lignite*, a keyword related hierarchically with *Coal* (its father).

Geology: This third cluster contains 128 metadata records from which 92 contain *Geology* and 34 *Marine geology* (narrower term of *Geology*). The remaining two records contain keywords with terms related to *Geology*. One of them contains *Earth Science* (broader term of *Geology*) and the other one contains *Mineralogy* (sibling term of *Geology*, i.e. having *Earth Science* as broader term).

Chemistry, substances and processes: This cluster contains 53 metadata records with keywords from the *Chemistry, substances, and processes* branch of GEMET.

Parameter & Others: This cluster of 32 metadata records is heterogeneous. It contains 14 metadata records from the *Parameter* branch of GEMET, however, the other 18 metadata records have no relation with them or between them.

Using the result obtained, the generated classification from the whole collection of metadata would be *Geology* (contained in 151+92 records), *Natural gas*, *Earth Science*, *Coal* and *Chemistry, substances and processes*. Each one being less relevant than the previous one (it appears fewer times).

The obtained results also show that inside the set of metadata records about *Geology* an important part of them is more specialized (they are also about *Natural gas* and *Earth Science*).

The expansion through the GEMET hierarchy in conjunction with clustering techniques allows detecting semantic aggregations not directly visible. For example, the cluster *Chemistry, substances, processes* is not detected when no hierarchy relations are considered. In addition, this technique also separates in different clusters records not sharing enough keywords. An example of this occurs in the *Geology* cluster. There are records that contain the terms *Geology* or *Earth Science*, also present in the *Natural gas* cluster, but do not contain the *Natural gas* term. In order to distinguish them from the set of records that always include *Natural gas* two different clusters have been created.

This separation causes that the two clusters share part of name, given that both contain the *Geology* term in most of their metadata. This can cause confusion when accessing the information, given that two subsets are said to be about the same matter. When this happens, the solution adopted is to include all the elements of the more specific cluster inside the most general considering the *Natural gas, Geology, Earth Science* as a subset of the *Geology* cluster.

Another problem found in the obtained results is that the *K-means* algorithm assigns very heterogeneous records to the smallest cluster (in the example, the *Parameter* cluster), because they can not be classified in the rest of clusters. The result is that the smallest cluster is quite useless. Therefore, the name obtained for the *Parameter* cluster is not adequate as it contains many heterogeneous elements. In order to remark this heterogeneity, the generic term *Others* has been added to the cluster name.

Keywords as free text and fuzzy clustering. The same $C(NxM)$ matrix generated for the previous example was used with the fuzzy *C-means* algorithm asking for five clusters ($K=5$) with $m=2$. But due to the low degree of cluster membership of the records in two of the clusters, the experiment was redone with $K = 4$. In the results obtained, these two clusters were joined producing a new cluster whose main keyword was *parameter* with a probability of 0.5. There were no changes in the other clusters. The following results were obtained; they include the cluster name and the number of records that have been assigned to each cluster with the highest probability:

Natural gas, Earth Science (151 records): The occurrences of *Natural gas* and *Earth science* keywords are contained in metadata records near the cluster center, with a mean probability of being in the cluster of 0.99, *Geology* is the following with a probability of 0.64.

Coal (117 records): The occurrences of *coal* keyword are contained in metadata records near the cluster center, with a mean probability of being in the cluster of 0.8.

Marine geology (128 records): The occurrences of *marine geology* are contained in metadata records near the cluster center, with a mean probability of being in the cluster of 0.75.

Others (85 records): The other two clusters are not adequate given that they
contain more or less the same keywords but with low probability of belonging
to the clusters.

In order to generate the name of each cluster (its main themes), the concept
names present in the records with the highest cluster membership were selected.
Depending on the requirements of the systems, concepts in records with a lower
degree could be also included (e.g. *Geology* in the first cluster). In this situation,
the membership degree could be displayed to indicate that not all the categories
represent the collection in the same way.

4.4   Comparison of Results

The results obtained in each situation are quite different. Figure 2 displays an
skeleton of the GEMET hierarchy and how the clusters of each approach are
located in this thesaurus hierarchy. Regarding the first approach, it has been
shown that hard clustering does not work, and that fuzzy clustering produces
very general results. In the second approach, fuzzy clustering produces more
specific results than hard clustering. However, all the results are valid. Each one
provides a vision from a different perspective of the collection: ranging from the
more general vision of the first approach to the more specific vision of the last
experiment.

The first approach only works correctly if the keyword section has been created
with terms of a single thesaurus. But, given that in practice metadata contain
terms from a wide range of thesauri, or even keywords randomly typed by the
users, the second approach is more flexible. This second approach expands the

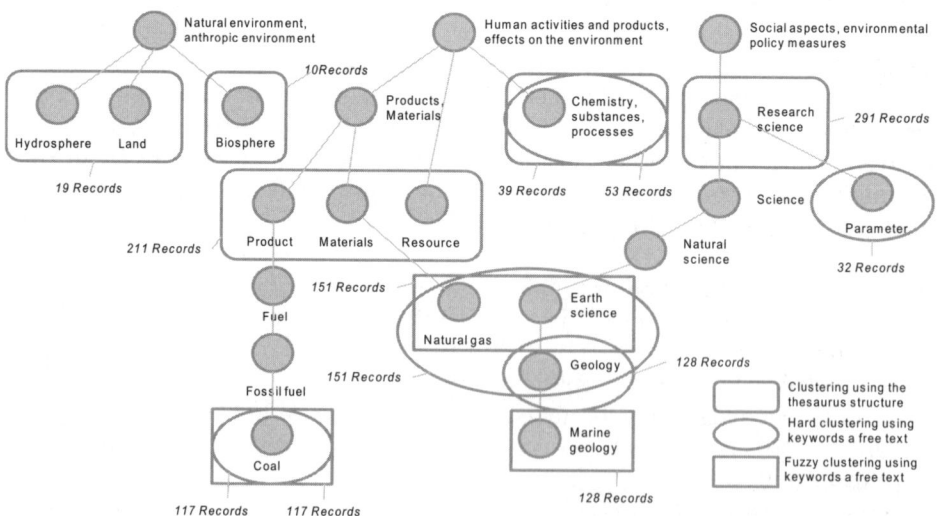

Fig. 2. Clusters generated with the different approaches proposed

keywords contained in the metadata (to improve the results) using the thesauri used as source. Nevertheless, it also works properly with several thesauri or even with keywords not contained in a controlled vocabulary.

In general, hard clustering has the advantage of being more simple but fuzzy clustering enables the identification of records with doubtful allocation (they have similar probability for two or more clusters). This is a clear advantage of fuzzy with respect to hard clustering. Fuzzy clustering detects a cluster not only on the basis of the number of times a keyword appears in a metadata record, but also taking into account that a keyword can not be found in other clusters. An example of this is the treatment of the *Geology* concept in the second approach. When hard clustering is applied, *Geology* was as important for the cluster theme as the rest of the main keywords. However, with fuzzy clustering it is shown that the records containing *Geology* are less centered in the cluster than the other ones.

5   Conclusions

This paper has shown some techniques to automatically generate a classification of a collection of metadata records using the elements of their keywords section. These techniques are based on hard (*K-means*) and fuzzy (*C-means*) clustering and have into account the hierarchical structure of the concepts contained in the Knowledge Organization Systems (KOS) used to select these keywords.

Two different approaches have been analyzed. In the first approach, the keywords in the metadata records picked up from a single thesaurus have been encoded as a numerical value that maintains the inner structure of the thesaurus. Then, this encoding has been used as the property for the definition of the thematic clusters. In the second approach, all the keywords in the metadata collection have been considered as free text. Additionally, in order to improve the results, when it has been recognized that a term belongs to a KOS, the terms in the hierarchy of ancestors have been added to the set of keywords. Then, these sets of expanded keywords have been clustered using hard and fuzzy clustering techniques. This last approach aimed at avoiding the deficiencies of the first approach that assumed the use of a single selected vocabulary.

For the experiments, the USGS metadata collection corpus has been used. This collection has been processed to identify terms from the GEMET thesaurus. Then, each approach has been tested with hard and fuzzy clustering processes to compare the obtained results, and to show the advantages and disadvantages of each technique. The results obtained have been quite different, ranging from general to specific classifications. Depending on the specific needs of the final application where the classification is integrated, the use of one or another will be more appropriate.

Further work will be focused on two different aspects: the improvement of clustering techniques, and the integration of these techniques within the services offered by an SDI.

With respect to the improvement of clustering techniques, we expect to adjust automatically the number of clusters. For instance, clustering algorithms requiring no previous knowledge on the number of clusters (e.g., Extended Star [26]) could be executed to estimate the initial number of clusters before applying the approaches presented in this paper. Additionally, we will study the effect of introducing modifications in the way to compute the relevance of an expanded keyword in the second approach. The modifications will take into account the distance from the ancestor to the original term in the "broader/narrower" hierarchy.

Last, concerning the integration of clustering within SDI services, future work will tackle, among other issues, the use of clustering for the improvement of the human computer interaction. For users more interested in browsing than in performing blind searches, thematic clustering will help them in this browsing task. Besides, in a distributed network of geographic catalogs, thematic clustering will help to filter those catalogs that are clearly related to user queries, and discard those not related at all. In fact, the names of the clusters obtained as a result of the techniques proposed in this paper may serve as the metadata that describes the whole collection of records in a repository.

Acknowledgements

This work has been partially supported by the University of Zaragoza through the project UZ 2006-TEC-04.

References

1. Nebert, D. (ed.): Developing Spatial Data Infrastructures: The SDI Cookbook v.2.0. Global Spatial Data Infrastructure (GSDI) (2004), http://www.gsdi.org
2. Baeza-Yates, R., Ribeiro-Neto, B.: Modern Information Retrieval. ACM Press, Addison Wesley, New York (1999)
3. Demšar, U.: A visualization of a Hierarchical Structure in Geographical metadata. In: Proceedings of the 7th AGILE Conference on Geographic Information Science, Heraklion, Greece, pp. 213–221 (2004)
4. Gruber, T.: A translation approach to portable ontology specifications. ACM Knowledge Acquisition, Special issue: Current issues in knowledge modeling 5(2)(KSL 92-71), 199–220 (1993)
5. Schlieder, C., Vögele, T., Visser, U.: Qualitative spatial representation for information retrieval by gazetteers. In: Montello, D.R. (ed.) COSIT 2001. LNCS, vol. 2205, pp. 336–351. Springer, Heidelberg (2001)
6. Schlieder, C., Vögele, T.: Indexing and browsing digital maps with intelligent thumbnails. In: SDH 2002. Spatial Data Handling 2002, Ottawa, Canada, pages 12 (2002)
7. International Organization for Standardization (ISO): Information technology – SGML applications – Topic Maps. ISO/IEC 13250, International Organization for Standardization (1999)
8. Boulos, M.N.K., Roudsari, A.V., Carson, E.R.: Towards a Semantic Medical Web: HealthCyberMap's Dublin Core Ontology in Protégé-2000. In: Fifth International Protégé Workshop, SCHIN, Newcastle, UK (2001)

9. Lacasta, J., Nogueras-Iso, J., Tolosana, R., López, F.J., Zarazaga-Soria, F.J.: Automating the Thematic Characterization of Geographic Resource Collections by Means of Topic Maps. In: Proceedings of the 9th AGILE International Conference on Geographic Information Science, Visegrád, Hungary (in press)

10. Pepper, S., Moore, G. (ed.): XML Topic Maps (XTM) 1.0. Technical report (2001), http://www.topicmaps.org

11. Steinbach, M., Karypis, G., Kumar, V.: A comparison of document clustering techniques. In: Proceedings of the KDD Workshop on Text Mining, Boston, USA, pp. 1–20 (2000)

12. Dubes, R.C., Jain, A.K.: Algorithms for Clustering Data. Prentice-Hall, Englewood Cliffs (1988)

13. Kaufman, L., Rousseeuw, P.J.: Finding Groups in Data: an Introduction to Cluster Analysis. John Wiley & Sons, Chichester (1990)

14. Torra, V., Miyamoto, S., Lanau, S.: Exploration of textual document archives using a fuzzy hierarchical clustering algorithm in the GAMBAL system. Information Processing and Management 41, 587–598 (2005)

15. Juan, A.: Reconeixement de formes. Technical report, Universidad Politecnica de Valencia (2005)

16. Krowne, A., Halbert, M.: An Evaluation of Clustering and Automatic Classification For Digital Library Browse Ontologies. Metacombine project report (2004), htttp://metacombine.org

17. Podolak, I., Demšar, U.: Discovering structure in geographical metadata. In: Proceedings of the 12th conference in Geoinformatics, Galve, Sweden, pp. 1–7 (2004)

18. Fisher, D.H.: Knowledge acquisition via incremental conceptual clustering. Machine Learning 2, 139–172 (1987)

19. Albertoni, R., Bertone, A., Demšar, U., Martino, M.D., Hauska, H.: Knowledge extraction by visual data mining of metadata in site planning. In: Knowledge Extraction by Visual Data Mining of Metadata in Site Planning, Espoo, Finland, pp. 119–130 (2003)

20. Ball, G., Hall, D.: ISODATA, A novel method of data analysis and pattern classification. NTIS AD699616, Standford Research Institute, Standford, California (1965)

21. OCLC: Dewey Decimal Classification System, 22nd edn. Online Computer Library Center (2003)

22. Berry, M., Drmac, Z., Jessup, E.: Matrices, Vector Spaces, and Information Retrieval. SIAM Review 41, 335–362 (1999)

23. Nogueras-Iso, J., Zarazaga-Soria, F.J., Muro-Medrano, P.R.: Geographic Information Metadata for Spatial Data Infrastructures - Resources, Interoperability and Information Retrieval. Springer, Heidelberg (2005)

24. Federal Geographic Data Committee (FGDC): Content Standard for Digital Geospatial Metadata, version 2.0. Document FGDC-STD-001-1998, Metadata Ad Hoc Working Group (1998)

25. European Environment Agency (EEA): GEneral Multilingual Environmental Thesaurus (GEMET). Version 2.0. European Topic Centre on Catalogue of Data Sources (ETC/CDS) (2001), http://www.mu.niedersachsen.de/cds/

26. Gil-García, R.J., Badía-Contelles, J.M., Pons-Porrata, A.: Extended Star Clustering Algorithm. In: Sanfeliu, A., Ruiz-Shulcloper, J. (eds.) CIARP 2003. LNCS, vol. 2905, pp. 480–487. Springer, Heidelberg (2003)

The Need for Web Legend Services

Bénédicte Bucher, Elodie Buard, Laurence Jolivet, and Anne Ruas

COGIT Laboratory, Institut Géographique National,
2 avenue Pasteur, 94 165 Saint Mandé Cedex, France
{Benedicte.Bucher, Elodie.Buard, Laurence.Jolivet,
Anne.Ruas}@ign.fr

Abstract. This paper addresses the issue of designing efficient on-demand maps in services oriented architectures. Some crucial steps of an on-demand map design process are not supported by today's architectures and may greatly benefit from knowledge that has been formalised by cartographers, like the definition of styles to draw geographical data on the map. Our proposal aims at integrating part of cartographic expertness, mostly semiologic rules, in Web mapping services oriented architectures. It comprises a set of Web services dedicated to facilitating the definition of accurate legends with respect to user objectives and data.

Keywords: Legend design, cartography, semiology, Web services.

1 Introduction

Maps on the web or on smartphones should be efficient and attractive both to support user decision based on a growing number of criteria related to the geographic space and to capture user attention during a possibly compulsive browsing activity. To use a striking formula from the literature, a good map should tend to be a 'beautiful evidence' ensuring 'wonder of the spectacle' and 'accuracy of the expression' as stated in [1]. This paper addresses the issue of drawing such maps on the Web.

Designing accurate on demand maps on the Web requires more than knowing what data services exist, what data processing services exist and how to interact with these services. It also requires some classical cartographical expertness. The accuracy of a map refers to the adequacy between the expected functions of this map and the actual user tasks it affords. This affordance is based on user seeing, reading and possibly interacting with the graphical signs that make up the map. Accuracy is thus closely related to selecting the relevant features –e.g. a navigation map needs to show roads-, the relevant representation for each feature –e.g. a navigation map does not need to have a surface representation of roads, the network representation is enough-, and the relevant styles –e.g. a navigation map needs to portray roads in a highlighted way so that the user easily visualises network information-. Cartographic servers propose predefined styles to portray available features. Yet, on-demand mapping may require the creation of new styles. This need emerges when a map combines data stemming from different sources and for which available styles are not consistent when used together. For instance, if the style for roads in a topographic data product is a red line with a black border and the

J.M. Ware and G.E. Taylor (Eds.): W2GIS 2007, LNCS 4857, pp. 44–60, 2007.

style for dangerous area in a fire danger data product is a red filling, the usage of a similar hue for these two types of features (roads and dangerous area) will visually create an association between them that does not exist in the intended message. For these reasons, the overlay of topographic data and risk data on the French portal for national geospatial information, which is technically possible, had to be turned off; the portal administrators detected colour inconsistencies in both legends. This need for creating new styles also emerges when the map has very specific objectives that have not been foreseen by the provider of cartographic layers. For instance, if a user wants to locate a firm with its logo on a topographic layer where the default style for the church feature has the same colour as the logo, he will need to change the church feature style. Correctly combining existing styles or creating new styles to draw accurate maps requires skills in graphical language. Using a graphical language is not quite as intuitive as speaking in a natural language. The mapmaker needs to be both a scientist and an artist [4]. As stated by the cartographer Max Eckert, quoted by the authors of [4]: 'the dictates of science will [..] impart to the map a fundamentally objective character in spite of all subjective impulses.' This science is made up of rules (read clues) formalised by cartographers about the usage of graphical variables to convey a message related to a portion of geographic space [2][3]. One may argue that knowing cartographic rules is not mandatory to draw good maps and that a person with sheer artistic dispositions and no specific knowledge of cartography may possibly create an efficient map. Artists indeed distinguish themselves by their capacity in exploring new paths of expression that reach a resonance in people perception. But there are still other people who are not especially gifted in graphical expression, or who lack the time to devote themselves to creating the maps they need. Besides, once styles have been created and the map is drawn, evaluating its accuracy should rely on objective criteria. The author of the map is not empowered to evaluate the adequacy between his map and his message, because he knows the message and will intuitively focus on specific aspects of the map. Thus, cartographic science, if not for the styles definition process, is very relevant for evaluation.

Yet, the fast evolution of GI software has led from a situation where cartographic experts were always involved in the process of map design to a situation where drawing a map is rather a matter of modular architecture and interoperability. Nowadays, cartographers are not systematically involved in the design process of the maps that will soon appear on our Web browser or mobile phones. In 'geovisualization' projects like [5], the respect of cartographic rules has been identified as an objective and project managers will allocate time and expertness to reaching this objective. Other projects put the accent on other relevant and crucial issues (algorithmic, interoperability, networking) and do not explicitly mention semiology in their objectives. Typically, in the field of data portrayal on the web, the standard concept to describe a map, the ISO/OGC MapContext element associated with Styled Layer Descriptors, is the specification of a data retrieval and portrayal process [6]. Somebody who does not know how to draw a map can not rely on MapContext to express his need. So far, cartographic expertness has often been present in web mapping projects; there are persons who are involved in the development or evaluation of the application and whose initial formation or personal interest has granted them some cartographic knowledge and skills. Who can guarantee that this will always be the case? In the context of Web2.0 mapping, 'geohackers' who seldom have a cartographic culture can produce maps based on simple programmatic interface like the Google API. Graphical user interfaces and programmatic interfaces are not enough, expert know-how is also needed to use them and produce 'beautiful evidences'.

The section 2 of this paper illustrates what cartographic inconsistencies are. It simply recalls principles earlier stated in the book 'How to lie with maps' [7].

Section 3 presents a proposal under elaboration at the COGIT laboratory of the French national mapping agency (IGN) to improve the accuracy of maps in services oriented architectures. It focuses on legend design for static maps. Besides, we concentrate on map design processes that are based on a composition of Web services. In this context, our proposal consists in a set of Web legend services specifications. These rely on standard models and enrich them to represent a user map during the design process.

2   Maps Accuracy

This section extends upon the notion of map accuracy.

2.1   Good Maps

Fig. 1 displays examples of maps drawn by cartographers. Upper maps are two French topographical maps centred at the same location but at two different levels of

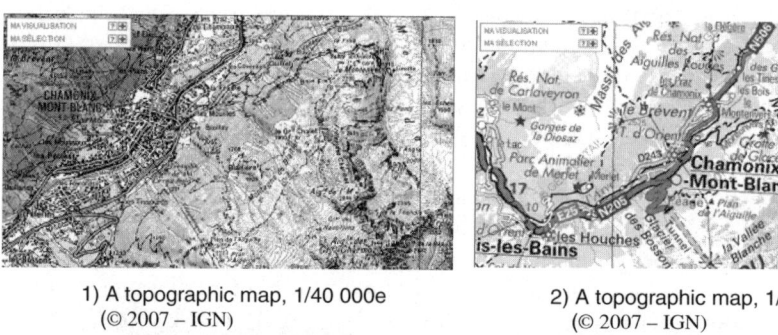

1) A topographic map, 1/40 000e
(© 2007 – IGN)

2) A topographic map, 1/100 000e
(© 2007 – IGN)

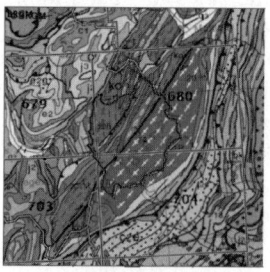

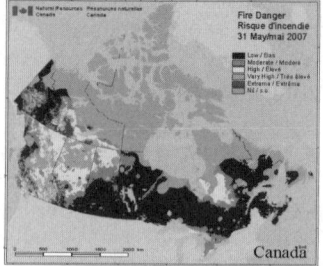

3) A geological map
(© 2007 – BRGM)

4) A fire danger map
(http://cwfis.cfs.nrcan.gc.ca/)

Fig. 1. Examples of 'good' maps drawn by experts using cartographical rules

details. All graphical variables have been used to build a legible representation of relevant topographical information at the desired level of detail. For instance, in the map *1)*, the location variable has been used to render density at first sight. The reader first sees densities of things. He then reads these things, i.e. look inside each portion of space with a homogeneous density to study what objects are there. The map *3)* is colourful and is not legible to anyone. It is dedicated to be read by experts in geology who are familiar with these styles. Its objective is not to convey a graphical message understood by any user. The map *4)* is a fire danger map dedicated to citizens. It is legible by anyone thanks to a good usage of graphical variables. The blue colour is used to denote a moderate danger and the red colour a high danger, which are commonsense connotations of these colours.

1) A map of wikipedia articles
www.placeopedia.com
(© 2007 – IGN)

2) A map from
Google mobile

1bis) A map of wikipedia articles
www.placeopedia.com
(© 2007 – IGN)

Fig. 2. Series of maps which styles have been generated based on the same generic style (the Google map API style). These styles ensure the drawing of good maps (aesthetic and efficient). Though the 1bis) style reaches a critical threshold of information content.

Maps produced in SOA architecture and on the Web are often series of maps, i.e. a set of maps that adopt the same styles to portray geographic features. Defining a set of styles for a series is complex because the result does not only depend on the styles but also a lot on the data. Designing a 'good' set of styles for a series ideally requires foreseeing graphical signs on maps that will result from applying the styles to the data and foreseeing possible graphical interactions between signs on each map belonging to the series. In the context of the Web, Google has proposed a generic style illustrated in Fig 2. The Google mobile style for contextual information is now a familiar style because it is the 'plan' style of the Google map API. The style for thematic information is a red pin with a number, and a popup window, both with a shadow. This style enhances accuracy; dynamic popup and shadows are attached to user information only, which clearly distinguishes them from contextual information.

2.2 Bad or Could-Be-Better Maps

Numerous pitfalls can be expected to deceive the neophyte map maker. The reader may confer [7] for a most exhaustive and illustrated list. We just list a few of them and how they can be observed on maps currently produced.

Fig. 3 shows a caricatural example we created: two maps which expected function is to warn of a danger close to a river bend. The author of the first map drew a simple and efficient map. Its reader correctly interprets the graphical signs into objects, properties and relationships in the reality. On the contrary, the author of the second map created an innovative legend. He selected colours based on his own taste only. The reader of the second map misinterpreted the graphical signs and did not get the intended message. For instance, since the circles and the line have similar properties of alignment, he thought it was one object of the reality. He first thought the purple sign was an arrow indicating the road direction and then perceived a notion of 'mystery' conveyed by the purple colour and guessed it was the location of a treasure. Drawing the legend on the side of the map may disambiguate the meaning of some signs, typically the meaning of the purple triangle. Yet, there is always some graphical information that does not have an explanation in the legend graphics and that must be self-understanding. For instance, relationships between graphical objects (the red sign is on the blue object bend, the grey houses are aligned with the river, the houses are similar) are not explained in the legend caption.

As stated by Monmonier, users tend to put too many colours on the map without knowing their role in visualisation. The geojoey maps, shown in fig. 4 and fig. 5, illustrate the pitfall of using colour to satisfy 'personal artistic needs' while forgetting that the first property of colour on a map is to add sense to the representation. Colours are meaningful on their own. For example red means danger or relevance, dark green means poison, light green means hope and life. Consequently, the immediate interpretation of the map in fig. 4 is: red is a very negative experience (or ill rated, or spicy), green is a very positive (or best rated) and blue are neutral. It takes quite a time browsing and testing this site to understand the styles. There are actually five styles: red means the latest entry, orange the second latest, yellow the third latest, green the fourth and the remaining are blue. In fig.4, the orange and yellow symbols are hidden.

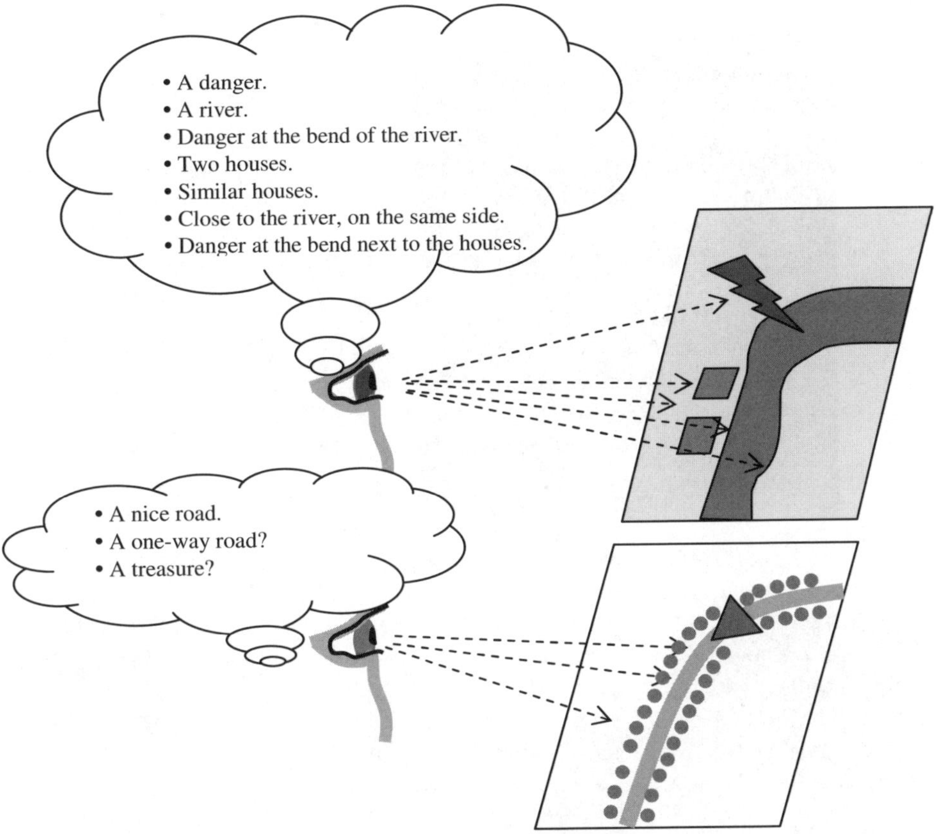

• A danger.
• A river.
• Danger at the bend of the river.
• Two houses.
• Similar houses.
• Close to the river, on the same side.
• Danger at the bend next to the houses.
• A nice road.
• A one-way road?
• A treasure?

Fig. 3. Schematized visualization processes of two maps that share the same objective (warning users of a danger close to a river bend)

1
votes
No
votes
yet
3
votes

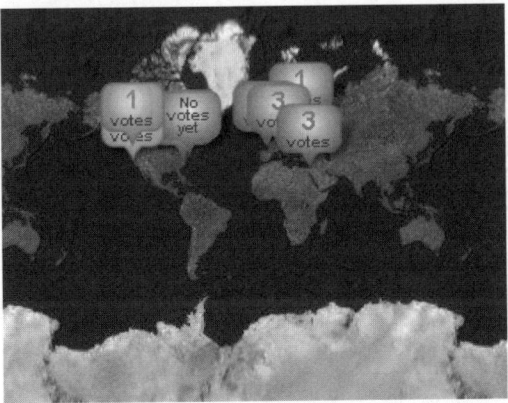

Fig. 4. A map from www.geojoey.com: 'newest blog entries' in the whole world

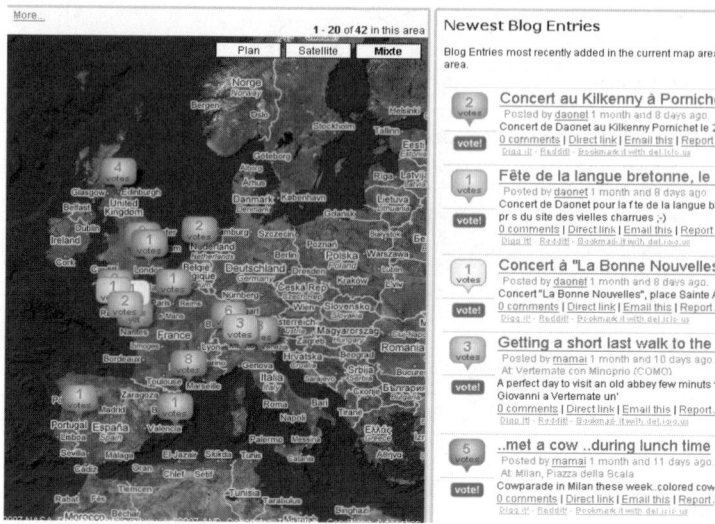

Fig. 5. Another map from www.geojoey.com

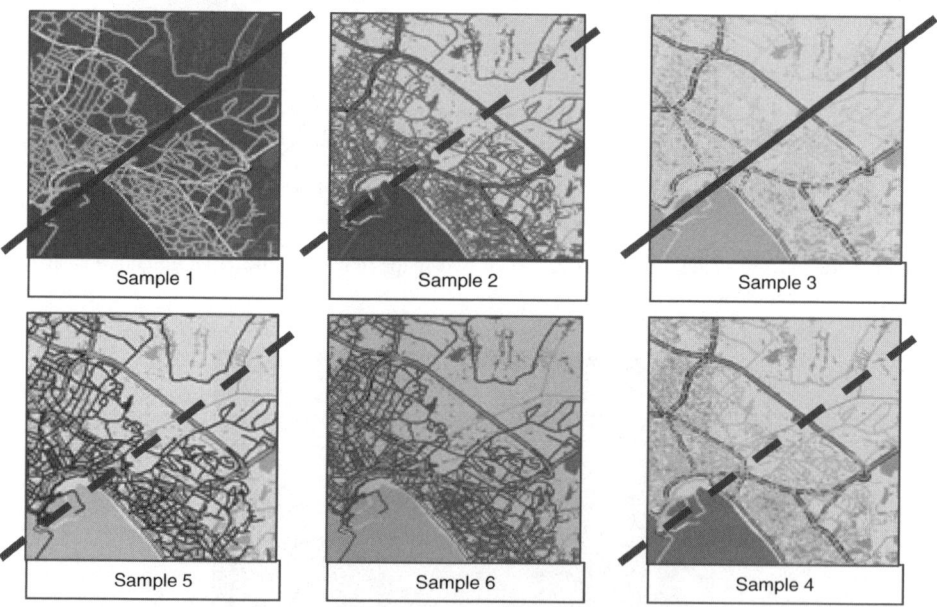

Fig. 6. Samples of good (sample 6) and bad maps (all other samples) built to study user perception of maps legibility and aesthetics. The qualification of 'good' or 'bad' is based on the assumption that the objective of the map is to display topographical information.

Even more importantly, relationships between graphical signs, and in particular between their colours, reveal relationships between represented objects. In the geojoey example, a variation in colour hues suggests that rendered items belong to different semantic classes whereas they are of the same nature ('blog entry'). More generally, misusing relationhips between colours to render relationships between classes of objects is a most common pitfall. For instance, in fig. 6 *sample 5*, the colour hues of the see and road network are very similar so that the reader may think there is a relationship between the purple network and the see.

Another pitfall is to fail drawing information in a legible way and to overwhelm the reader with graphical information. On the map in fig. 6 *sample 1,* the see and the background have too high colour values with respect to their surfaces, which prevents the reader from an accurate reading of other graphical elements. The map on *sample 6*, with lower colour values, is more legible.

Last, managing relative colour contrasts is a difficult task. In fig. 6 the maps *sample 3* and *sample 4* are rather aesthetic but they are illegible because the road network is not enough contrasted with the background. Using dark borders for the road style may mend this.

3   Embedding Cartographic Expertness in Web Legend Services

Part 2 of this paper has illustrated the importance and difficulty of drawing accurate maps. This part describes our proposal to embed assistance to draw accurate maps in Web mapping architectures.

3.1   Analysis of Required Capacities

There does not exist a generic method to draw an accurate map with respect to user objectives. As stated in the introduction of this paper, using a graphical language to convey a message related to a portion of the earth refers to the expertness of cartographers. This expertness has been formalised by many authors. First cartographic rules were adapted to static paper maps, like these from [2]. Cartography has evolved to integrate the potential of the digital media, i.e. to integrate the existence of new graphical variables and the possibility to automate some operations involved in visualization tasks [2][8]. Cartography has also evolved in its foundations. The usage of cognitive science enriches the foundations of this empirical science. [9] observe and analyse the eye movements of map readers to capture visual operations and tasks effectively performed by maps readers. Importantly, despite these evolutions, cartographical rules are often seen as 'hypothetic rules' rather than 'systematic rules', i.e. rules that may not always apply. More specifically, a rule is often to be used relatively to a context and rules may be conflicting. There exist cartographic methods but they are adapted to limited contexts like the portraying of thematic field. It is possible to transmit these methods to users for instance in a tutorial like the ESRI guide written by [10] for drawing good thematic maps.

If cartographic rules do not form a ready-to-use method to define styles, there is still an important recurring analysis pattern in cartography that we build our approach on. This analysis pattern consists in decomposing a map after 'abstraction' levels.

- – The first level is that of concrete variables. These are properties that distinguish one map from another independently of any reader. For instance [2] gives a typology of graphical variables for static maps. Screen maps have more variables related to movement and to available operations.

- The second level is what may be called 'abstract properties of a map'. These are generic properties related to the interpretation of a map. Examples of such properties are: the domain of reality shown on the map, the relationships between them and the number of communication levels of the map.
- The last level is that of map functions or user tasks.

In the remaining of this section 3.1, we enumerate required capacities to draw an accurate map, based on this decomposition, as well as existing services that provide them. This is summarized in fig. 7.

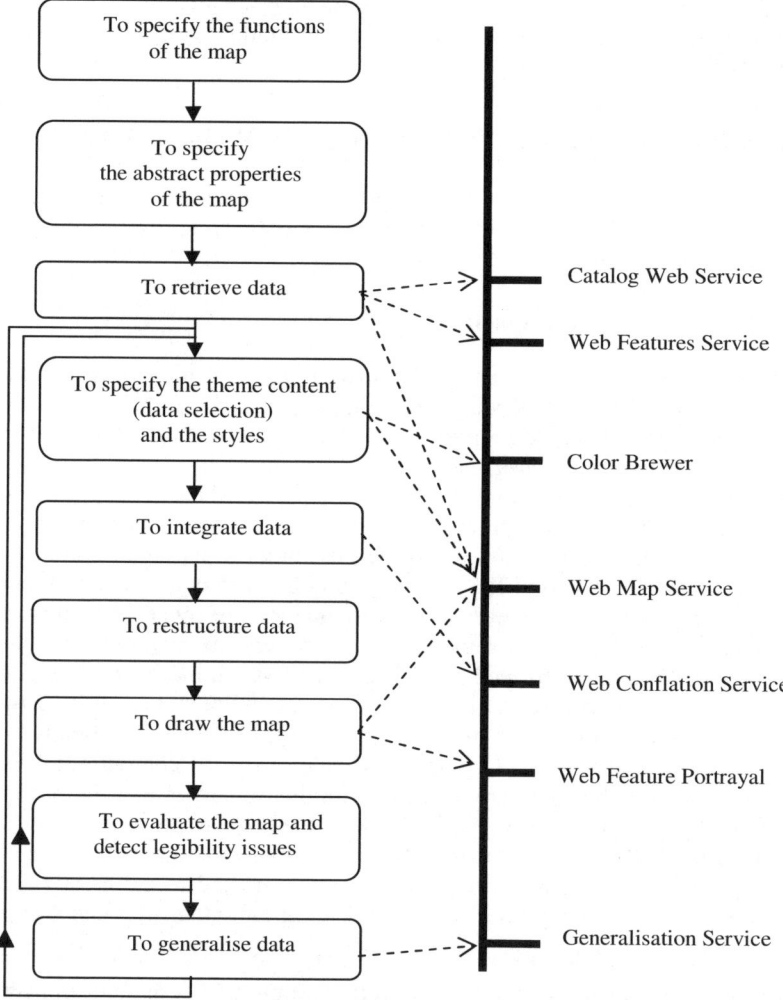

Fig. 7. The main capacities required for on-demand mapping (on the left) and the available services (or services specifications) to support them (on the right)

First capacity is to specify map functions, like to warn from a specific danger or to support navigation. These may be very generic expressions like an inventory map, a topographic map, a communication map or an analysis map. There is no service, in today interoperable SOA, to support this capacity.

The next capacity is to specify the map abstract properties. The abstract properties are for example the types of information to be displayed on the map, which are sometimes called 'themes' and the relationships between these themes. This step can follow the specification of the map function. For instance, an inventory map is a map to read and a travel planning map should contain roads, airport, public transport, hotels. This step can also come first for instance if the map maker expresses his need as a map about a specific town on which green parks should be seen at first sight and pollution sources should be perceived only during a more careful reading. This capacity is not supported as such in today interoperable SOA. But the specification of information belonging to a theme may rely on ontologies involved in cataloguing services.

The next step is the specification of the remaining legend items to produce a map. These are the concrete variables of the map. It is decomposed into several capacities.

The first capacity is to retrieve data that will provide the required information. This is supported by existing catalogue services specifications (WCS).

Another capacity is to organise retrieved data for portrayal and possibly refining some abstract properties like the type of relationships between elements within a theme. This may require integrating data acquired from two different sources and transforming the data to derive the classes of objects that will be portrayed with the same style.

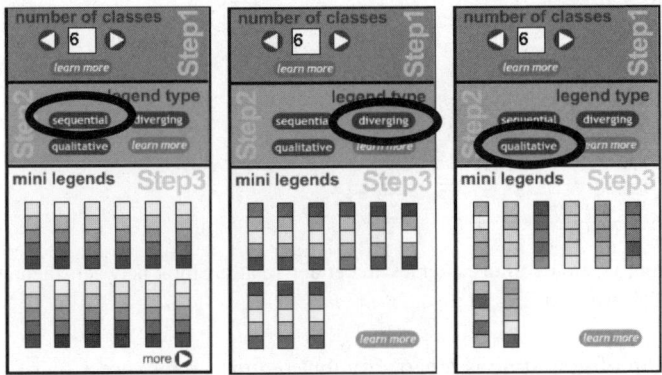

Fig. 8. Color schemes generated by the ColorBrewer to render different kind of relationships between classes of features: sequential relationships (e.g.: classes of density of inhabitants), diverging relationships (e.g.: deserted to overpopulated densities), qualitative (ex: the dominant religion). Source of snapshots: http://www.personal.psu.edu/cab38/ColorBrewer/ColorBrewer.html

Another capacity is to define styles to portray data. As explained in the introduction, the user may rely on existing predefined styles. If he needs to create new ones, it is advised to specify abstract properties of his map. These abstract properties will stand as constraints on the definition of concrete variables. Most cartographic rules do not relate an abstract property to a concrete variable but rather trace the opposite way. Examples of rules are 'colour, shape and orientation are adapted to render classification', 'size is adapted to render quantities', 'shape is not adapted to render quantities'. The ColorBrewer of [11] generates colour schemes according to the number of classes in user data and to the type of relationships between these classes. Fig. 8 shows three set of schemes obtained from the ColorBrewer with different queries. First set of schemes are adapted to portray 6 classes of features with 'sequential relationships'. Sequential relationships are mainly rendered by a variation of colour value. It is not a service so far but its implementation into a geotool library is a first step towards that.

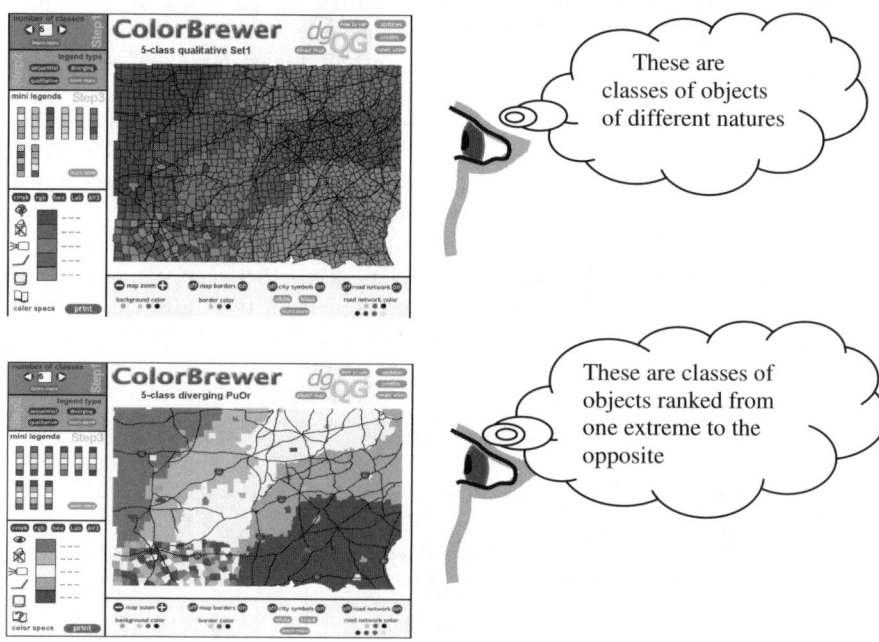

Fig. 9. Selecting accurate color schemes to render relationships between thematic categories based on the ColorBrewer

The creation of new style is a capacity that requires the expression of creativeness. Facilitating user creativeness necessitates understanding his vocabulary, his taste, his creation strategies and adapting the map design process to them. Current man-machine interaction techniques exist to share knowledge that is difficult to formalise (such as taste or creation strategy). These are man-machine dialogue techniques. A PhD work is on-going within COGIT laboratory to support on-demand map design through dialogue techniques with an accent on facilitating the expression of the user taste and creativeness [12][13].

Once data and styles are defined, the map must be drawn. This capacity is to be supported by the OGC Web Feature Portrayal Services (WFP) specifications, provided the drawing process is specified in a MapContext element.

Next step is evaluating the map. The set of graphical signs that will appear on the map and their interactions cannot be completely anticipated based on styles only. For instance, if there is a large portion of sea on the map the colour used to fill it will appear more intense than if there is a smaller portion. This step is not supported in today's architectures.

Last, the author of the map may need to generalise his data to improve the legibility of his map. Label placement is also tuned after all other features have been drawn. Generalisation web services are being developed to provide this capacity [14][15].

3.2 Web Legend Services

To support capacities described in the preceding section that are not supported by existing services, we propose three services corresponding to an increasing specification of a map. Importantly, we focus on services that may be stateless. A

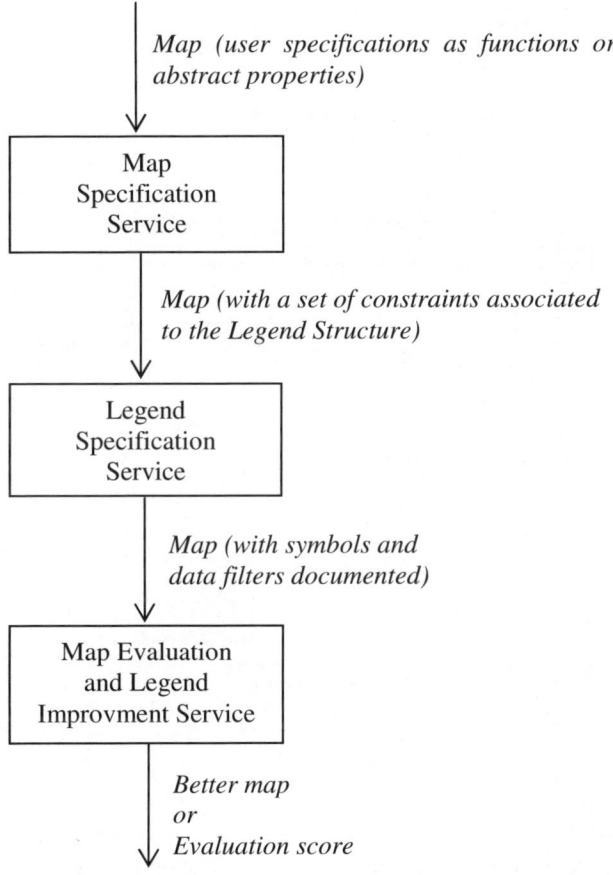

Map (user specifications as functions or abstract properties)

Map
Specification
Service

Map (with a set of constraints associated to the Legend Structure)

Legend
Specification
Service

Map (with symbols and data filters documented)

Map Evaluation
and Legend
Improvment Service

*Better map
or
Evaluation score*

Fig. 10. Global suite of services dedicated to embed cartographic expertness in legend design

stateless service is a service that does not keep session information, which makes it easier to optimise the service performances. These services are summarized in f. Inputs and outputs are of one same datatype: a *Map*. This concept of *Map* entails the classical items of a map (features and styles) but plus some metadata. User specifications are gathered in this *Map* object, as well as the cartographic solution under construction. Fig. 11 illustrates the main lines of our *Map* conceptual model. Whereas OGC *MapContext* is dedicated to describe a map drawing process, the *Map* is dedicated to support iterative on-demand map specification, definition, drawing, evaluation and improving.

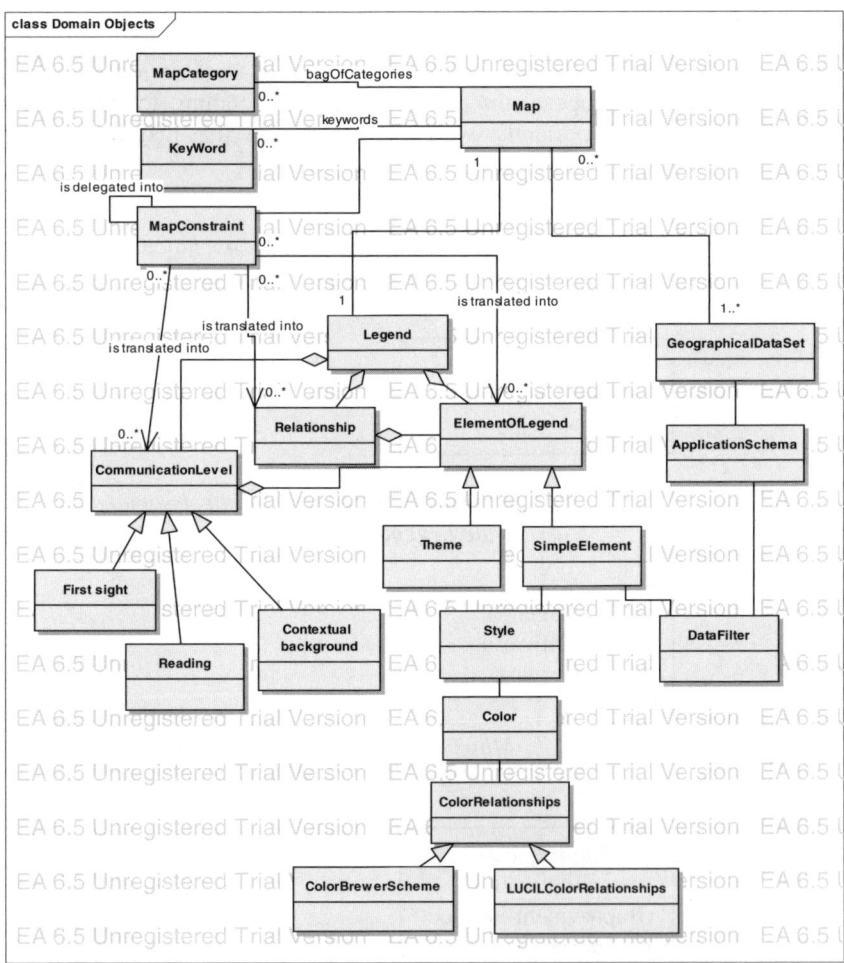

Fig. 11. Main lines of our conceptual model to represent a map and its objectives. The grey area delimits concepts shared with OGC MapContext. The most important elements are the legend structural elements that are used to translate and organise the map constraints: Relationships, CommunicationLevels and Elements.

3.2.1 Map Specification Service

The first item of our proposal is the Map Specification Service. This service assists the user in defining relevant abstract properties of his map, independently from available data and from styles (capacities described in section 3.1.1 and 3.1.2). The user may use keywords and categories to express his need for a map. A *Map* may have several *MapCategory*, for instance it may be altogether a 'map to read' and 'an inventory map' and a 'biological content map'. It may also have several keywords like 'prestigious' or 'children'.

This service takes a *Map* –after the model illustrated Fig. 10- and returns a *Map*. It parses *MapCategory* and *Keyword* elements in the input Map and translates them into a more tractable form: a set of *MapConstraints*. A *MapConstraint* is a constraint that can be associated to legend structural items. There are several types of legend structural items: elements (a data filter associated with a drawing style), including themes, relationships between elements, communication levels. For instance, if the user specifies that he wants to draw a 'map to see', the service will generate several constraints related to this category. One constraint will be the existence of a CommunicationLevel 'first sight' that groups all elements to be seen at first sight.

Locating map constraints on structural items of the legend is a key aspect of this proposal. It serves several objectives:

- Constraints are expressed with the same vocabulary.
- Structural items are organised into a hierarchy that will be used to prioritise constraints.
- Constraints are indirectly related to graphical signs on the map. This will be exploited during map evaluation. Graphical signs and their interactions will be analysed and interpreted in terms of 'de facto structure of the map' that can be compared to the 'specified structure of the map'.

3.2.2 Legend Specification Service

The second item of our proposal is the Legend Specification Service. This service assists the user in defining a complete legend to draw a map according to the user objectives. This service also takes a *Map* as an input and returns a *Map*. In the return *Map*, all information necessary to draw the map is specified: data queries and styles. A crucial aspect in this service is compromising between the constraints of the input *Map*. Indeed, as explained in section 2.1, cartographic rules need to be interpreted in the context of the current map and it is sometimes necessary to compromise between them. As mentioned in the section before, our approach is to associate constraints with a structural item and to try and satisfy constraints from the higher structural item to the lower. This approach is possible as long as structural items are ordered in a hierarchy.

Each structural item is related to user constraints and to cartographic constraints. Examples of user constraints on a CommunicationLevel 'first sight' are: it should exist, it should include the road theme. A structural item may also be associated to cartographic constraints. Examples of cartographical constraints associated to a CommunicationLevel 'first sight' are: the set of graphical signs belonging to this level should be limited, the level of detail of this level should be coarse. An example of cartographic constraints associated to an association relationship between elements of legend is: 'styles should include symbols with similar hues'.

Some of these cartographic constraints can only be interpreted during map evaluation, i.e. once styles have been defined and applied to geographical data. Others can be used to guide the definition of styles. Practically, some constraints can be interpreted based on relationships between colours implemented in the Geotools ColorBrewer library and in the Geoxygene LUCIL library (presented in the next section).

The *Map* object, once specified, may be automatically translated into a ISO/OGC *MapContext* and this object can be sent to a Web Feature Portrayal Service. So far, we use the geotools library to interpret a MapContext object into a gif image or a SVG document.

3.2.3 Map Evaluation and Legend Improvement Service

The last item of this proposal is a Map Evaluation Service. It assists the user in evaluating the adequacy between the map graphics and his objectives. The input of this service is a *Map* object and the data. The service checks if relationships expressed

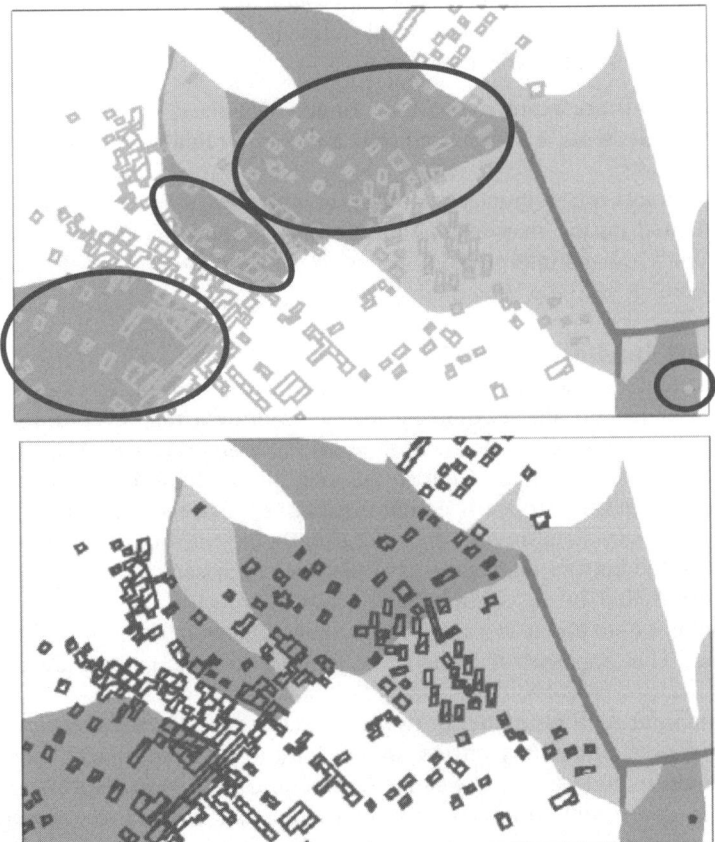

Fig. 12. A result of the LUCIL application that automatically evaluates and improves colour contrasts in a map

in the legend structure are conveyed by the graphics obtained by applying the style to the data. A Map Evaluation and Legend Improvement Service is already under prototyping at our laboratory. It is built upon [16] work to automatically evaluate and improve colour contrasts in risk maps.

The initial application, embedded within a proprietary GIS, is being translated into an application within the GEOXYGENE platform, an open-source java platform implementing ISO/OGC standards developed within COGIT. This includes a palette of colours for which contrast relationships have been calculated based on theoretical measures and on user tests. Fig. 12 presents a result of this application. The application evaluates the initial map (above map) and detects low contrasts occurring when houses portrayed in grey are located on a dark green area. These low contrasts are associated to objects on the map. The application aggregates them to obtain a contrast score for the corresponding elements of legend. It selects the element of legend with the worst score, here the buildings, and propose a new colour. The proposal of the new colour is constrained by the relationships of this element. For instance, if this element (i.e. the set of buildings drawn on the map) is to be associated with other elements (for instance a set of churches), then it is important to keep a similar hue.

This application is a stand alone application. To build a Map Evaluation and Legend Improvement Service based on this application, we have concentrated on deploying two methods of this application. One method calculates contrasts score for two given themes on a map. The other method detects contrast inconsistencies on a map and proposes to change a colour in the legend to improve contrast. We are currently analyzing the preconditions of these methods and developing factories to yield their required input structure from our *Map* objects.

4 Conclusion

This paper has analysed the limits of on-demand map design in today's Web mapping services oriented architecture. It presents a proposal to integrate relevant cartographic rules during some important steps of on-demand map design process. The stake is to produce on-demand maps that convey the intended message, thanks to an expert usage of graphical variables, and that do not fail to convey this message, or convey wrong messages, due to unfortunate interactions between graphical signs on the map.

Our approach integrates into the dominant paradigm of services oriented architecture by consisting in a suite of three stateless services. When developing modular services it is important to provide sharable interfaces for these services, which means at least to use standard types for inputs and output. Doing this ensures that they will be easier to chain with other services that use similar standard types. Yet, there is no standard model to describe user specification of a map (or user need for a map). Our proposal includes a formal model to represent in a tractable way user specifications of his map. User specifications are expressed as constraints associated to explicit structural items of the legend: relationships, themes and properties.

Current works consist in implementing these services.

Acknowledgments

The authors wish to thank Sébastien Mustière, Guillaume Touya and Julien Gaffuri from the COGIT laboratory for their help during the writing of this paper and for fruitful discussions on this topic.

References

1. Tufte, E.: Beautiful Evidence. Graphic Press (2006)
2. McMaster, R.B., Shea, S.: Generalization in digital cartography, p. 18. Association of American Geographers publications, Washington D.C. (1992)
3. Bertin, J.: Semioloy of Graphics: Diagrams, Networks, Maps. University of Wisconsin Press, Madison (1983)
4. MacEachren, A.-M.: How maps work. Guilford Publications, New York (1995)
5. Andrienko, N., Andrienko, G.: Intelligent Visualisation and Information Presentation for Civil Crisis. In: Suares, J., Markus, B. (eds.) AGILE 2006, 9th AGILE Conference on Geographical Information Science, Proceedings, Visegrad, Hungary, pp. 291–298 (2006)
6. Sonnet, J. (ed.).: Open Geospatial Consortium (OGC): Web Map Context Documents. OGCTM Implementation Specification (2005)
7. Monmonier, M.: How to lie with maps. University of Chicago Press, Chicago (1991)
8. Kraak, M.J., Brown, J. (eds.): Web Cartography. Developments and Prospects. Taylor & Francis, New York (2001)
9. Fabrikant, S.I., Montello, D.R., Rebich, S.: The Look of Weather Maps. In: Raubal, M., Miller, H.J., Frank, A.U., Goodchild, M.F. (eds.) GIScience 2006. LNCS, vol. 4197, pp. 269–271. Springer, Heidelberg (2006)
10. Mitchell, A.: The ESRI Guide to GIS Analysis, vol. 1, Geographic patterns & relationships. USA (1999)
11. Brewer, C., Hatchard, G., Harrower, M.: ColorBrewer in print: a catalog of color schemes for maps. Cartography and Geographic Information Science 30(1), 5–32 (2003)
12. Domingues, C., Bucher, B.: Legend design based on map samples. In: Raubal, M., Miller, H.J., Frank, A.U., Goodchild, M.F. (eds.) GIScience 2006. LNCS, vol. 4197, Springer, Heidelberg (2006)
13. Christophe, S., Bucher, B., Ruas, A.: A dialogue application for creative portrayal. In: The XXIIIe International Cartographic Conference, ICC, Moscow (to be published)
14. Chesneau, E., Ruas, A., Bonin, O.: Colour Contrasts Analysis for a better Legibility of Graphic Signs on Risk Maps. In: International Cartography Conference ICC 2005, Proceedings, La Coruna, Spain (2005)
15. Harrower, M., Bloch, M.: MapShaper.org: A Map Generalization Web Service. IEEE Computer Graphics and Applications 26(4), 22–27 (2006)
16. Burghardt, D., Neun, M., Weibel, R.: Generalization Services on the Web - A Classification and an Initial Prototype Implementation. In: CaGIS, vol. 32(4) (2005)

Supporting Range Queries on Web Data Using *k*-Nearest Neighbor Search

Wan D. Bae[1], Shayma Alkobaisi[1], Seon Ho Kim[1], Sada Narayanappa[1],
and Cyrus Shahabi[2],[*]

[1] Department of Computer Science, University of Denver, USA
{wbae,salkobai,seonkim,snarayan}@cs.du.edu
[2] Department of Computer Science, University of Southern California, USA
shahabi@usc.edu

Abstract. A large volume of geospatial data is available on the web through various forms of applications. However, access to these data is limited by certain types of queries due to restrictive web interfaces. A typical scenario is the existence of numerous business web sites that provide the address of their branch locations through a limited "nearest location" web interface. For example, a chain restaurant's web site such as McDonalds can be queried to find some of the closest locations of its branches to the user's home address. However, even though the site has the location data of all restaurants in, for example, the state of California, the provided web interface makes it very difficult to retrieve this data set. We conceptualize this problem as a more general problem of running spatial range queries by utilizing only *k*-Nearest Neighbor (*k*-NN) queries. Subsequently, we propose two algorithms to cover the rectangular spatial range query by minimizing the number of *k*-NN queries as possible. Finally, we evaluate the efficiency of our algorithms through empirical experiments.

1 Introduction

Due to the recent advances in geospatial data acquisition and the emergence of diverse web applications, a large amount of geospatial data have become available on the web. For example, numerous businesses release the locations of their branches on the web. The web sites of government organizations such as the US Postal Office provide the list of their offices close to one's residence. Various non-profit organizations also publicly post a large amount of geospatial data for different purposes.

---

[*] The author's work is supported in part by the National Science Foundation under award numbers IIS-0324955 (ITR), EEC-9529152 (IMSC ERC) and IIS-0238560 (PECASE) and in part by unrestricted cash gifts from Microsoft and Google. Any opinions, findings, and conclusions or recommendations expressed in this material are those of the authors and do not necessarily reflect the views of the National Science Foundation.

J.M. Ware and G.E. Taylor (Eds.): W2GIS 2007, LNCS 4857, pp. 61–75, 2007.

Unfortunately, access to these abundant and useful geospatial data sets is only possible through their corresponding web interfaces. These interfaces are usually designed for one specific query type and hence cannot support a more general access to the data. For example, the McDonalds web site provides a restaurant locator service through which one can ask for the five closest locations from a given zip code. This type of web interface to search for a number of "nearest locations" from a given geographical point (e.g., a mailing address) or an area (e.g., a zip code) is very popular for accessing geospatial data on the web. It nicely serves the intended goal of quick and convenient dissemination of business location information to potential customers. However, if the consumer of the data is a computer program, as in the case of web data integration utilities (e.g., wrappers) and search programs (e.g., crawlers), such an interface may be a very inefficient way of accessing the data. For example, suppose a web crawler wants to access the McDonalds web-site to retrieve all the restaurants in the state of California. Even though the site has the required information, the interface only allows the retrieval of five locations at a time. Even worse, the interface needs the center of the search as input. Hence, the crawler needs to identify a set of *nearest location* searches that both covers the entire state of California (for completeness) and has minimum overlap between the result sets (for efficiency).

In this paper, we conceptualize the aforementioned problem into a more general problem of supporting spatial range queries using k-Nearest Neighbor (k-NN) search. Besides web data integration and search applications, the solution to this more general problem can be beneficial for other application domains such as sensor networks when sensors have a limited number of channels to communicate with their nearest neighbors. Assuming that we have no knowledge about the distribution of the data sets and no control over the value of k, we propose two algorithms: the Quad Drill Down (QDD) and Dynamic Constrained Delaunay Triangulation ($DCDT$) algorithms. The efficiencies of the algorithms are empirically compared.

The remainder of this paper is organized as follows: In Section 2, the problem is formally defined. Our proposed algorithms are discussed in Section 3. Next, the algorithms are experimentally evaluated in Section 4. The related work is presented in Section 5. Finally, Section 6 concludes the paper.

2   Problem Definition

A range query in spatial database applications returns all objects inside a query region R. A k-Nearest Neighbor (k-NN) query returns the k closest objects to a query point q. More formal definitions for the range query and the k-NN query can be found in [1,2].

Our focus is on finding all objects within a given rectangular query region R using a minimum number of k-NN searches as supported in various web applications. The complexity of this problem is unknown. Therefore only approximation results are provided. In reality, there can be more than one web site involved in the given range query. Then, the general problem is to find all objects within the query region R using k-NN searches on each of these web sources and to integrate the results from the sources into one final query result. To make the problem more general, no assumptions are made about the data set, i.e., neither the number of objects in the data set nor the data distribution is known.

Figure 1 (a) and (b) illustrate the challenges of this general problem through examples. Suppose that we want to find the locations of all the points in a given region (rectangle) and the only available interface is a k-NN search with a fixed k, $k = 3$ in (a) and $k = 6$ in (b). Hence, the only parameter that we can vary is the query points for the k-NN searches. Given a query point q, the result of the k-NN search is a set of k nearest points to q. It defines a circle centered at q with radius equal to the distance from q to its kth nearest neighbor (i.e., the 3rd closest point in (a)). The area of this circle, covering a part of the rectangle, determines the covered region of the rectangle. To *complete* the range query, we need to identify a set of input locations corresponding to a series of k-NN searches that would result in a set of circles covering the entire rectangle.

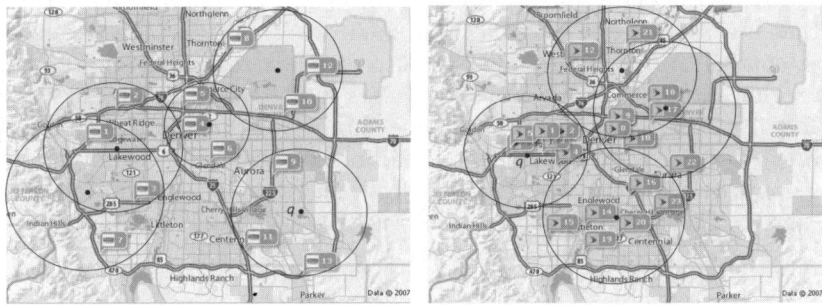

(a) A range query using 3-NN search (b) A range query using 6-NN search

Fig. 1. Range queries on the web data

The optimization objective of this problem is to perform as few k-NN searches as possible. This is because each k-NN search results in communication overhead between the client and the server in addition to the actual execution cost of the k-NN search at the server. Hence, the circles should have as few overlaps as possible. In addition, the fact that we do not know the radius of each circle prior to the actual evaluation of its corresponding k-NN query makes the problem even harder. Hence, the *sequence* of our k-NN searches become important as well. To summarize, the optimal solution should find a *minimum set* of query locations that minimizes the number of k-NN searches to completely cover the given query region. In many web applications, the client has no knowledge about the data set at the server and no control over the value of k (i.e., k is fixed depending on the web applications). These applications return a fixed number of k nearest objects to the user's query point. Considering a typical value of k in real applications, which ranges from 5 to 20 [3], multiple queries are evaluated to completely cover a reasonably large region.

Without loss of generality, and to simplify the discussion, we made the follow-ing assumptions: 1) a k-NN search is supported by a single web source, 2) the web source supports only one type of k-NN search while there can be multiple k-NN searches based on the value of k, 3) the parameter to the k-NN search is the query point q, 4) the value of k is fixed, i.e., the user has no control over k.

3   Range Queries Using k-NN Search

Our approach to the range query problem on the web is as follows: 1) divide the query region R into subregions, $R = \{R_1, R_2, ..., R_m\}$, such that any $R_i \in R$ can be covered by a single k-NN search, 2) obtain the resulting points from the k-NN search on each of the subregions, 3) combine the partial results to provide a complete result.

The main focus is how to divide R so that the total number of k-NN searches can be minimized. We consider the three following approaches: 1) a naive ap-proach: divide the entire query region R into equi-sized subregions (grid cells) such that any cell can be covered by a single k-NN search. 2) a recursive ap-proach: conduct a k-NN search in the current query region, divide the current query region into subregions with same sizes if the k-NN search fails to cover R, and call k-NN search for each of these subregions. Repeat the process until all subregions are covered. 3) a greedy and incremental approach: divide R into subregions with different sizes. Then select the largest subregion for the next k-NN search. Check the covered region by the k-NN search and select the next largest subregion for another k-NN search.

Table 1. Notations

Notation	Description
R	a query region (rectangle)
P_{ll}	the lower-left corner point of R
P_{ur}	the upper-right corner point of R
q	the query point for a k-NN search
R_q	half of the diagonal of R; the distance from q to P_{ll}
C_{Rq}	circle inscribing R; the circle of radius R_q centered at q
P_k	the set of k nearest neighbors obtained by a k-NN search ordered ascendingly by distance from q
r	the distance from q to the farthest point p_k in P_k
C_r	k-NN circle; the circle of radius r centered at q
C_r'	tighter bound k-NN circle; the circle of radius $\epsilon \cdot r$ centered at q

The notations in Table 1 are used throughout this paper and Figure 2 illus-trates our approach to evaluate a range query using k-NN searches. If $r > R_q$, then the k-NN search must have returned all the points inside C_{R_q}. Therefore, we can obtain all the points inside the region R after pruning out any points of

P_k that are outside R but inside C_r. Then, the range query is complete. Otherwise, we need to perform some additional k-NN searches to completely cover R. For example, if we use 5-NN search for the data set shown in Figure 2, then the resulting C_r is smaller than C_{R_q}. For the rest of this section, we focus on the latter case that is a more realistic scenario. Note that it is not sufficient to have $r \geq R_q$ in order to retrieve all the points inside R when q has more than one k^{th} nearest neighbor since one of them will be randomly returned as q's k^{th}-NN. Hence, we use $r > R_q$ as a terminating condition in our algorithms.

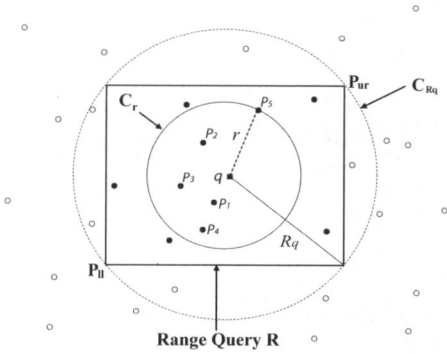

Fig. 2. A range query using 5-NN search

In a naive approach, R is divided into a grid where the size of cells is small enough to be covered by a single k-NN search. However, it might not be feasible to find the exact cell size with no knowledge of the data sets. Even in the case that we obtain such a grid, this approach is inefficient due to large amounts of overlapping areas among k-NN circles and wasted time to search empty cells (consider that most real data sets are not uniformly distributed). Thus, we provide a recursive approach and an incremental approach in the following subsections.

3.1   Quad Drill Down (QDD) Algorithm

We propose a recursive approach, the Quad Drill Down (QDD) algorithm, as a solution to the range query problem on the web using the properties of the quad-tree. A quad-tree is a tree whose nodes either are leaves or have four children, and it is one of the most commonly used data structures in spatial databases [11]. We adopt the partitioning idea of the quad-tree but we do not actually construct the tree. The main idea of QDD is to divide R into equal-sized quadrants, and then recursively divide quadrants further until each subregion is fully covered by a single k-NN search so that all objects in it are obtained.

Algorithm 1 describes QDD. First, a k-NN search is invoked with a point q which is the center of R. Next, the k-NN circle C_r is obtained from the k-NN result. If C_r is larger than C_{R_q}, it entirely covers R, then all objects in R are

Algorithm 1. QDDrangeQuery(P_{ll}, P_{ur})

1: $q \leftarrow$ getCenterOfRegion(P_{ll}, P_{ur})
2: $R_q \leftarrow$ getHalfDiagonal(q, P_{ll}, P_{ur})
3: $Knn\{\} \leftarrow$ add kNNSearch(q)
4: $r \leftarrow$ getDistToK^{th}NN($q, Knn\{\}$)
5: **if** $r > R_q$ **then**
6:      $result\{\} \leftarrow$ pruneResult($Knn\{\}, P_{ll}, P_{ur}$)
7: **else**
8:      clear $Knn\{\}$
9:      $P_0 \leftarrow (q.x, P_{ll}.y)$
10:     $P_1 \leftarrow (P_{ur}.x, q.y)$
11:     $P_2 \leftarrow (P_{ll}.x, q.y)$
12:     $P_3 \leftarrow (q.x, P_{ur}.y)$
13:     $Knn\{\} \leftarrow$ add QDDrangeQuery(P_{ll}, q) // q1
14:     $Knn\{\} \leftarrow$ add QDDrangeQuery(P_0, P_1) // q4
15:     $Knn\{\} \leftarrow$ add QDDrangeQuery(P_2, P_3) // q2
16:     $Knn\{\} \leftarrow$ add QDDrangeQuery(q, P_{ur}) // q3
17:     $result\{\} \leftarrow Knn\{\}$
18: **end if**
19: return $result\{\}$

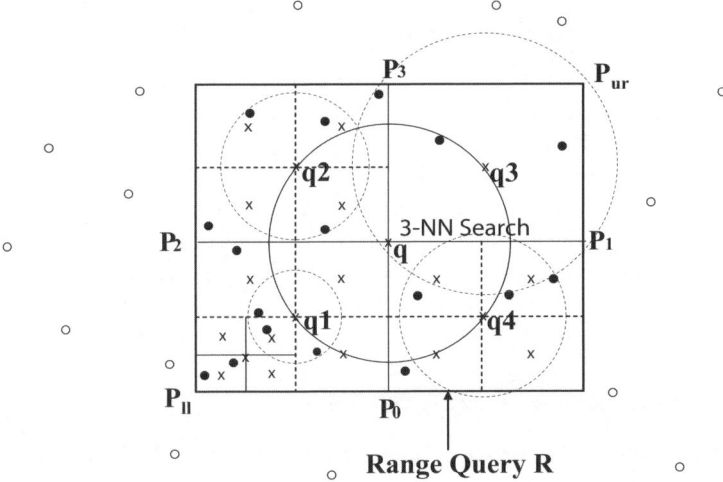

Fig. 3. A QDD Range Query

retrieved. Finally, a pruning step is necessary to retrieve only the objects that are inside R − a trivial case. However, if C_r is smaller than or equal to C_{R_q}, the query region is partitioned into four subregions by equally dividing the width and height of the region by two. The previous steps are repeated for every new subregion. The algorithm recursively partitions the query region into subregions until each subregion is covered by a k-NN circle. The pruning step eliminates those objects that are inside the k-NN circle but outside the subregion. An example of QDD is illustrated in Figure 3 (when $k=3$), where twenty one k-NN calls are required to retrieve all the points in R.

3.2 Dynamic Constrained Delaunay Triangulation ($DCDT$) Algorithm

In this section, we propose the Dynamic Constrained Delaunay Triangulation ($DCDT$) algorithm − a greedy and incremental approach to solve the range query on web data using the Constrained Delaunay Triangulation (CDT)[1]. $DCDT$ uses triangulations to divide the query range and keeps track of covered triangles by k-NN circles using the characteristics of CDT. $DCDT$ greedily selects the largest uncovered triangle for the next k-NN search while QDD follows a pre-defined order. To overcome redundant k-NN searches on the same area, $DCDT$ includes a propagation algorithm to cover the maximum possible area within a k-NN circle. Hence, no k-NN search will be wasted because a portion of a k-NN circle is always added to the covered area of the query range.

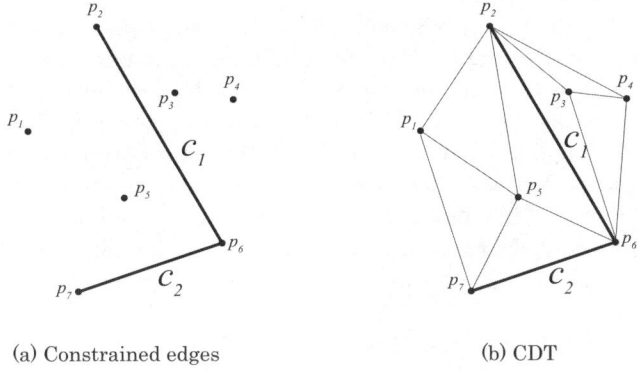

(a) Constrained edges   (b) CDT

Fig. 4. An example of Constrained Delaunay triangulation

Given a planar graph G with a set of vertices, P, a set of edges, E, and a subset of edges, C, in the plane, CDT of G is a triangulation of the vertices of G that includes C as part of the triangulation [12], i.e., all edges in C appear as edges of the resulting triangulation. These edges are referred to as constrained edges and they are not crossed (destroyed) by any other edges of the triangulation. Figure 4 illustrates a graph G and the corresponding CDT. Figure 4 (a) shows a set of vertices, $P = \{P_1, P_2, P_3, P_4, P_5, P_6, P_7\}$, and a set of constrained edges, $C = \{C_1, C_2\}$. Figure 4 (b) shows the result of CDT that includes the constrained edges of C_1 and C_2 as part of the triangulation.

---

[1] The terms, CDT and $DCDT$, are used for both the corresponding triangulation algorithms and the data structures that support these algorithms in this paper.

We define the following data structures to maintain the covered and uncovered regions of R:

- $pList$: a set of all vertices of G; initially $\{P_{ll}, P_{rl}, P_{lr}, P_{ur}\}$
- $cEdges$: a set of all constrained edges of G; initially empty
- $tList$: a set of all uncovered triangles of G; initially empty

$pList$ and $cEdges$ are updated based on the current covered region by a k-NN search. R is then triangulated (partitioned into subregions) using CDT with the current $pList$ and $cEdges$. For every new triangulation, we obtain a new $tList$. $DCDT$ keeps track of covered triangles (subregions) and uncovered triangles; covered triangles are bounded by constrained edges (i.e., all three edges are constrained) and uncovered triangles are kept in $tList$, which are sorted in descending order by the area of the triangles. $DCDT$ chooses the largest uncovered triangle in $tList$ and calls k-NN search using the centroid (which always lies inside the triangle as opposed to the center) of the triangle as the query point. Our algorithm uses a heuristic approach – to pick the largest triangle from the list of uncovered triangles. The algorithm terminates when no more uncovered triangles exist. For the rest of this section, we describe the details of $DCDT$ shown in Algorithm 2 and Algorithm 3.

$DCDT$ invokes the first k-NN search using the center point of the query region R as the query point q. Figure 5 (a) shows an example of a 3-NN search. If the resulting k-NN circle C_r completely covers R $(r > R_q)$, then we prune and return the result – a trivial case. If C_r does not completely cover R, $DCDT$ needs to use C_r' (a little smaller circle than C_r) for checking covered region from this point for further k-NN searches. We have previously discussed the possibility that the query point q has more than one k^{th} nearest neighbor and one of them is randomly returned. In that case, $DCDT$ may not be able to retrieve all the points in R if it uses C_r. Hence we define $C_r' = \epsilon \cdot C_r$, where $0 < \epsilon < 1$ ($\epsilon = 0.99$ in our algorithm).

$DCDT$ creates an n-gon inscribed into C_r'. Choosing the value of n depends on the tradeoff between computation time and the coverage of area; $n = 6$ (a hexagon) is used in our experiments as shown in Figure 5 (b). All vertices of the n-gon are added into $pList$. To mark the n-gon as a covered region, the n-gon is triangulated and the resulting edges are added as constrained edges into $cEdges$ ($getConstrainedEdges()$ line 15 of Algorithm 2). The algorithm constructs a new triangulation with the current $pList$ and $cEdges$, then a newly created $tList$ is returned ($constructDCDT()$ line 16 of Algorithm 2).

$DCDT$ selects the largest uncovered triangle from $tList$ for the next k-NN search. With the new C_r', $DCDT$ updates $pList$ and $cEdges$ and creates a new triangulation. For example, if an edge lies within C_r', then $DCDT$ adds it into $cEdges$; on the other hand, if an edge intersects C_r', then the partially covered edge is added into $cEdges$. Figure 6 shows an example of how to update and maintain $pList$, $cEdges$ and $tList$. $\triangle_1$ ($\triangle_{max}$) is the largest triangle in $tList$, therefore $\triangle_1$ is selected for the next k-NN search. $DCDT$ uses the centroid of $\triangle_1$ as the query point q for the k-NN search. C_r' partially covers $\triangle_1$: vertices v_5 and v_6 are inside C_r' but vertex v_3 is outside C_r'. C_r' intersects $\triangle_1$ at k_1 along $\overline{v_3 v_6}$ and

---

Algorithm 2. DCDTrangeQuery(P_{ll}, P_{ur})

1: $pList\{\} \leftarrow \{P_{ll}, P_{rl}, P_{lr}, P_{ur}\}$;
2: $cEdges\{\} \leftarrow \{\}$;
3: $tList\{\} \leftarrow \{\}$;
4: $Knn\{\} \leftarrow \{\}$; all objects retrieved by k-NN search
5: $q \leftarrow$ getCenterOfRegion(P_{ll}, P_{ur})
6: $Knn\{\} \leftarrow$ add kNNSearch(q)
7: $r \leftarrow$ getDistToK^{th}NN($q, Knn\{\}$)
8: $C_r \leftarrow$ circle centered at q with radius r
9: $R_q \leftarrow$ getHalfDiagonal(q, P_{ll}, P_{ur})
10: $\epsilon = 0.99$
11: **if** $r \leq R_q$; C_r does not cover the given region R **then**
12:     $C_r^{'} \leftarrow$ circle centered at q with radius $\epsilon \cdot r$
13:     $N_g \leftarrow$ create an n-gon inscribing $C_r^{'}$ with q as the center
14:     $pList\{\} \leftarrow$ add {all vertices of N_g}
15:     $cEdges\{\} \leftarrow$ getConstrainedEdges({all vertices of N_g})
16:     $tList\{\} \leftarrow$ constructDCDT($pList\{\}$, $cEdges\{\}$)
17:     **while** $tList$ is not empty **do**
18:         mark all triangles in $tList$ to unvisited
19:         $\triangle_{max} \leftarrow$ getMaxTriangle()
20:         $q \leftarrow$ getCentroid($\triangle_{max}$)
21:         $Knn\{\} \leftarrow$ add kNNSearch(q)
22:         $r \leftarrow$ getDistToK^{th}NN($q, Knn\{\}$)
23:         $C_r^{'} \leftarrow$ circle centered at q with radius $\epsilon \cdot r$
24:         checkCoverRegion($\triangle_{max}, C_r^{'}, pList\{\}, cEdges\{\}$)
25:         $tList\{\} \leftarrow$ constructDCDT($pList\{\}$, $cEdges\{\}$)
26:     **end while**
27: **end if**
28: $result\{\} \leftarrow$ pruneResult($Knn\{\}, P_{ll}, P_{ur}$)
29: return $result\{\}$

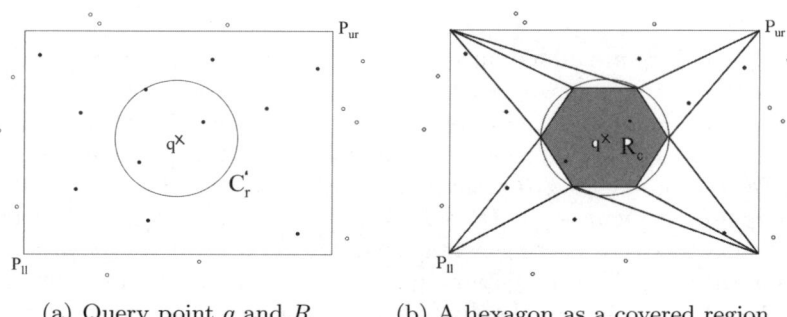

(a) Query point q and R (b) A hexagon as a covered region

Fig. 5. An example of first k-NN call in query region R

at k_2 along $\overline{v_3 v_5}$, hence, k_1 and k_2 are added into $pList$ (*getCoveredVertices()*
line 3 of Algorithm 3). Let R_c be the covered area (polygon) of $\triangle_1$ by $C_r^{'}$
(Figure 6 (c)). Then R_c is triangulated and the resulting edges, $\overline{k_1 k_2}$, $\overline{k_1 v_6}$,

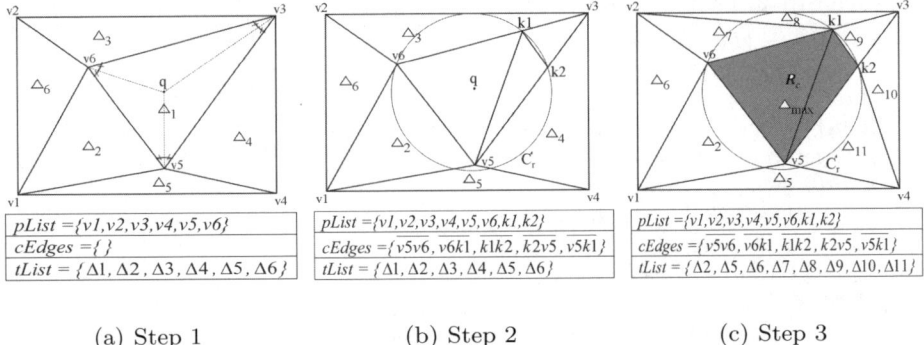

(a) Step 1   (b) Step 2   (c) Step 3

Fig. 6. Example of pList, cEdges and tList: covering $\triangle_{max}$

Algorithm 3. checkCoverRegion($\triangle$, C_r', $pList\{\}$, $cEdges\{\}$)

1: **if** $\triangle$ is marked unvisited **then**
2:     mark $\triangle$ as visited
3:     $localPList\{\} \leftarrow$ getCoveredVertices($\triangle, C_r'$)
4:     $localCEdges\{\} \leftarrow$ getConstrainedEdges($localPList$)
5:     **if** $localCEdges$ is not empty **then**
6:         $pList\{\} \leftarrow$ add $localPList\{\}$
7:         $cEdges\{\} \leftarrow$ add $localcEdges\{\}$
8:         $neighbors\{\} \leftarrow$ findNeighbors($\triangle$)
9:         **for** every $\triangle_i$ in neighbors$\{\}$, $i = 1, 2, 3$ **do**
10:             checkCoverRegion($\triangle_i, C_r', pList\{\}, cEdges\{\}$)
11:         **end for**
12:     **end if**
13: **end if**

$\overline{k_2 v_5}$, $\overline{v_5 v_6}$ and $\overline{k_1 v_5}$ are added into $cEdges$. The updated $pList$ and $cEdges$ are used for constructing a new triangulation of R (Figure 6 (c)).

As shown in Figure 6 (c), the covered region of $\triangle_1$ can be a lot smaller than the coverage area of C_r'. In order to maximize the covered region by a k-NN search, $DCDT$ propagates to (visits) the neighboring triangles of $\triangle_1$ (triangles that share an edge with $\triangle_1$). $DCDT$ marks the covered regions starting from $\triangle_{max}$, and recursively visiting neighboring triangles ($findNeighbors()$ line 8-11 of Algorithm 3). Figure 7 shows an example of the propagation process. In Figure 7 (a), $\triangle_{max}$ is completely covered and its neighboring triangles $\triangle_{11}$, $\triangle_{12}$ and $\triangle_{13}$ (1-hop neighbors of $\triangle_{max}$) are partially covered in Figure 7 (b), (c) and (d), respectively. In Figure 7 (e), the neighboring triangles of $\triangle_{max}$'s 1-hop neighbors, i.e., 2-hop neighbors of $\triangle_{max}$, are partially covered. Finally, based on the returned result of $checkCoverRegion()$ shown in Figure 7 (e), $DCDT$ constructs a new triangulation as shown in Figure 7 (f). Algorithm 3 describes the propagation process.

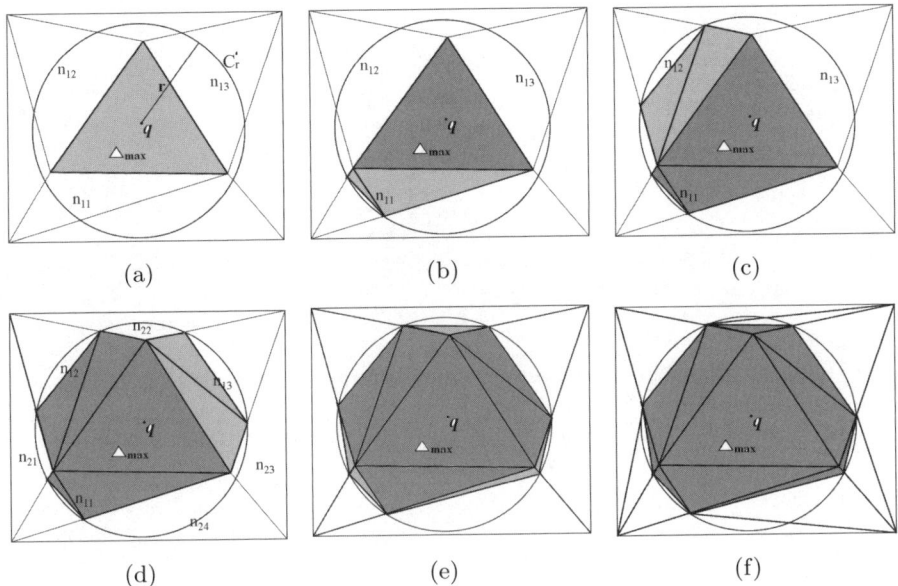

Fig. 7. Example of checkCoverRegion()

Notice that $DCDT$ covers a certain portion of R at every step, and therefore the covered region grows as the number of k-NN searches increases. On the contrary, QDD discards the returned result from a k-NN call when it does not cover the whole subregion, which results in re-visiting the subdivisions of the same region with 4 additional k-NN calls (see line 5-16 in Algorithm 1). Also note that the cost of triangulation is negligible as compared to k-NN search because triangulation is performed in memory and incrementally.

4   Experiments

In this section, we evaluate QDD and $DCDT$ using both synthetic and real GIS data sets. Our synthetic data sets use both uniform (random) and skewed (exponential) distributions. For the uniform data sets, (x, y) locations are distributed uniformly and independently between 0 and 1. The (x, y) locations for the skewed data sets are independently drawn from an exponential distribution with mean 0.3 and standard deviation 0.3. The number of points in each data set varies: 1K, 2K, 4K, and 8K points. Our real data set is from the U.S. Geological Survey in 2001: Geochemistry of consolidated sediments in Colorado in the US [13]. The data set contains 5,410 objects (points).

Our performance metric is the number of k-NN calls required for a given range query. In our experiments, we varied a range query size between 1% and 10% of the entire region of the data set. Different k values between 5 and 50 were used. Each size of range queries was conducted for 100 trials and the average values

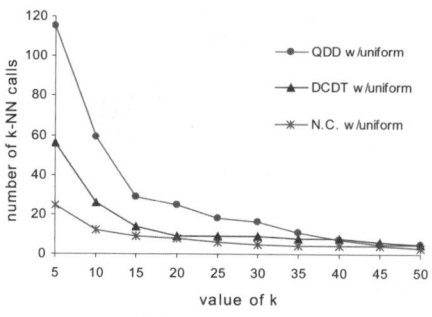 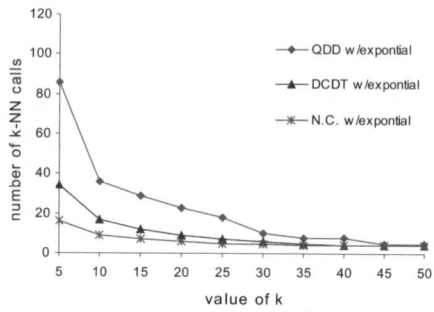

(a) Uniformly distributed data set   (b) Exponentially distributed data set

Fig. 8. Number of k-NN calls for 4K synthetic data set with 3% range query

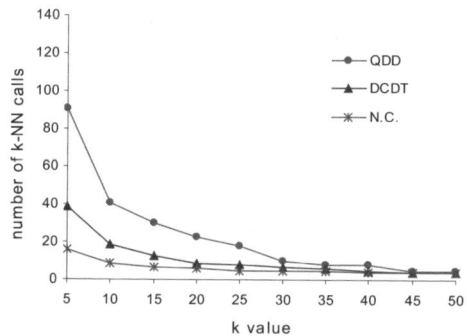

Fig. 9. Number of k-NN calls for real data set with 3% range query

were reported. We present only the most illustrative subset of our experimental results due to space limitation. Similar qualitative and quantitative trends were observed in all other experiments.

First, we compared the performance of QDD and $DCDT$ with the theoretical minimum, i.e., the necessary condition (N.C.) $\lceil \frac{n}{k} \rceil$, where n is the number of points in R. Figure 8 (a) and (b) show the comparisons of QDD, $DCDT$ and N.C. with uniformly distributed and exponentially distributed 4k synthetic data sets, respectively. For both $DCDT$ and QDD, the number of k-NN calls rapidly decreased as the value of k increased in the range 5-15, and 5-10, for the uniformly and exponentially distributed data, respectively. Then, it slowly decreased as the value of k became larger, approaching to those of N.C. when k is over 35. The results for the real data set have similar trends to those of the exponentially distributed synthetic data set (Figure 9). Note that k is determined not by the client but by the web interface and a typical k value in real applications ranges between 5 and 20 [3]. In both synthetic and real data sets, $DCDT$ needed a significantly smaller number of k-NN calls compared to QDD. For example, with

Table 2. The average percentage reduction in the number of k-NN calls (%)

data distribution	query size	$k=5$	10	15	20	25	30	35	40	45	50
uniform	3%	51.3	55.9	51.7	52.0	50.0	43.8	27.3	0	0	0
	5%	56.3	53.5	56.9	56.3	55.6	52.5	54.0	43.8	35.7	0
	10%	56.2	53.6	48.2	45.0	37.5	30.0	22.2	22.2	25.0	20.0
exponential	3%	60.5	58.8	58.6	56.9	51.1	40.0	37.5	37.5	20.0	20.0
	5%	68.4	67.2	68.8	65.7	60.7	58.3	50	41.7	25.0	20.0
	10%	69.4	59.7	56.5	56.7	55.8	34.4	46.9	46.4	35.0	31.3

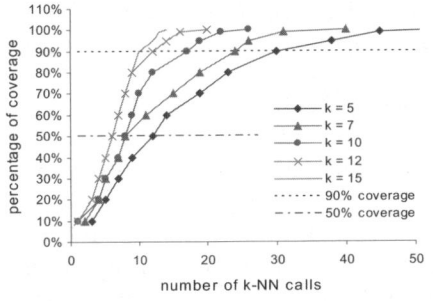
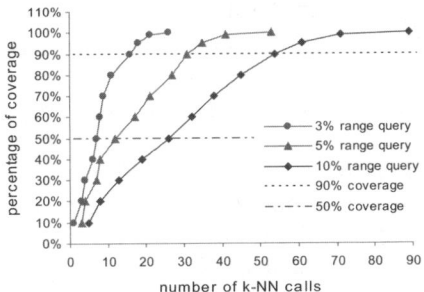

(a) 3% range query with 4K synthetic data set   (b) 4K synthetic data set with $k=10$

Fig. 10. Percentage of coverage

the exponential data set (4K), $DCDT$ showed 55.3% of average reduction in the required number of k-NN searches compared to QDD. As discussed in Section 3, $DCDT$ covers a portion of the query region with every k-NN search while QDD revisits the same subregion when a k-NN search fails to cover the entire subregion. $DCDT$ still required more k-NN calls than N.C. On the average, $DCDT$ required 1.67 times more k-NN calls than N.C. However, the gap became very small when k was greater than 15.

Next, we conducted range queries with different sizes of query region: 3%, 5%, and 10% of the entire region of data set. Table 2 shows the average percentage reduction in the number of k-NN calls between $DCDT$ and QDD for 4K uniformly and exponentially distributed synthetic data sets. On the average, $DCDT$ resulted in 50.4% and 58.2% of reduction over QDD for the uniformly and exponentially distributed data set, respectively. The results show that the smaller the value of k is, the greater the reduction rate.

In Figure 10, we plotted the average coverage rate versus the number of k-NN calls required for $DCDT$ on 4K synthetic uniformly distributed data set. Figure 10 (a) shows the average coverage rate of the $DCDT$ algorithm while varying the value of k. For example, when $k=7$, 50% of the query range can be covered by first 8 k-NN calls while the entire range is covered by 40 k-NN calls. The same range query required 24 calls for 90% of coverage and this number is approximately 60% of the total required number of k-NN searches. Figure 10 (b) shows the average coverage rate of $DCDT$ while varying query size. For 5% range query, 12 and 31 k-NN searches were required for 50% and 90% coverage, respectively.

5   Related Work

Some studies have discussed the problems of data integration and query processing on the web [4,5]. A query processing framework for web-based data integration was presented in [4]. It also implemented the evaluation of query planning and statistics gathering modules. The problem of querying web sites through limited query interfaces has been studied in the context of information mediation systems in [5].

Several studies have focused on performing spatial queries on the web [6,3,7]. In [6], the authors demonstrated an information integration application that allows users to retrieve information about theaters and restaurants from various U.S. cities, including an interactive map. Their system showed how to build applications rapidly from existing web data sources and integration tools. The problem of spatial coverage using only the k-NN interface was introduced in [3]. The paper provided a quad-tree based approach. A quad-tree data structure was used to check the complete coverage. Our work defines a more general problem of range queries on the web and provides two different solutions with detailed experiments and comparisons. The problem of supporting k-NN query using range queries was studied in [7]. The idea of successively increasing query range to find k points was described assuming statistical knowledge of the data set. This is the reverse of the problem we are focusing on.

The Delaunay triangulation and Voronoi diagram based approaches have been studied for various purposes in wireless sensor networks [8,9,10]. In [9], the authors used the Vornonoi diagram to discover the existence of coverage holes, assuming that each sensor knows the location of its neighbors. The authors in [10] proposed a wireless sensor network deployment method based on the Delaunay triangulation. This method was applied for planning the positions of sensors in an environment with obstacles. It retrieved the location information of obstacles and pre-deployed sensors, then constructed a Delaunay triangulation to find candidate positions of new sensors. CDT was dynamically updated while covering the query region with k-NN circles.

6   Conclusions

In this paper, we introduced the problem of evaluating spatial range queries on the web data by using only k-Nearest Neighbor searches. The problem of finding a minimum number of k-NN searches to cover the query range appears to be hard even if the locations of the objects (points) are known.

The Quad Drill Down (QDD) and Dynamic Constrained Delaunay Triangulation ($DCDT$) algorithms were proposed to achieve the completeness and efficiency requirements of spatial range queries. For both QDD and $DCDT$, we showed that they can cover the entire range even with the most restricted environment where the value of k is fixed and on which users have no control. The efficiencies of these two algorithms were compared each other as well as with the necessary condition. $DCDT$ provides a better performance than QDD.

We plan to extend our approaches for the cases in the presence of known data distribution and more flexible k-NN interfaces, for example, when users can change the value of k.

Acknowledgments

The following people provided helpful suggestions during this work: Dr. Petr Vojtěchovský and Brandon Haenlein.

References

1. Gaede, V., Günther, O.: Multidimensional access methods. ACM Computing Surveys 30(2), 170–231 (1998)
2. Song, Z., Roussopoulos, N.: k-Nearest Neighbor search for moving query point. In: Jensen, C.S., Schneider, M., Seeger, B., Tsotras, V.J. (eds.) SSTD 2001. LNCS, vol. 2121, pp. 79–96. Springer, Heidelberg (2001)
3. Byers, S., Freire, J., Silva, C.: Efficient acquisition of web data through restricted query interface. In: Poster Prceedings of the World Wide Web Conference, pp. 184–185 (2001)
4. Nie, Z., Kambhampati, S., Nambiar, U.: Effectively mining and using coverage and overlap statistics for data integration. IEEE Transactions on Knowledge and Data Engineering 17(5), 638–651 (2005)
5. Yerneni, R., Li, C., Garcia-Molina, H., Ullman, J.: Computing capabilities of mediators. In: Proceedings of SIGMOD, pp. 443–454 (1999)
6. Barish, G., Chen, Y., Dipasquo, D., Knoblock, C.A., Minton, S., Muslea, I., Shahabi, C.: Theaterloc: Using information integration technology to rapidly build virtual application. In: Proceedings of ICDE, pp. 681–682 (2000)
7. Liu, D., Lim, E., Ng, W.: Efficient k Nearest Neighbor queries on remote spatial databases using range estimation. In: Proceedings of SSDBM, pp. 121–130 (2002)
8. Sharifzadeh, M., Shahabi, C.: Utilizing Voronoi cells of location data streams for accurate computation of aggregate functions in sensor networks. GeoInformatica 10(1), 9–36 (2006)
9. Wang, G., Cao, G., Porta, T.L.: Movement-assisted sensor deployment. In: Proceedings of IEEE INFOCOM, pp. 2469–2479 (2003)
10. Wu, C., Lee, K., Chung, Y.: A Delaunay triangulation based method for wireless sensor network deployment. In: Proceedings of ICPADS, pp. 253–260 (2006)
11. Samet, H.: Data structures for Quadtree approximation and compression. Communications of the ACM 28(9), 973–993 (1985)
12. Chew, L.P.: Constrained Delaunay triangulations. Algorithmica 4(1), 97–108 (1989)
13. USGS mineral resources on-line spatial data (2001), http://tin.er.usgs.gov/

XFormsGI –
Extending XForms for Geospatial and Sensor Data

Jürgen Weitkämper and Thomas Brinkhoff

Institute for Applied Photogrammetry and Geoinformatics (IAPG)
FH Oldenburg/Ostfriesland/Wilhelmshaven (University of Applied Sciences)
Ofener Str. 16/19, 26121 Oldenburg, Germany
{juergen.weitkaemper,thomas.brinkhoff}@fh-oow.de

Abstract. User interaction is mostly based on forms. However, forms are restricted to alphanumerical data – the editing of geospatial data is not supported. This statement does not only hold for established form standards like HTML forms, but also for the new W3C standard XForms. XForms defines a declarative framework for using forms and processing form data. It can be used with a variety of XML host languages. In this paper, we present an extension to XForms (called XFormsGI) that provides control elements for entering new geometries, for modifying the properties of existing objects as well as for selecting geometric entities. As XForms, XFormsGI is declarative, thus avoiding script extensions. XFormsGI supports coordinate systems. Additionally, it can be used for capturing and displaying sensor data like GPS or for the measurements of distances or temperatures. A prototypical implementation has been integrated into a mobile SVG client running on a PDA.

Keywords: XForms, SVG, Geographic Information Systems (GIS), Location-Based Services (LBS), User Interaction, Sensors.

1 Introduction

Nearly all software applications require human interaction for editing data or for displaying the results of a computation. The most important approach for such user interaction is based on *forms*. Forms are applied for depicting alphanumerical data as well as for requesting input from the user. They are not only used by desktop applications, but also by web and mobile applications. For web applications, *HTML forms* [10] are typically utilized.

As mentioned before, forms only offer display and input capabilities for alphanumerical data. In case of Geographic Information Systems (GIS) and other GI applications, however, it must be possible to display and capture geospatial data by graphical user interaction. Especially for web and mobile applications, it would be desirable to be able to describe the geospatial input facilities of the application declaratively in analogy to forms. Without such means, additional scripts and/or

J.M. Ware and G.E. Taylor (Eds.): W2GIS 2007, LNCS 4857, pp. 76–93, 2007.

applets must be used for implementing the presentation and input of geospatial data. The SPARK framework [3] is an example for such an approach that leads to a conceptional break between the handling of alphanumeric data and of geospatial data.

The new W3C standard *XForms* [14] allows describing the user manipulation of alphanumeric data in the context of XML documents. Like XHTML [16] should take the place of HTML, XForms is supposed to replace HTML forms. But in addition, it is able to be integrated into other XML standards. In case of graphical and geospatial XML documents, *Scalable Vector Graphics (SVG)* [12] and the *Geography Markup Language (GML)* [8] could be host languages for XForms. Like HTML forms, XForms does not support interaction with geospatial data. But XForms is suitable to be extended for geospatial data. In contrast to HTML forms, XForms separates the visual representation of the user interface from the underlying data model. A single device-independent XML form definition has the capability to work with a variety of standard or proprietary user interfaces.

In this paper, we propose an extension that is conform to XForms. It allows for editing geometric entities like georeferenced points, lines and areas. Our extension is called *XFormsGI*. Its main characteristics are:

- XFormsGI provides control elements for entering new geometries, for modifying the properties of existing objects as well as for selecting geometric entities.
- XFormsGI is declarative, avoiding complicated and confusing script extensions.
- XFormsGI takes spatial reference systems into account.

We will show in the following that the concept of XFormsGI is not only suitable for capturing geospatial data, but can also be used for *sensor data* like GPS positions or the measurements of distances or temperatures. This is especially useful for supporting mobile GIS applications and location-based services (LBS). In this case, metadata like quality, number of visible GPS satellites, the moving direction are processed in addition to the pure sensor data. For validating the concept of XFormsGI, a prototype has been implemented and integrated into a mobile SVG client running on a PocketPC PDA.

The rest of the paper is organized as follows: After presenting the essentials of XForms, the concept and control elements of XFormGI for geospatial data are introduced. The fourth section deals with supporting sensor data. Section 5 presents an application of our approach and Section 6 gives an overview on related work. The paper concludes with a summary and an outlook to future work.

2 The XForms Standard

In this section, we give a brief outline of the XML-based standard XForms.

XForms [14] has its origins in HTML forms and will replace the current forms in future versions of XHTML. The use of XForms is not restricted to (X)HTML but can be embedded in arbitrary XML applications. XForms overcomes several of the limitations of HTML forms. Notably, it separates clearly the purpose from the presentation of a form: "XForms are comprised of separate sections that describe what the form does, and how the form looks. (…) XForms is designed to gather instance data, serialize it into an external representation, and submit it with a protocol." [15].

The functionality of a form is described in one or more *model elements*, whereas its presentation is represented by a set of *control elements*.

2.1 The XForms Model

The *model element* describes several properties of the form:

- An XForms form collects its data – the so-called *instance data* – as child elements of its `instance` element. The instance data will be displayed by control elements.
- The structure of the instance data, i.e. the set of attributes, is defined by the `schema` element. The *schema* is also used to constrain the domain of data.
- The *data flow* to and from the server that is processing the form data is given by the `submission` element.
- An optional *model binding expression* (`bind` element) establishes the relation between the control elements and their instance data

Figure 1 gives an overview on the cooperation between the components of XForms.

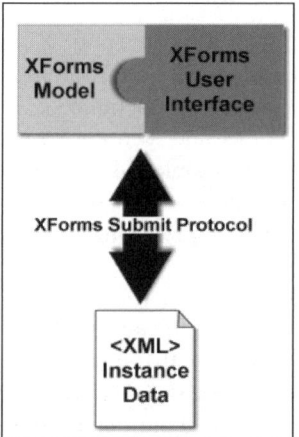

Fig. 1. Overall structure of XForms [15]

Figure 2 depicts the essential parts of an XForms model by a small example document.

The URI `http://www.w3.org/2002/xforms` is the namespace of XForms. In our examples, we will use `xforms` as namespace prefix.

```
<xforms:model id="form1" schema="xxx.xsd">
   <xforms:submission  method = "post"
                       id     = "send"
                       action = "http://..." />
   <xsd:schema>
      ...
      <xsd:simpleType name="status">
         <xsd:restriction base="xsd:string">
            <xsd:enumeration value="prof"/>
            <xsd:enumeration value="stud"/>
            <xsd:enumeration value="other"/>
         </xsd:restriction>
      </xsd:simpleType>
   </xsd:schema>
   <xforms:instance>
      <data>
         <name>Wacker</name>
         <firstname>Willi</firstname>
         <status>prof</status>
      </data>
   </xforms:instance>
</xforms:model>
```

Fig. 2. Example of the model element

2.2 Controls

The form presented to a user is defined by a set of *control elements* defining the principal appearance and interaction of the form. They are embedded in a so-called *host language*. The host language is the XML application that contains the form. The most prominent example is XHTML. In case of graphical applications, the Scalable Vector Graphics (SVG) is a suitable host language.

In contrast to HTML forms, there exists no element which serves as container for all controls of one model. Instead, all controls referencing instance data of the same model element are considered a unit. They may be spread over the host document. There are several predefined controls like input fields, buttons, checkboxes, file upload elements, etc. Each control may have a *label element* as child which is placed automatically by the client.

Figure 3 depicts an example of an XForms `trigger` control with an embedded label usually displayed as a button.

```
<xforms:trigger>
        <xforms:label>Submit</xforms:label>
</xforms:trigger>
```

Fig. 3. Example of an XForms trigger

Data Binding

The relation between a control element of the form and its instance data is established by the XForms binding mechanism. Each control refers to a model by a `model` attribute that may be omitted if the document only contains a single model. For specifying the element storing the data within the instance element, there exist mainly two alternatives:

- using the `ref` attribute to specify an XPath pointing to a child element of `instance` or
- using the `bind` attribute to reference a `bind` element in the model which in turn uses e.g. an XPath expression for defining the location of the data element.

In contrast to HTML, the XForms specification allows the definition of repeating structures. This permits the multiple insertion of input from a single set of controls as a sequence into the instance data.

Transmission of Data

An important functionality of forms is the transmission of data to a server that performs the processing of the input data or that provides data to be displayed. For this purpose, the `submit` element exists as a control element. When the *submit control* is activated by the user, the instance data is collected and transmitted. This is performed according to the protocol and address specified in the `submission` element of the corresponding model (see Figure 2).

Embedding of XForms into SVG

As mentioned before, SVG is a suitable host language for graphical applications. The `foreignObject` element of SVG can be used for embedding XForms controls. Figure 4 depicts some XForms controls embedded into SVG.

```
<model id="form1" >
  ...
  <instance>
    <daten>
      <name>Müller</name>
      <firstname>Willi</firstname>
      <status>prof</status>
    </daten>
  </instance>
</model>

<foreignObject  x="0" y="110" width="100" height="20">
    <input model="form1" ref="/daten/name">
      <label>Name:</label>
    </input>
</foreignObject>
```

Fig. 4. Example of an XForms fragment that can be embedded in an SVG document

Figure 5 shows the corresponding form displayed by the implemented prototype running on a PocketPC PDA.

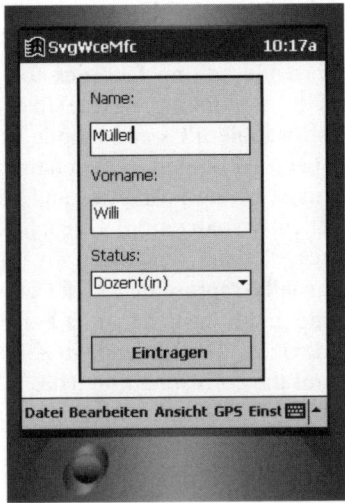

Fig. 5. Display of an XForm by the PDA prototype

2.3 Event Handling

XForms defines several *events* sent on state transitions of the document, controls or models. These events can either directly be routed to XForms actions like `reset` (resetting the form), `setfocus` (setting the focus to a specified control element), etc. or can be processed by *scripts*. XForms itself defines no method for script-based event handling. The definition of such facilities is a responsibility of the host language.

Figure 6 shows the usage of the action `reset` in order to model the behavior of the reset button known from HTML forms; such a control is not defined by XForms.

```
<xforms:trigger>
        <xforms:label>Reset</xforms:label>
        <xforms:reset ev:event="DOMActivate" model="themodel"/>
</xforms:trigger>
```

Fig. 6. Example of the usage of the action `reset`

3   XFormsGI

In the following we describe an extension of XForms for capturing geospatial and sensor data. This extension will be called *XFormsGI*.

For XFormsGI we introduce the namespace URI `http://www.fh-oldenburg.de/xformsgi`. In the subsequent examples the namespace prefix `xfmgi` is used.

3.1 Controls

Several XFormsGI controls provide editing facilities for geospatial data. Geometric editing is usually performed by a pointing device like a mouse or track ball for desktop computers or a stylus for tablet PCs and mobile devices. Like in XForms, the details of interaction are not specified but left to the client implementation.

XFormsGI controls share most of the properties and behavior of XForm controls. One difference concerns the client visualization: To each XForms control, there exists a corresponding user interface element visualized by the client. For example, an XForms `input` element is usually represented as an edit box that also displays its instance data. In contrast, the main task of an XFormsGI control is to capture geometric data by pointing devices. To this end, it neither requires a direct visual representation nor the display of the corresponding data.

The controls introduced by XFormsGI are summarized in Table 1.

Table 1. XFormsGI controls

Control	Description
input	Definition of a new base geometry.
construct	Like `input`. Additionally, from the base geometry a geometric entity of the host language is constructed based on a template.
select	Selection of elements for further processing.
modify	Modification of geometric properties of existing elements, mostly used in conjunction with a `select` control.
sensor	Definition of sensors like GPS (see Section 4).

State of XFormsGI Controls
Another difference between XForms and XFormsGI concerns the start and termination of an input. An XForms control like the text input element can be activated any time by the user to add, remove or modify one or more characters. If the user chooses to switch to another input field, the first one is not affected. In contrast, geometric interaction in general consists in a sequence of distinct steps that have to be performed in a certain order and that cannot be interrupted easily. Thus, an XFormsGI control needs to keep track of a *state* which consists of at least two possible values `stopped` or `running`. In order to start an input sequence, the state of the control has to be switched from stopped to running. This can be performed either by a trigger like a button or implicitly when the user picks an element. Selection and modification operations are typical examples where implicit activation is often used: Whenever an element is picked, the object is, e.g., added to the selection set or it is marked by a frame with "drag grapples" to allow stretching, rotation or displacement.

In order to represent the state, all XFormsGI controls have an attribute `state` which can have one of the following values:

- `stopped`, when the input is currently deactivated (default value of `state`),
- `running`, when input is currently active, and
- `start-implicit`, when the input sequence starts whenever the user picks certain elements.

This attribute can be modified by scripts in order to start an input sequence.

A further difficulty arises from the fact that some interaction sequences not only need a start trigger but also a stop trigger. Consider as example the construction of a line string with a previously unknown number of points. If a closed ring should be constructed, the user can pick the first point again to terminate the construction. If it needs not being closed, a separate trigger is required to terminate the input sequence.

In order to support different types of termination, two specific *XFormsGI events* are introduced in addition to the standard XForms events: `xformsgi-constructionstep` and `xformsgi-constructionterminated`. The first one is sent after every major construction step. In case of capturing a line string, it is sent any time a new point is defined. A further example is the input of a circle. After the construction of the center an `xformsgi-constructionstep` is raised to allow a special center mark to be set. During the processing of the `xformsgi-constructionstep` event, the geometric data already captured is available in the instance data.

The `xformsgi-constructionterminated` is raised after completion of the input sequence. Then, all data are stored in the corresponding instance data element.

Construction by the `input` Element

The `input` element allows capturing standard geometries. Its `type` attribute defines the geometry type. Possible values are `point`, `circle`, `line-segment`, `rect`, `polygon`, and `polyline`. For convenience, the `trigger` control activating the input can be placed as child to the `input` element.

Figure 7 illustrates an input control for capturing a line constructor.

```
<xfmgi:input
      type   = "line-segment"
      model  = "#geomodel"
      ref    = "#linedata "
      state  = "stopped">
   <xforms:trigger>
      <xforms:label>Start input</xforms:label >
   </xforms:trigger>
</xfmgi:input>
```

Fig. 7. Example of the usage of the `input` control

In contrast to the start trigger, there is no direct support for a stop trigger. One possible solution would be to switch the trigger from "Start input" to "Terminate input" by a script. Alternatively a client implementation could add the "Terminate input" automatically to the menu or context menu when a geometric input is running.

Construction by the `construct` Element

The `construct` control is introduced as a convenience to facilitate the generation of geometrical entities of the host language by user input sequences. In contrast to the `input` element, which stores the input only in the instance data, the construction control generates a new entity after completion of the input sequence. The new entity is constructed according to a *template* specified by the `template` attribute and placed as new child to the element of the host language denoted by the `parent` attribute of `construct`.

Figure 8 illustrates this by using SVG as host language. The `template` attribute refers to the `image` element of SVG.

```
<defs>
    <image   id       = "img-template"
             width    = "60"
             height   = "40"
             xlink:href = "tisch.svg" />
</defs>
<xfmgi:construct
     type     = "point"
     template = "#img-template"
     parent   = "#new-elements"
     state    = "stopped">
     <xforms:trigger>
        <xforms:label>Start input</xforms:label >
     </xforms:trigger>
</xfmgi:construct>
```

Fig. 8. Example `construct` element referring to an SVG element as template

How the constructed geometry data is applied to the newly generated geometric entity depends on the types of both, the `construct` and the template element. Consider the example of an SVG `image` element. If the `construct` element has the type `point`, the x and y coordinates of the image are set to the coordinates of the constructed point. For a `polyline`, the image is placed on every vertex, and finally, if it is of type `rect`, the image is scaled and positioned to fit inside the constructed rectangle.

Selection

The `select` control allows the selection of elements by the user. Subsequently the user may query information about the selected elements or modify them.

Selected elements are usually highlighted by changing some of their style attributes or by adding decorations like a bounding box. Often a bounding box additionally displays some "grapples" where the user can drag the bounding box to change the size, position or orientation of the element. How the highlighting is performed, is – in correspondence to the XForms specification – out of the scope of XFormsGI.

By default, only elements having an id may be selected. It is task of the host language to control the selectivity of their elements. This can be done by (extensions to) the styling language (e.g. to Cascading Style Sheets) used in the host language like it is proposed in [2].

The main attributes of the select control are described in Table 2.

Table 2. Attributes of the select element

Attribute	Description
ref	ref refers to the "selection set" in the instance data. The selection set is represented as a space-separated list of the identifiers of the selected elements.
type	type has alternatively the value single or multiple. In the first case, at most one element can be selected. In the second case, the selection set may contain an arbitrary number of elements.
modifier	Space-separated list of IDREFs referring to modify controls; explained in Section 0.

Figure 9 depicts an example for the select control.

```
<xfmgi:select
    type        = "single "
    model       = "#geomodel"
    ref         = "/data/selected-elements"
    modifier    = "#modify-placement #modify-delete
    state       = "start-implicit"
/>
```

Fig. 9. Example select element

Modification

The modify control can be used either standalone or in conjunction with a select element to allow the modification of an existing graphical entity. Its main attributes are described in Table 3.

Table 3. Attributes of the modify element

Attribute	Description
type	type defines the kind of modification that can be applied. It has one of the following values: move-vertex, move-edge, move, rotate, resize, homogeneous-resize, or delete.
applyto	applyto has alternatively the value selected , when the selected entity or entities should be modified, or, in the case of a stand-alone modifier, it consists of a space-separated list of IDREFs of the elements that are subject to the modification operation.
description	A description of the operation displayed to the user.
ref	Reference to the instance data element that stores the modified data.

Figure 10 shows an example of a selection control with three corresponding modify controls. The modifier attribute of the select control allows the direct "routing" of a selected element to one or more modification operations. A client may display the bounding box with the appropriate "grapples" (e.g., for scale and rotation) to give a hint to the user which modifications are possible. Figure 10 also depicts the interpretation of those controls by our prototype implementation. The description attributes are taken to be shown in a popup menu allowing the user to choose one of the three modifications.

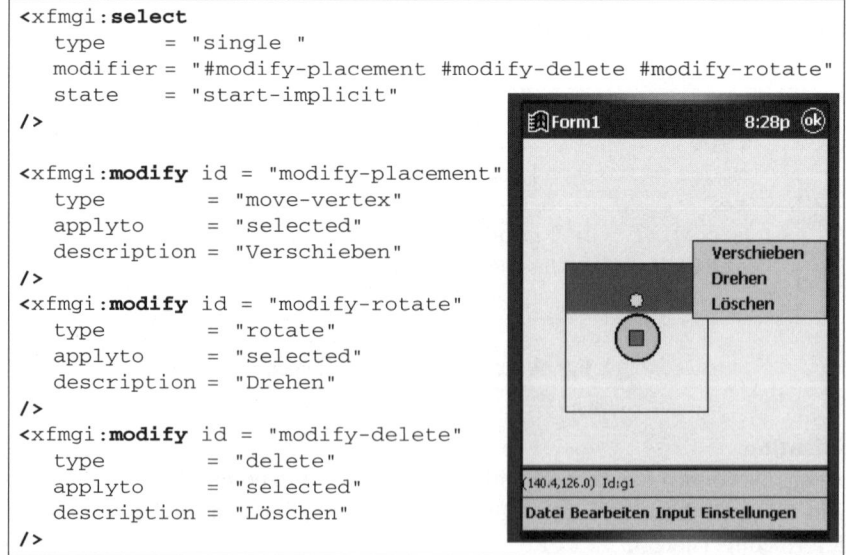

```
<xfmgi:select
    type     = "single "
    modifier = "#modify-placement #modify-delete #modify-rotate"
    state    = "start-implicit"
/>

<xfmgi:modify id = "modify-placement"
    type        = "move-vertex"
    applyto     = "selected"
    description = "Verschieben"
/>
<xfmgi:modify id = "modify-rotate"
    type        = "rotate"
    applyto     = "selected"
    description = "Drehen"
/>
<xfmgi:modify id = "modify-delete"
    type        = "delete"
    applyto     = "selected"
    description = "Löschen"
/>
```

Fig. 10. Example of selection with corresponding modify elements

3.2 Data Binding

XFormsGI uses the same data binding mechanism as XForms. Since geospatial data refers to spatial reference systems, we have to extend the XFormsGI data binding accordingly.

Coordinate Systems

In the context of GIS or LBS, we have to consider three coordinate systems:

- GPS-coordinates defining the current position (typically referenced by WGS'84 coordinates),
- the application reference system (e.g., a projected coordinate system), and
- the coordinate system of the host language.

The relation between the three systems can be described by two transformations T_1 and T_2 illustrated in Figure 11.

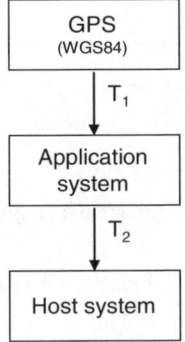

Fig. 11. Transformation chain

The transformations are used in the following ways: GPS data (e.g., used to position a cursor depicting the current location) is mapped into the host system by first applying T_1 and then T_2. Visual construction or modification of an element by the user takes place in the host coordinate system. When coordinates are displayed numerically to the user or sent to the server for further processing, these coordinates have to be transformed to the application system by the inverse of T_2.

The SVG standard proposes to include information about the underlying geographic reference system into an SVG document by adding a metadata element containing an RDF description [11] of the coordinate reference system (see Section 7.12 "Geographic Coordinate Systems" in [12]). The coordinate system itself has to be defined by the "OpenGIS Recommendation on the Definition of Coordinate Reference Systems" [9]. This proposal leads to rather lengthy XML fragments. Instead, we introduced a simple XFormsGI element to describe the two transformations. The application system is identified by its EPSG number (see [5]) which establishes T_1. The mapping T_2 to host coordinates is described by a `transform` attribute which has the same functionality as the SVG `transform` attribute.

```
<xfmgi:coordinatereferencesystem
    crs-id    = "epsg:31467"
    transform = "scale(1000,-1000) translate(-5308.5,-812)"
/>
```

Fig. 12. Example of the `coordinatereferencesystem` element

Data Representation

If a `coordinatereferencesystem` element is present in the document, the instance data collects two sets of corresponding coordinate representations. The data is stored in two child elements of the reference element: the `hostcoords` and the `appcoords` element. An example is depicted in Figure 13. The data itself is stored as element content by a space-separated sequence of coordinates.

```
<data>
    <xfmgi:hostcoords>27.2 -157.4<xfmgi:hostcoords>
    <xfmgi:appcoords>8.19 53.14<xfmgi:appcoords>
</data>
```

Fig. 13. Example of coordinate data representing a single point

The data representation according to the type of the geometric entity is straightforward, see Table 4.

Table 4. Data representation for selected geometric entities

Type	Data	Description
point	x y	Coordinates
line-segment	x_0 y_0 x_1 y_1	Coordinates of start and end point
circle	x y r	Coordinates of center, radius

4 Sensors

Especially for mobile applications, the collection of data by sensors connected to the device becomes more and more important. For this purpose it is essential that XFormsGI supports this type of data capturing.

The `sensor` element is used to declare a "data source" which reads data from measuring equipment like GPS-sensors, compasses, velocity sensors, gas sniffers, barcode readers etc. The attributes of the `sensor` control are described in Table 5.

Table 5. Attributes of the sensor element

Attribute	Description
`type`	Signifies the type of the sensor. The meaning of the other attributes depends on this value.
`source`	Either the name of the communication port used by the sensor – like e.g. COM1 and possibly further communication parameters – or the URL of a file containing emulation data. The source attribute may be omitted. In this case data is read from the standard port configured by the client.
`display-element`	Reference to host element used to visualize the readings
`cursor`	*For georeferenced sensors:* Optional. An IDREF to a host element used as a cursor to display the current location and track direction.
`speed-arrow`	*For moving georeferenced sensors:* Optional. An IDREF to a host element used to track speed (by scaling) and direction (may reference the same element as `cursor`).
`mode`	*For geo-sensors:* Optional. Any combination of following modes: `track-view`: keep the view centered on the current location; `track-angle`: cursor is rotated according to current heading; `track-velocity`: the speed arrow is scaled according to current velocity.

An example of a GPS sensor element is shown in Figure 14.

```
<xfmgi:sensor
    type        /  = "gps"
    ref            = "#gps-data"
    source         = "nmea2.txt"
    cursor         = "#gps-cursor"
    speed-arrow    = "#gps-speed"
    mode           = "trackview trackangle"
    state          = "stopped"
/>
```

Fig. 14. Example of `sensor` element

Additionally, metadata produced by the sensor may be placed in the instance data besides the coordinate information. For a GPS sensor, these may be height, date, and quality information like number of visible satellites, and so on. For the representation in the case of a GPS sensor, we choose the "way point element" `wpt` defined by the GPS exchange standard GPX, see [4]. Figure 15 shows an example of such instance data.

```
<xfm:instance>
   <gps-data>
      <xfmgi:hostcoords>8.19 -53.14</xfmgi:hostcoords>
      <xfmgi:appcoords>8.19 -53.14</xfmgi:appcoords>
      <gpx:wpt lat="8.190375" lon="53.132135">
         <gpx:time>4-04-07T15:23:32.809Z</gpx:time>
         <gpx:ele>62.100</gpx:ele>
         <gpx:sat>4</gpx:sat>
         <gpx:dhop>1.9</gpx:dhop>
      </gpx:wpt>
   </gps-data>
</xfm:instance>
```

Fig. 15. Example instance data for a GPS sensor

5   Application

We outlined the concept of XFormsGI in the previous sections. In this section, we will shortly present a prototype, which serves to validate the concepts and which implements a specific XFormsGI user interface. The second objective is motivated by the fact that XForms leaves the details of the user interface to the implementation. The prototype has been developed as a mobile application running on a PocketPC PDA. SVG is used as host language of XFormsGI. Apart from the visualization of spatial data, the purpose of this application is the support for redlining and the input and editing of simple graphical elements.

The dominating input device for PDAs is the stylus. Therefore, the prototype is designed for supporting this type of input. The `input` and `construct` elements as described in Section 3.1 are implemented for the geometric primitives point, line segment, circle and polyline. Compared to the usage of a traditional mouse pointer on a desktop computer, the stylus has several restrictions. The prototype takes these specific characteristics into account. For example, the stylus lacks buttons and its position is only tracked when it touches the screen. As a consequence, a PDA is not able to distinguish between touches signifying for example the construction of a new point and simple display of the movement of the current track position.

In addition, the manipulation of existing geometric elements is addressed by the prototype. For example, it is possible to select one or more elements for deletion or geometric manipulation. At the moment dragging, insertion and deletion of vertices of a polyline are implemented. A GPS sensor element can be declared as described in Section 4.

Figure 16 shows a screenshot of a graphical editor defined by XFormsGI controls.

The integration of GPS sensors has been used for a tourist application: a historic city walk for the City of Oldenburg. This application offers the user historic (SVG and raster) maps of Oldenburg, corresponding textual information and shows the current position of the user by using the sensor element of XFormsGI. Figure 17 illustrates this application.

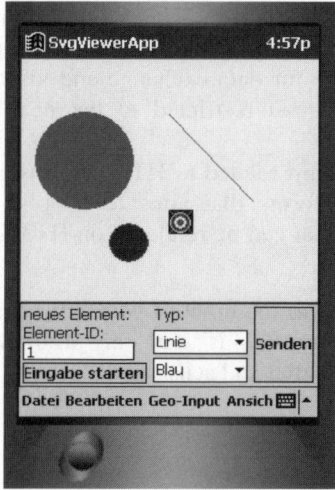

Fig. 16. Example of a complete graphical editor

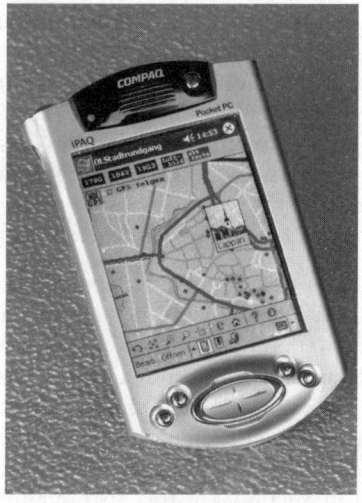

Fig. 17. Historic city walk application for the City of Oldenburg

6 Related Work

The following three standards are the currently most prominent frameworks for developing form-based desktop and web applications:

- XUL is the *XML User-Interface Language* [7] developed by Mozilla. It allows the development of platform-independent graphical user interfaces.
- XAML (*Extensible Application Markup Language*) [6] is an XML application defined by Microsoft that allows the description of user interface similar to

XForms for defining controls and to SVG for describing the visual aspects of the user interface. XAML is embedded into the broader framework .Net that provides the technical infrastructure for data exchange and visualisation. However, XAML is a proprietary standard and restricted to the next versions of the Windows operating system.

- *WebForms* [13] is a standard related to HTML forms and XForms. However, it is an extension to HTML forms that aims to simplify the task of transforming XForms into documents that can be rendered on HTML Web browsers that do not support XForms.

None of these standards support the input of geometric data. To the knowledge of the authors, the proposed XFormsGI is the only framework allowing descriptive definition for capturing geospatial and sensor data.

7 Conclusions

The presented XFormsGI framework allows the declarative definition of a graphical editor that needs no or only a small amount of scripting. Such an editor could be used, e.g., on a mobile device to collect field data. The representation by geographic coordinates is taken into account so that the seamless integration with a back-end application processing the captured data is possible.

For validating the concept of XFormsGI, a prototype has been implemented and integrated into a mobile SVG client [1] running on a PocketPC PDA.

A further feature of XFormsGI is the integration of sensors. In case of using SVG as host language for LBS, it is possible to make an SVG map location-aware by only adding two XFormsGI elements.

Several aspects of XFormsGI are still under development:

- In the current version, it is not possible to confine newly constructed points to predefined areas. This may lead to errors and inconsistent data. It would be desirable to have a means to define geometric input restrictions.
- In general, the construction of a new geometry is influenced by other geometries. As examples, consider the vertex of a building that has a certain distance from a given point, or the location of an accident on a road. In the simplest case, a new point could be snapped onto a given point. In analogy to the question of "selectibility" discussed in Section 3.1, this also requires extensions to the styling language used for the host document.
- In the context of sensors, a general method to map sensor measurements to element attributes of the host language is needed.

Acknowledgements

This work was funded by the Ministry for Research and Culture of Lower Saxony by the program "Praxisnahe Forschung und Entwicklung an niedersächsischen Fachhochschulen" and the German Federal Ministry for Education and Research by the program „FH³ Angewandte Forschung an Fachhochschulen im Verbund mit der Wirtschaft". This support is gratefully acknowledged.

References

1. Brinkhoff, T., Weitkämper, J.: Mobile Viewers based on SVG±geo and XFormsGI. In: Proceedings 8th AGILE Conference on Geographic Information Science, Estoril, Portugal, pp. 599--604 (2005)
2. Brinkhoff, T.: Towards a Declarative Portrayal and Interaction Model for GIS and LBS. In: Proceedings 8th AGILE Conference on Geographic Information Science, Estoril, Portugal, pp. 449–458 (2005)
3. Fettes, A., Mansfield, P.A.: Creating a Graphical User Interface in SVG. In: Proceedings 3rd Annual Conference on Scalable Vector Graphics, Tokyo, Japan (2004)
4. GPX: The GPS exchange format, http://www.topografix.com/gpx.asp
5. International Association of Oil & Gas Producers, Surveying & Positioning Committee: EPSG Geodetic Parameter Dataset, http://www.epsg.org/
6. Microsoft MSDN: XAML Overview, http://windowssdk.msdn.microsoft.com/library/default.asp?url=/library/en-us/wpf_conceptual/html/a80db4cd-dd0f-479f-a45f-3740017c22e4.asp
7. Mozilla.org: XML User Interface Language (XUL), http://www.mozilla.org/projects/xul/
8. Open Geospatial Consortium (OGC): OpenGIS Geography Markup Language (GML) Implementation Specification, Version 3.1.1, OGC 03-105r1 (2004)
9. Open Geospatial Consortium (OGC): Recommended XML/GML 3.1.1 encoding of common CRS definitions, OpenGIS Recommendation, OGC 05-011 (2005)
10. W3C: HTML 4.01 Specification, W3C Recommendation (December 24, 1999), http://www.w3.org/TR/html401
11. W3C: Resource Description Framework (RDF), http://www.w3.org/RDF/
12. W3C: Scalable Vector Graphics (SVG) 1.1 Specification, W3C Recommendation (January 14, 2003), http://www.w3.org/TR/2003/REC-SVG11-20030114/
13. W3C: Web Forms 2.0, W3C Member Submission (April 11, 2005), http://www.w3.org/Submission/web-forms2/
14. W3C: XForms 1.0 (Second Edition), W3C Recommendation (March 14, 2006), http://www.w3.org/TR/2006/REC-xforms-20060314
15. W3C: XForms - The Next Generation of Web Forms, http://www.w3.org/MarkUp/Forms/
16. W3C: XHTML 1.0 The Extensible HyperText Markup Language (Second Edition), W3C Recommendation, 26, revised 1 August 2002. (January 2000), http://www.w3.org/TR/xhtml1

Web Architecture for Monitoring and Visualizing Mobile Objects in Maritime Contexts

F. Bertrand[1], A. Bouju[1], C. Claramunt[2], T. Devogele[2], and C. Ray[2]

[1] University de La Rochelle, France
{alain.bouju,frederic.bertrand}@univ-lr.fr
[2] Naval Academy Research Institute, Brest Naval, France
{claramunt,devogele,ray}@ecole-navale.fr

Abstract. Recent advances in telecommunication and positioning systems, web and wireless architectures offer new perspectives and challenges to the management and visualization of mobile and geo-referenced objects. Amongst many research challenges still opened, the development of integrated sensor-based architectures necessary to the integration of mobile data has been often neglected. Real-time integration of sensor-based data, their management within a distributed system, and diffusion to different levels of services and interfaces to what should be considered as a collaborative web GIS constitutes the objective of the research presented in this paper. This paper introduces a modular and experimental web GIS framework applied to maritime navigation, and where mobile objects behave in a maritime environment. Different levels of services are developed, including a web-based wireless access and interface to a traffic monitoring application.

Keywords: mobile objects, web-based interface, maritime navigation.

1 Introduction

Over the past years location-based services and applications have been the object of considerable development. Although technological-driven by continuous advances in wireless-communications and positioning systems, the software and service dimen-sions should not be neglected as people are expecting and asking for more data and services, anywhere and anytime. As for many information-based systems, location-based services are at the crossroads of many engineering and scientific domains. The technological component is of course essential and supported by the essential progress in geo-positioning systems (Taylor, 2006). At a different level, also crucial as it concerns the way mobile data should be integrated within computing systems, the representation and manipulation of mobile objects has been one of the most active area of spatio-temporal database research. The problem has been considered as a data description and storage issue where the objective is to design appropriate physical indexes, query languages and prediction mechanisms for moving objects (Kollios *et al.*, 1999; Brinkhoff, 2000; Papadias *et al.*, 2003; Saltenis *et al.*, 2000, Trajcevski *et al.*, 2004, Teixeira and Güting,

J.M. Ware and G.E. Taylor (Eds.): W2GIS 2007, LNCS 4857, pp. 94–105, 2007.

2005). Recent work concerns the development of conceptual languages (Noyon *et al.*, 2007) and data mining algorithms (Sharifzadeh *et al.*, 2005) for the manipulation of trajectory at the interface level, and the search for trajectory patterns in space and time.

The development of location-based services is related to a great degree to emerging web and wireless geographical information systems. Similarities are multiple and have attracted many recent works. Early developments concern mobile applications for transportation systems, and where the objective is to provide accurate information and services to bus locations and arrival times (Peytchev and Claramunt, 2001; Bertolotto *et al.*, 2002). Web and wireless GIS have also rapidly evolved towards parallel but complementary directions: search for standardised platforms and exploration of novel mobile-based services. On the one hand, the integration of geographical information standards under the umbrella of the Open GIS consortium, web specifications imposed by the W3C endowment have lead to the definition of several web GIS architectures (Luaces *et al.*, 2004). On the other hand, additional services on mobile systems have been explored for neighbourhood retrieval systems (Ishikawa *et al.*, 2003), directional queries (Gariner and Carswell, 2003), diffusion of multi-resolution maps (Follin *et al.*, 2003), personalisation of geographical maps in mobile environments (Doyle *et al.*, 2004) and diffusion of three-dimensional geographical data (Guth and Thibaud, 2003), management of geographical data (Kang and Li, 2006).

The research presented in this paper proposes an integrated framework that can be considered as an intermediate layer of the works mentioned above in the sense it proposes a data integration framework completed by a distribution of services to the users. The framework is experimented in the context of maritime navigation, where ship trajectories generate incoming data, and where wireless communications constitute an heterogeneous data integration environment, and where different levels of services are experimented for different users, from local navigators to monitoring authorities.

The reminder of the paper is organised as follows. Section 2 introduces the main sensor-based techniques used in maritime navigation, and the data integration principles retained for our framework. Section 3 develops the modular platform developed so far for the monitoring of ship trajectories. Section 4 presents the interface and visualisation services implemented. Finally Section 5 concludes the paper and discusses further work.

2 From Sensor-Based Systems to an Architecture for Mobile Objects

2.1 Sensor-Based Systems

One of the main objectives of location-based services is to integrate a large number of mobile objects within a distributed architecture, and to offer different levels of manipulation and visualisation facilities to mobile end-users. Several aspects can be considered for the design of such a system:

- A geo-localisation system that delivers positioning information. The choice of such a location based-system is often driven by the telecommunication environment, that is, GPS, radio-based systems (GSM or WiFi) or inertial systems.
- A data format that supports the delivery of real-time geo-referenced data and additional contextual (e.g., ship properties) and spatio-temporal data (e.g., speed, direction, rate of turn). These formats might be either constrained by the sensor design, or preferably structured in a standard meta language (Botts *et al.*, 2006).
- A medium and a communication protocol. Aerial, terrestrial and maritime transportation systems use a large spectrum of wireless telecommunications for the integration of moving object trajectories. Although the choice of wireless communication system is immediate, the properties of the wireless network (e.g., WiFi, WiMax, GSM, and VHF) determine the extent in space of the communication capabilities.

Current systems used for real-time tracking and monitoring of mobile objects use different approaches, depending of the environment. For instance, the aerial domain is mainly based on VHF communication. The Automatic Dependent Surveillance-Broadcast (ADS-B) gives to an aircraft the ability to broadcast real-time location information in a wide area. Additional functionalities include a Traffic Alert and Collision Avoidance System (TCAS) and object identification facilities. The application of wireless communications to the terrestrial domain is also an active area. These are either based on radio or digital audio systems, and more recently on high speed communication protocols such as WiFi and WiMax. Recent works on vehicle-to-vehicle or vehicle-to-infrastructure communications are largely based on these protocols.

In the maritime and aerial domains, communications have been long developed on the basis of VHF communications. One of the most successful systems used in maritime navigation and positioning is the Automatic Identification System (AIS) system whose objective is to identify and locate vessels at distance. An AIS generally integrate a transceiver system, a GPS receiver and other navigational sensors on board such as a gyrocompass and a rate of turn indicator. An AIS transponder runs in an autonomous and continuous mode, regardless its location (e.g., open sea, coastal or inland areas). A position report from an AIS station fits into one of the numerous time slots established every 60 seconds on each frequency. AIS stations are continuously cross-synchronized in order to avoid overlapping of slot transmissions. The transmitted information to a range of 35 miles to surrounding ships and maritime Vessel Traffic Systems (i.e., maritime authorities) include data on the ship characteristics, its position, route, speed, and estimated arrival time to a given place. The International Maritime Organization has made the AIS a mandatory standard for the Safety of Life at Sea (SOLAS). Passenger ships and international voyaging ships with gross tonnage are now equipped with AIS.

2.2 Toward an Architecture for Mobile Objects in Maritime Environments

The objective of the platform developed so far is to provide an integrated and flexible system for the monitoring and tracking of mobile objects active in a given maritime environment, and delivery of several distributed components and services. Figure 1 presents the distributed architecture of our proposed platform that includes two permanent AIS transponders implemented in two important harbours located in North West France (listening mode only). The first AIS is located in the Brest harbour, the second one at the La Rochelle harbour, and a third one acts as a mobile receiver. One of the preliminary principle of this architecture is to separate the AIS receivers from the platform itself. This leads us to design a system where mobile objects considered are both AIS receivers and transmitters. The system is scalable as it can be easily extended to additional sites, either mobile or not. The Java-based platform is a computing system that includes an AIS management server, a PostGIS spatial database for data manipulation, and a Web server based on Apache and Tomcat. The distributed components of the platform are able to automatically get AIS data either through the Web server (i.e., using incremental HTTP queries) or directly through a serial port.

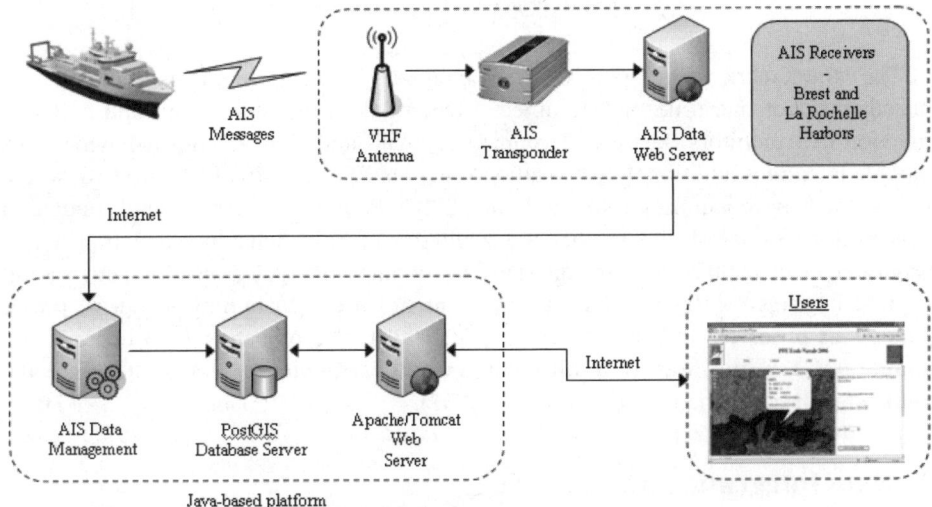

Fig. 1. Distributed architecture for maritime navigation

The functional components of the platform depicted in figure 2 rely on a modular approach, whose objective is to integrate heterogeneous positional and real-time data from different sources: public traffic data available from the Internet, and mobile ships sending information using the AIS or ad-hoc communication systems based on WiFi or WiMAX. These sensors generate real-time data (frames from different types) on maritime trajectories and vessel characteristics. Several parsers send mobile objects data to the PostGIS database through a "sensor data management module" which is able to manage incoming data from many distributed sites. Within this module, these sensor-based data are automatically integrated after a filtering and pre-processing process.

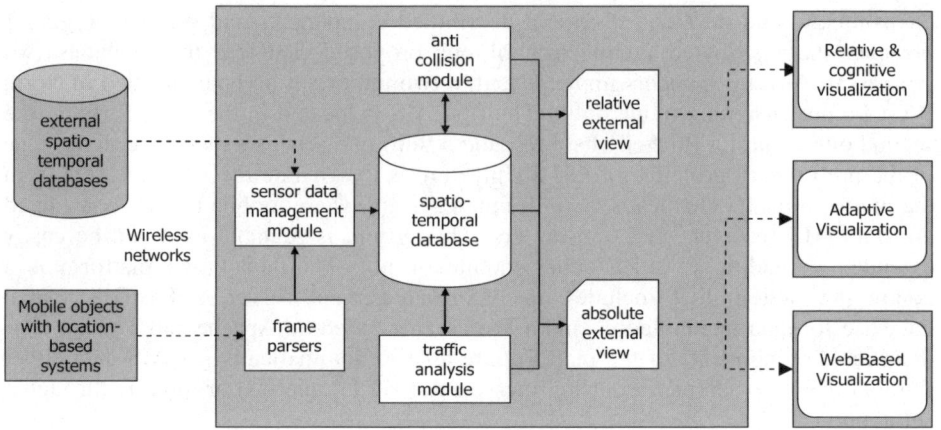

Fig. 2. Framework components

The framework developed so far integrates several modules tied to the management of maritime mobile objects. One of these modules is an anti-collision function that monitors the risk of running aground and evasive ship behavior. This module integrates additional visualization functionalities to provide the users different views on their trajectories (Noyon *et al.*, 2007; Petit *et al.*, 2006), and simulation capabilities to control and predict the evolution of ship behaviors and trajectories (Fournier *et al.*, 2003). The traffic analysis module still under development should explore intelligent inference mechanisms that can use data mining to derive traffic patterns as observed and studied in (Imfeld, 2000). The objective is to observe and understand maritime traffic at different levels of granularity and to integrate the human factor component (Chauvin *et al.*, 2007).

3   Data Integration and Filtering

The data integration level constitutes the central component of the platform. Data volumes are relatively important as for instance the AIS system located in the port of Brest can receive up to 50000 relevant AIS messages per day. A representative daily figure for the Brest harbor is 13817 positions recorded from 26 moving vessels, 13915 positions from 20 moving vessels for the La Rochelle harbor. Navigation data, whatever the kind of sensor, are stored within a PostgreSQL[1] database and its PostGIS[2] extension for spatial data management, and logically organised in different relational tables according to the conceptual model presented in Figure 3. The frequency of the data inputs varies according to the speed of the vessels and the density of the traffic within the maritime region considered.

---

[1] http://www.postgresql.org/
[2] http://postgis.refractions.net/

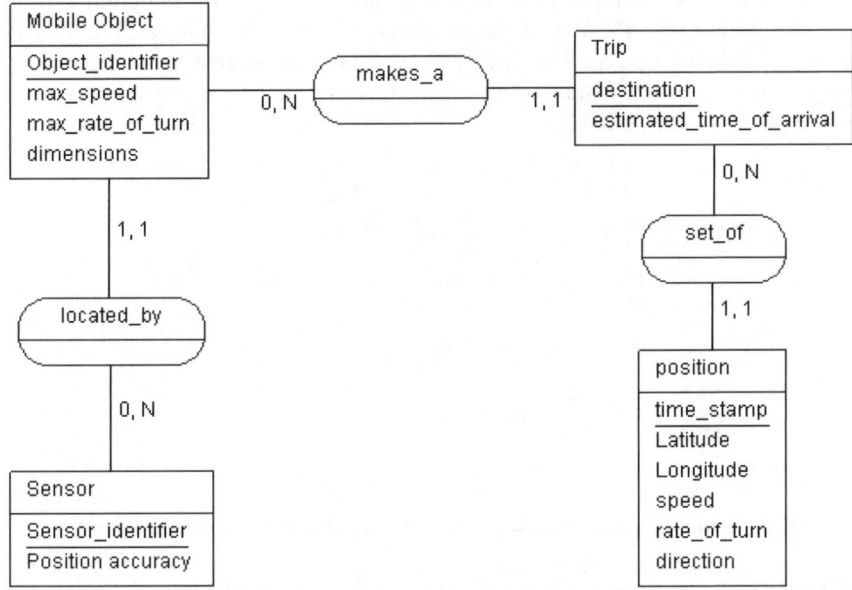

Fig. 3. Conceptual model for mobile objects

In order to reduce the volumes of trajectory data, a filtering process has been applied and implemented. A trajectory is defined as an ordered set of positions (P_i). A given position P_i is represented by a tuple [x, y, speed, direction, time]. A filtering process selects the key positions of a given trajectory. A position is considered as a key position when either the speed or the direction change significantly. The other positions are removed. Our filtering process is based on a line simplification algorithm initially introduced by Douglas and Peuker (1973), and that preserves directional trends in a line using a tolerance distance. In fact, it rapidly appears that this polyline-based approach is computationally efficient, and adapted to the main characteristics of maritime trajectories which are most often linear over long periods of time. The principles of this algorithm are as follows. The start and end points of a given polyline are connected by a straight line segment. Perpendicular offsets for all intervening end points of segments are then calculated from this segment, and the point with the highest offset is identified. If the offset of this point is less than the tolerance distance, then the straight line segment is considered adequate for representing the line in a simplified form. Otherwise, the point is selected, and the line is subdivided at this point of maximum offset. The selection procedure is then recursively repeated for the two parts of the line until the tolerance criteria is satisfied. Selected points are finally chained to produce a simplified line. This simplification algorithm can be customized for trajectory filtering. The perpendicular distance is derived as a spatio-temporal distance, and as follows:

$$d_T(T_i, T_j, t) = \left\| P_i(t) - P_j(t) \right\|^2$$

Let us considered the example presented in figure 4. The spatio-temporal distances between the points (P_2, P_3, P_4) of the trajectory T_i, and the points (P'_2, P'_3, P'_4), respectively, of the interpolated trajectory T_j taken at a same time are computed (note that these spatio-temporal distances are influenced by the speed and the direction of the mobile object).

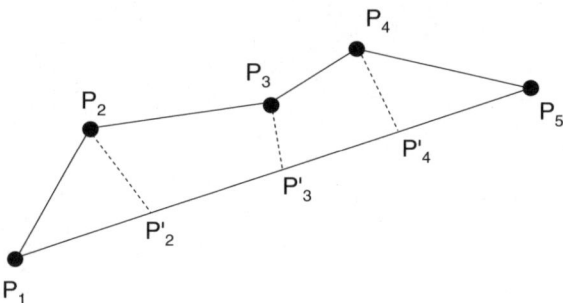

Fig. 4. Spatio-temporal distance between recorded and interpolated positions

A tolerance distance should be defined appropriately. When the tolerance is too large, too many positions are removed and the interpolated positions are likely to be inappropriate. Conversely, when the tolerance is too small, the data volumes generated are too large. These tolerance values are experimentally defined depending on the sensor accuracy and characteristics of the maritime region considered. The experiments made on the maritime region of Brest showed that the filtering process generally reduce the volume of trajectory data by more than 90% when considering tolerances of either 5 or 10 meters. The performance of the filtering process is also likely to increase for large ships, and decrease for small ships due to the characteristics of their navigation.

4 Visualisation

The main services offered so far to the users concern the visualisation of real-time maritime traffic and collision detection. Several visualization services introduced in our previous works have been implemented and integrated within the framework (cf. Figure 2). The first one relies on a conventional cartographical view provided to an adaptive GIS (Petit *et al.*, 2006). An adaptive GIS is defined as a generic and context-aware GIS that automatically adapts its interface according to its context. Such a context is defined by (i) the properties and location of the geographical data being manipulated (e.g. maritime traffic data); (ii) the underlying categories that reflect different user profiles (e.g. port authorities); and (iii) the characteristics of the computing system, including Web and wireless techniques. The second one takes a rather egocentric view that takes into account the relative view of the environment as it is perceived by a mobile user (Noyon *et al.* 2007). The objective is to provide a complementary view to the usual absolute vision of space. Trajectories are characterized from the perception of a moving observer, and where relative positions and relative velocities are the basic primitives. This allows for a more intuitive identification of elementary trajectory configurations, and their relationships with the regions that compose the environment.

The additional services offered by our platform concern a web-based visualisation system. This module has been designed with a Java Server Page (JSP)[3] approach, compliant to an application server like Tomcat, and to external access to our spatial-temporal navigation database using a Java Database Connectivity (JDBC) interface. In order to guaranty maximum portability of the web interface, the GoogleMaps API[4] has been retained including satellite images in background. The interface and the maritime navigation data associated are refreshed on a temporal basis predefined at the interface and user levels. Usual manipulation functionalities have been implemented, e.g., pan, zoom in, zoom out, point and select operations. The web interface also allows integration of additional geographical data. Figure 5 shows an example of interface output where additional information on a given vessel are presented to the user (e.g., latitude, longitude, speed, direction, and time of last position). The example presented in this figure shows the trajectories of two mobile objects, and additional contextual traffic information.

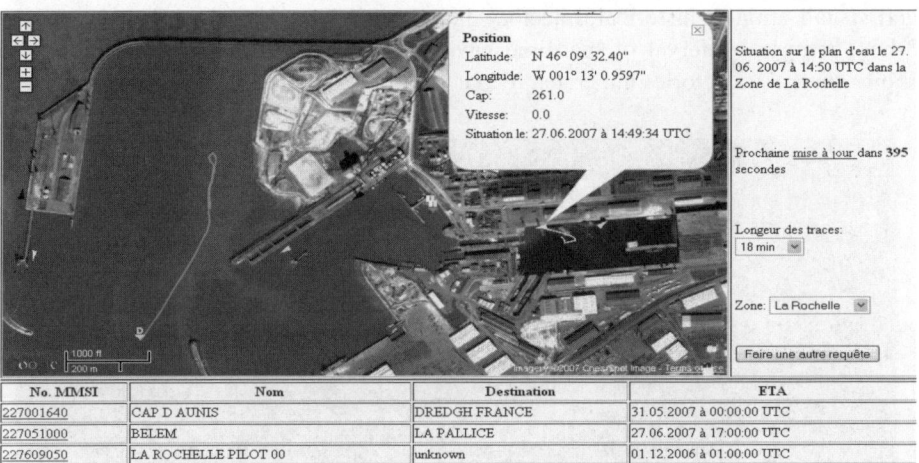

No. MMSI	Nom	Destination	ETA
227001640	CAP D AUNIS	DREDGH FRANCE	31.05.2007 à 00:00:00 UTC
227051000	BELEM	LA PALLICE	27.06.2007 à 17:00:00 UTC
227609050	LA ROCHELLE PILOT 00	unknown	01.12.2006 à 01:00:00 UTC

Fig. 5. Web based visualization interface

Additional contextual objects predefined by the user can be integrated at the interface level (e.g., waiting area, access channel). Figure 6 shows the example of a vessel on a waiting area (displayed as a square), a vessel that leaves an access channel (to the left), and a vessel leaves abnormally the same channel at it goes out of the limit of that channel (to the top).

---

[3] http://java.sun.com/products/jsp/
[4] http://www.google.com/apis/maps/

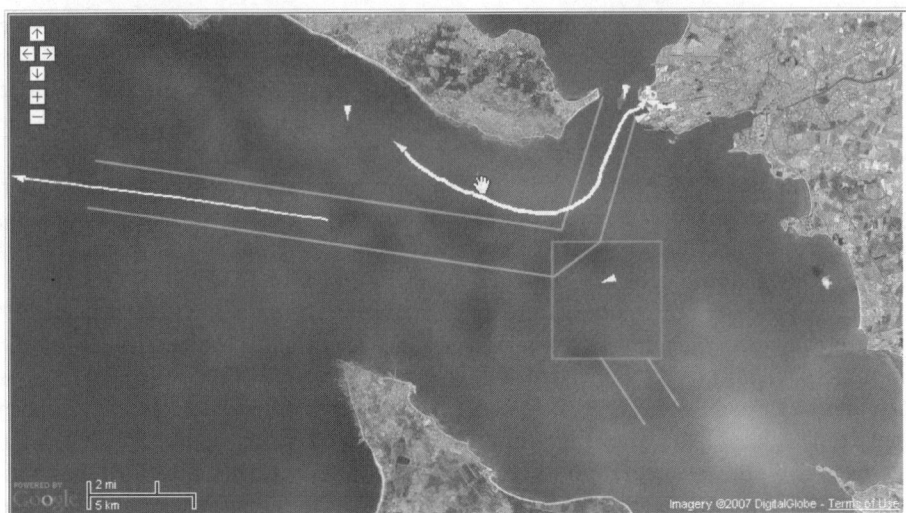

Fig. 6. Trajectories behaviours

Others service provided by the web interface are oriented towards the manipulation and visualisation of historical trajectory data. Figure 7 presents the visualisation of a 3-hour temporal interval of maritime navigation on the Brest harbour. This figure shows several trajectories integrated by our system.

Fig. 7. Example of trajectory analysis in the Brest harbour

Such views can be useful for the analysis of navigation traffic as patterns and trends analysis are particularly useful for safety purpose, or the study of specific events and incidents in a given maritime area. Additional functionalities have been also developed for the manipulation and analysis of trajectories (Noyon *et al.*, 2007).

5 Conclusion

Although current maritime systems are relatively successful, thanks to the development of vessel traffic systems and automated information systems, there is still a need for additional services that favor the diffusion of maritime navigation data. The research presented in this paper introduces a modular data integration platform and web interface to monitor and visualize mobile objects in a maritime environment. This architecture allows for the integration of various sensor-based data, and provides several visualization services to either vessels considered as clients or monitoring authorities. The interface developed can be used either in real-time for navigation monitoring and control, or for the analysis of maritime navigation behaviors.

Further work concerns the integration within our framework of Open Geospatial Consortium recommendations on the encoding of sensor data. In particular, the forthcoming Sensor Model Language (SensorML) based on XML provides an information model and encoding level that might facilitate the integration of additional sensors and the flexibility of our platform. Other aspects still to consider concern the development of additional services between the different sensor nodes of the architecture, and exploration of novel functionalities for detecting trends and patterns in maritime trajectories.

Acknowledgement

The authors wish to thanks the undergraduate students who contributed to the development of the platform.

References

Bertolotto, M., O Hare, G.M.P, Strahan, R., Brophy, A., Martin, A.N., McLoughin, E.: Bus catcher: a context sensitive prototype system for public transportation users. In: Huang, B., et al. (eds.) WISE 2002 Workshops. Proceedings of the 3rd International Conference on Web Information Systems Engineering Workshops, pp. 64–72. IEEE Computer Society, Singapore (2002)

Brinkhoff, T.: Generating network-based moving objects. In: Günther, O., Lenz, H.-J. (eds.) Proceedings of the 12th International Conference on Scientific and Statistical Database Management, pp. 253–255. IEEEE Computer Society, Berling, Germany (2000)

Chauvin, C., Clostermann, J.P., Hoc, J.M.: Situation Awareness and the decision-making process in a dynamic situation: avoiding collisions at sea. Journal of Cognitive Engineering and Decision Making (JCEDM), p. 33 (to appear, 2007)

Botts, M., Robin, A., Davidson, J., Simonis, I.: OpenGIS® Sensor Web Enablement Architecture Document, OpenGIS® Discussion Paper n OGC 06-021r1, Version: 1.0, p. 69 (2006)

Douglas, D., Peucker, T.: Algorithms for the reduction of the number of points required to represent a digitized line or its caricature. The Canadian Cartographer 10(2), 112–122 (1973)

Doyle, J., Han, Q., Weakliam, J., Bertolotto, M., Wilson, D.C.: Developing Non-proprietary Personalized Maps for Web and Mobile Environments. In: Kwon, Y.-J., Bouju, A., Claramunt, C. (eds.) W2GIS 2004. LNCS, vol. 3428, pp. 181–194. Springer, Heidelberg (2005)

Follin, J.-M., Bouju, A., Bertrand, F., Boursier, P.: Management of multi-resolution data in a mobile spatial information visualization system. In: Bertolotto, M., et al. (eds.) Proceedings of the 2nd Workshop on Web and Wireless GIS, WISE workshops, pp. 92–99. IEEE press, Singapore (2003)

Fournier, F., Devogele, T., Claramunt, C.: A role-based multi-agent model for concurrent navigation systems. In: Gould, M., et al. (eds.) Proceedings of the 6th AGILE Conference on Geographic Information Science, Presses Polytechniques et Universitaires Romandes, pp. 623–632 (2003)

Gardiner, K., Carswell, J.: Viewer-based directional querying for mobile applications. In: Bertolotto, M., et al. (eds.) Proceedings of the 2nd Workshop on Web and Wireless GIS, WISE workshops, pp. 83–91. IEEE press, Singapore (2003)

Guth, P., Oertel, O., Bénard, G., Thibaud, R.: Pocket panorama: 3D GIS on a handheld device. In: Bertolotto, M., et al. (eds.) Proceedings of the 2nd Workshop on Web and Wireless GIS, WISE workshops, pp. 100–111. IEEE press, Singapore (2003)

Imfeld, S.: Time, points and space, Towards a better analysis of wildlife data in GIS, Unpublished PhD report, University of Zurich, Switzerland (2000)

Ishikawa, Y., Tsukamoto, Y., Kitawaga, H.: Implementation and evaluation of an adaptive neighborhood information retrieval system for mobile users. In: Bertolotto, M., et al. (eds.) Proceedings of the 2nd Workshop on Web and Wireless GIS, WISE workshops, pp. 25–33. IEEE press, Singapore (2003)

Kollios, G., Gunopulos, D., Tsotras, V.J.: On indexing mobile objects. In: Proceedings of the 18th ACM Symposium on Principles of Database Systems, pp. 261–272. ACM Press, Philadelphia, Pennsylvania (1999)

Kang, H.-Y., Kim, J.-S., Li, K.-J.: Indexing Moving Objects on Road Networks in P2P and Broadcasting Environments. In: Carswell, J.D., Tezuka, T. (eds.) Web and Wireless Geographical Information Systems. LNCS, vol. 4295, pp. 227–236. Springer, Heidelberg (2006)

Li, X., Claramunt, C., Ray, C., Lin, H.: A semantic-based approach to the representation of network-constrained trajectory data. In: Kainz, W., Reidl, A., Elmes, G. (eds.) SDH 2006. Proceedings of the 12th International Symposium on Spatial Data Handling, pp. 451–464. Springer, Vienna, Austria (2006)

Luaces, M.R., Brisaboa, N.R., Parama, J.R., Rios Viqueira, J.R.: A Generic Framework for GIS Applications. In: Kwon, Y.-J., Bouju, A., Claramunt, C. (eds.) W2GIS 2004. LNCS, vol. 3428, pp. 94–109. Springer, Heidelberg (2005)

Noyon, V., Claramunt, C., Devogele, T.: A relative representation of trajectories in geographical spaces, Geoinformatica. Springer, Heidelberg (to appear, 2007)

Papadias, D., Zhang, J., Mamoulis, N., Tao, Y.: Query processing in spatial network databases. In: Freytag, J.F., et al. (eds.) Proceedings of 29th International Conference on Very Large Data Bases, pp. 802–813. Morgan Kaufmann, San Francisco (2003)

Petit, M., Ray, C., Claramunt, C.: A contextual approach for the development of GIS: Application to maritime navigation. In: Carswell, J.D., Tezuka, T. (eds.) Web and Wireless Geographical Information Systems. LNCS, vol. 4295, pp. 158–169. Springer, Heidelberg (2006)

Peytchev, E., Claramunt, C.: Experiences in building decision support systems for traffic and transportation GIS. In: Aref, W.G. (ed.) Proceedings of the 9th International ACM GIS Conference, pp. 154–159. ACM Press, Atlanta (2001)

Saltenis, S., Jensen, C.S., Leutenegger, S.T., Lopez, M.A.: Indexing the positions of continuously moving objects. In: Chen, W., Naughton, F., Bernstein, P.A. (eds.) Proceedings of ACM SIGMOD International Conference on Management of Data, pp. 331–342. ACM Press, Dallas, Texas (2000)

Sharifzadeh, M., Azmoodeh, F., Shahadi, C.: Change detection in time series data using wavelet footprints. In: Bauzer Medeiros, C., Egenhofer, M.J., Bertino, E. (eds.) SSTD 2005. LNCS, vol. 3633, pp. 127–144. Springer, Heidelberg (2005)

Taylor, G., Blewitt, G.: Intelligent Positioning: GIS-GPS, Unification Wiley, p. 200 (2006)

de Almeida, V.T., Güting, R.H.: Supporting uncertainty in moving objects in network databases. In: Proceedings of the 13th annual ACMGIS International Workshop on Geographic Information Systems, pp. 31–40. ACM Press, New York (2005)

Trajcevski, G., Wolfson, O., Hinrichs, K., Chamberlain, S.: Managing uncertainty in moving objects databases. ACM Transactions on Database Systems 29(3), 463–507 (2004)

A Theoretical Grounding for Semantic Descriptions of *Place*

Alistair J. Edwardes and Ross S. Purves

Department of Geography, University of Zurich
Winterthurerstrasse 190, 8057 Zurich, Switzerland
{aje,rsp}@geo.unizh.ch

Abstract. This paper is motivated by the problem of how to provide better access to ever enlarging collections of digital images. The paper opens by examining the concept of *place* in geographic theory and suggests how it might provide a way of providing keywords suitable for indexing of images. The paper then focuses on the specific challenge of how places are described in natural language, drawing on previous research from the literature that has looked at eliciting geographical concepts related to so-called basic levels. The authors describe their own approach to eliciting such terms. This employs techniques to mine a database of descriptions of images that document the geography of the United Kingdom and compares the results with those found in the literature. The most four most frequent basic levels encountered in the database, based on fifty possible terms derived from the literature, were road, hill, river and village. In co-occurrence experiments terms describing the elements and qualities of basic levels showed a qualitative accordance with expectations for example, terms describing element of beaches included shingle and sand and, those for beach qualities included being soft and deserted.

Keywords: place, semantics, image retrieval, basic levels, concept ontology, natural language description.

1 Introduction

Needs for geographic information increasingly encompass sources of data that do not fit easily with conventional systems for spatial data handling. In addition, techno-logies supporting these needs such as geographic information retrieval, collaborative web-maps and location-based services suppose quite different modes of use to more conventional Geographic Information Systems (GISystems). In particular, they seek to bring access to geographic information much closer to the ways in which people engage with the world and each other in everyday life [13][20]. This move from the universal and objective to the more individual and subjective necessitates a change in how geography is presented in GISystems. One consideration in achieving this shift is in how the idea of *place* might be articulated in GISystems [17][18][11]. Fisher and Unwin [14] note of this problem, "GI theory articulates the idea of absolute Euclidean spaces quite well, but the socially-produced and continuously changing notion of

J.M. Ware and G.E. Taylor (Eds.): W2GIS 2007, LNCS 4857, pp. 106–120, 2007.

place has to date proved elusive to digital description except, perhaps, through photography and film." (p. 6).

Curiously, this notion of *place* as being encapsulated by images has received relatively little attention in the field of image retrieval where indexing and retrieving a photograph is predicated on the ability to access a description of its contents in some form. Often such a description will have a strong geographical basis related to aspects of *where* the photograph was taken [2][3][26] such as by naming locations and features, events and activities, identifying types of geographical forms or environmental scene, or annotating particular characteristics of the physical or socio-economic landscape depicted.

In the field of image retrieval, which is generally recognised to lag behind the current state of the art in text retrieval, two main techniques exist to index the contents of a photograph, text-based and content-based [15].

Text-based image retrieval uses textual descriptions associated with image. These might be drawn from:

1. Descriptive terms or phrases chosen freely by an indexer [1].
2. Classification codes or controlled keyword lists manually assigned by an indexer [5].
3. The name of the image file or text associated with the photograph, e.g. the contents of a web-page [23].

The main issue with the first method is that terms selected can be highly subjective, prone to misspellings and inter-annotator agreement is often low [22]. One method to overcome such shortcomings has been through the deployment of annotation games where players gain points when they provide the same terms for a given image [1]. A possible criticism of this approach however is that it might bias annotation towards the use of terms that are essentially the 'lowest common denominators', for example objects and dominant colours, and so miss out on more complex semantics. The second method can ameliorate such factors by using a controlled list of terms and a more systematic approach to indexing. However, such word lists will usually be extensive, for example the Getty Images collection is indexed using 12,000 keywords with 45,000 synonyms [5], and are usually designed for specialised purposes e.g. Medicine or Art History. This means it is a time-consuming and labour intensive process to index images. In addition retrieval may be dependent on how familiar a user is with the indexing system and its terminology. The third method has the problem that there is often only a limited semantic overlap [23] between the contents of an image and the related text, which can be misleading to image retrieval. For example, Balasubramanian *et al.* [3] note that for most document genres on the web the text does not describe the image, but rather the image is used as support for the text.

Content-based image retrieval uses image processing techniques to try to match images to primitive features, for example based on colour, shape and texture [10]. A number of systems have been made commercially available; however their take-up has been minimal owning to a mismatch between the needs of users for higher level semantics and the capabilities of such systems that operate at a more primitive level. This difference is generally termed *the semantic gap*. Smeulders *et al.* [27] define this as:

> *"The semantic gap is the lack of coincidence between the information that one can extract from the visual data and the interpretation that the same data have for a user in a given situation."* (p.1353)

In attempting to bridge the semantic gap, content based techniques will often classify primitive level features against higher level semantic concepts (e.g. 'car' or 'mountain') defined in a controlled vocabulary (for example LSCOM http://www.lscom.org/). There is therefore a need in both text and content-based retrieval techniques for a structured set of concepts which can be used to index photographs.

The work described here focuses on text-based image retrieval and deals with the question of how people describe photographs in natural language, and in particular photographs whose contents are particularly geographic such as landscapes. The motivation of this work is the development of new technologies for indexing and retrieval of photographs using geographical aspects related to the setting in which they were taken. We hypothesise that, by describing *where* a picture was taken, we will also be able to describe *what* is in the picture on many occasions. Furthermore, we assume that by mining geographic data at a given location we will be able to formulate a useful description of *place*. Ultimately, the aim of this work will be to develop a *concept ontology* that structures a set of geographical descriptors that can be used to describe photographs and that can be derived from spatial data sources.

In this paper we firstly set out a theoretically solid foundation for exploring ways of describing *place* by reviewing literature on how the problem has been dealt with in different domains. Secondly, we examine a rich collection of georeferenced images to empirically explore how, in a given context, *place* is described. We then discuss how these results can be employed to help develop a *concept ontology* based on the use of basic levels [25].

2 Describing Images Through *Place*

2.1 Image Description

Shatford [26] considered the general issue of how people classify images, their contents and meaning. Shatford simplified the terminology of art historian Panofsky into three levels, *Generic Of, Specific Of* and, *About*. In relation to these she specified four facets, *who, what, where* and *when*, to generate the Panofsky-Shatford facet matrix (see also [2]) (Table 1).

The 'Where?' facet in Table 1 is particularly interesting for geographers. Here, *"Specific Of"* can be seen as referring to toponyms and entries in a gazetteer and *"Generic Of"* represents different types of topographic and architectural features. *"About"* then essentially introduces the concept of *place* and associates meaning with geographic locations. Hence, it is interesting to investigate the extent to which *place* related *About* concepts are used by people when describing images.

Table 1. The Pansofsky-Shatford facet matrix (Shatford, [26], p. 49)

Facets	Specific Of	Generic Of	About
Who?	Individually named persons, animals, things	Kinds of persons, animals, things	Mythical beings, abstraction manifested or symbolised by objects or beings
What?	Individually named events	Actions, conditions	Emotions, Abstractions manifested by actions
Where?	Individually named geographic locations	Kind of place geographic or architectural	Places symbolised, abstractions manifest by locale
When?	Linear time; dates or periods	Cyclical time; seasons, time of day	Emotions or abstraction symbolised by or manifest by

2.2 Place

It is often stated that Geography is the study of *Space* and *Place* [14]. These seemingly simple categories underlie many of the differences among geographers both in terms of what they study and how they go about it. The two concepts are often examined with respect to a continuum of viewpoints that range from the particular and the experiential at one end, to the abstract and the universal at the other [7]. *Place* is most distinct at the start, relating geography to human existence, experiences and interaction [29]. At the other end is the more detached, abstract and objective view of *space*, ultimately represented as geometry, which provides a means to think about, describe and encode the world in a logical way. As such, *Space* and *Place* are often emphasised as a ways of understanding (epistemology) rather than an as ontological categories such as a geographic location [9].

A number of researchers have sought to characterise the concept of *place*. Relph [24] highlighted elements of the physical setting (including location and physical appearance), the activities performed in places, and the meanings places have for people. Canter [6] identified four facets of place, *functional differentiation* (activities), *place objectives* (individual, social and cultural aspects of place), *scale of interaction* (e.g. room, home, neighbourhood city), and *aspects of design* (physical characteristics of a place). Gustafson [16] developed a somewhat different model encompassing, *self* (e.g. personal meanings and self-identification), *others* (social relations and norms), and *the environment* (the physical characteristics and natural conditions).

2.3 Natural Language Descriptions of *Place*

Work has also sought to understand how places can be described using natural language. Such studies have tended to take one of three approaches:

1. Collection of words from a dictionary validated by experts or experiment participants e.g. [12]

2. Collection of terms from documents relevant to a domain validated by experts or experiment participants e.g. [19]
3. Elicitation of terms directly from participants in an experiment

Results from the third approach are particularly relevant to this work. For example, Craik [8] sought to characterise the personality 'traits' of a landscape by eliciting adjectival qualities from participants in response to photographic stimuli of natural landscapes. Tversky and Hemenway [30], looked at how people categorise environmental scenes using *basic levels* [25]. They identified different types of basic-level scene by considering the informativeness of a word for a scene in respect to the different characteristics, attributes (qualities), types of related activities and, component parts. Their resultant basic level scene types have since been employed by researchers interested in content based image retrieval. For example, Vogel and Schiele [32] perform statistical analysis of local detail in images (amounts of *sky*, *water*, *grass*, *trunks*, *foliage*, *field*, *rocks*, *flowers* and *sand*) to categorise images into basic-level scene categories and transitional states between them. The work of Smith and Mark [28] also falls into this type of approach. They drew on the methodology of Battig and Montague [4] for eliciting category norms (similar to basic levels) from participants, and sought to identify 'folk' categories related to geography. They showed that, for example, that "mountain" was the most common norm identified for the category "a geographic feature." Most recently, Van Overschelde *et al.* [31] have also repeated Battig and Montague's work, also finding "mountain" as the most common norm for "a natural earth formation."

3 Experiments in Data Mining Descriptions of Places

A common goal in the work described in the previous section has been to identify vocabularies of concepts related to *place*. An obstacle for many of these studies is that they require human subject experimentation which is both costly and time consuming to perform. In this work we attempt to overcome this problem by analysing a database of textual descriptions of photographs which have a strong geographic component.

In order to explore the potential of this resource for developing such a vocabulary (which we will term a *concept ontology*), we present two experiments that employ simple data mining techniques to identify significant concepts. In the first we consider relative term frequencies (as an indicator of basic level categories) in order to compare the database as a resource with human subject testing reported in the literature. In the second we investigate how the resource can be employed to obtain sets of concepts that help characterise the semantics of basic level types of places.

3.1 Data

In these experiments we used data obtained from Geograph (www.geograph.org.uk). Geograph is a project with the aim to collect "geographically representative photographs and information for every square kilometre of the UK and the Republic of Ireland."

Fig. 1. Typical Geograph image – the associated caption reads: **Gatliff Trust Hostel on Berneray:** Picture taken from the beach on Berneray of the historic Gatliff Trust Hostel. Visited in the 1990s, shortly before the causeway linking Berneray to North Uist was built.

The project allows contributors to submit photographs representing individual 1km grid squares, and after moderation these images are uploaded together with descriptive captions to a publicly available web site. Figure 1 shows a representative example. Since a key aim of the project is to "…show at close range one of the main geographical features within the square" and to produce an associated description, we contend that the dataset may well be a digital representation of *place*. The database underlying Geograph is available under a Creative Commons licence. We obtained a snap shot taken on the 24th of February 2007, consisting of 346270 images taken by a total of 3659 individuals.

3.2 Methods

We performed two experiments using the Geograph data. In the first of these we looked at counts of terms identified as being basic levels or category norms in the experiments of Battig and Montague, Smith and Mark and Van Overschelde *et al.* (Table 2). Since we expected that some of these terms may vary according to the domain in which the original experiments were carried out (the US) we added appropriate synonyms for a UK context, (see [21] for a discussion). For example, we included the terms *loch* (Gaelic for lake) and *glen* (Gaelic for valley). We counted the total number of occurrences of each of these terms in the image descriptions in the Geograph database.

Table 2. Set of seed terms used for Experiment 1, sources are 1: Battig and Montague, 2: Smith and Mark, 3: Van Overschelde *et al.* and, 4: Added by the authors

Term	Source	Term	Source	Term	Source
Atlas	2	Glacier	3	Plain	1,3
Beach	3	Glen	4	Plateau	1,3
Building	2	Globe	1	River	1,2,3
Canyon	1, 3	Gorge	4	Road	1
Cave	1, 3	Grass	3	Rock	1,2,3
City	2	Highway	1	Sea	1
Cliff	1, 3	Hill	1,2,3	Sky	4
Clouds	4	Island	1,3	Stream	1,3
Coast	4	Lake	1,2,3	Street	1
Compass	1	Land	1	Town	1
Continent	1	Loch	4	Tree	4
Country	1	Map	1	Valley	1,2,3
County	1	Motorway	4	Village	4
Delta	1	Mountain	1,2,3	Volcano	1,3
Desert	2,3	Ocean	1,2,3	Waterfall	3
Elevation	1	Park	1	Woodland	4
Field	4	Peninsula	1	Woods	4
Forest	1				

In the second experiment we wished to explore how basic levels can be characterised. We chose four *basic levels* identified in the first experiment and explored the co-occurrence of these basic levels with two wordlists containing terms about: *elements* and *adjectives*. The list of elements attempted to describe the component parts of a scene as suggested by Tversky and Hemenway [30] and consisted of a list of 540 terms obtained from dictionary analysis using WordNet (http://wordnet.princeton.edu/). The list of adjectives employed the Landscape Adjective Check List of Craik and was aimed at capturing more abstract, *About* type concepts relating to *place*. In this experiment we searched for descriptions in which both a basic level and a term co-occurred. A description was counted only once however many times a pair of terms co-occurred within it, although in practice most descriptions contained only one or two sentences. We retrieved not only the count of co-occurrences of terms with basic levels, but also the count of occurrences of both the term and the basic level, in order to explore whether the frequency of a co-occurrence was simply explained by a term being very common.

4   Results

Table 3 compares the top rankings of basic level terms described in the literature with the ranking from the Geograph database using the seed terms given in Table 2. For the Battig and Montague and Van Overschelde *et al.* experiments participants had been asked to name 'natural earth formations' for the Smith and Mark experiment we unioned all terms elicited by all their experimental categories (e.g. 'a kind of geographical feature', 'something geographic' etc.) and ranked them according to the total number of responses.

Table 3. Top 10 terms from Battig and Montague (B&M), Top 20 from Smith and Mark (S&M), Van Overschelde *et al.,* (vanO) and this work (Geograph)

B&M	VanO	S&M	Geograph
Mountain	Mountain	Mountain	Road
Hill	River	River	Hill
Valley	Ocean	Lake	River
River	Volcano	Ocean	Village
Rock	Lake	Hill	Building
Lake	Valley	Country	Park
Canyon	Hill	Sea	Street
Cliff	Rock	City	Valley
Ocean	Canyon	Continent	Field
Cave	Plateau	Valley	Loch
	Tree	Plain	Land
	Plain	Plateau	Town
	Cave	Map	Forest
	Glacier	Road	Map
	Grand Canyon	Island	Sea
	Island	Desert	Woodland
	Stream	Peninsula	Tree
	Cliff	State	Beach
	Desert	Volcano	Country
	Beach	Forest	Glen

Table 4. Correlation between the relative rankings of each study using only the terms found by Battig and Montague

	B&M	S&M	VanO	Geograph	with subs
Mountain	1	1	1	6	7
Hill	2	6	6	1	1
Valley	3	5	5	3	3
River	4	2	2	2	2
Rock	5	7	7	5	6
Lake	6	3	4	4	5
Canyon	7	8.33	8	10	10
Cliff	8.5	8.33	10	7	8
Ocean	8.5	4	3	9	4
Cave	10	8.33	9	8	9
Spearman's Rank Correlation					
	B&M	S&M	VanO	Geograph	with subs
B&M	1	0.625	0.597	0.7	0.555
S&M		1	0.968	0.463	0.555
vanO			1	0.345	0.539
Geograph				1	0.818
With subs					1
significance level at $P_{0.05} = 0.648$					

In Table 4, the top ten terms of Battig and Montague are used to consider the rank obtained from the other experiments. The Geograph data are first ranked according to the raw occurrences of terms and secondly after substituting the terms sea for ocean, loch for lake and, gorge for canyon, since these were considered to be synonyms for these terms in British English. Correlations between the rankings are also shown with the ranking of Smith and Mark being significantly correlated with that of van (p < 0.05) and the Geograph ranking (without substitution) significantly correlated with that of Battig and Montague (p < 0.05).

Tables 5 and 6 show the total number of counts of elements and adjectival terms co-occurring with four basic level types of place (beach, village, hill and mountain). The terms were ranked by the frequency of the co-occurrence with respect to the occurrence of the term (e.g. shingle appears 254 times in total, of which 43% of occurrences are found in conjunction with beach).

Table 5. Element terms associated with basic-levels. N is the total number of occurrences of element term, n the no. occurrences of the basic level scene term and *Freq* the number of co-occurrences/N.

Elements						
Term	*Freq.*	*N*		*Term*	*Freq.*	*N*
Beach		n=2824		**Village**		n=12707
Shingle	0.43	254		Pub	0.15	3205
Sand	0.17	1037		Shop	0.13	1464
Cliff	0.10	1124		Inn	0.12	1892
Headland	0.10	524		Church	0.10	16157
Bay	0.08	2770		Housing	0.07	2820
Sea	0.08	3725		Edge	0.07	6541
Rock	0.06	1914		Cottage	0.06	2966
Coast	0.04	2548		Main Road	0.05	6735
Shore	0.04	1298		Village green	0.05	12707
Island	0.03	2645		Stone	0.05	4817
Hill		n=16232		**Mountain**		n=1256
Fort	0.30	1124		Peak	0.03	823
Top	0.21	6086		Summit	0.03	3427
Summit	0.18	3427		Ridge	0.02	3263
Horizon	0.14	2047		Moorland	0.02	1484
Ridge	0.13	3263		Quarry	0.01	2273
Sheep	0.10	2659		Stream	0.01	2478
Valley	0.08	8563		Sheep	0.01	2659
Side	0.08	11369		Forest	0.01	4542
Trees	0.08	7753		Top	0.01	6086
Track	0.08	8331		Path	0.01	6477

Table 6. Adjectival terms associated with basic-levels. N is the total number of occurrences of adjectival term, n the no. occurrences of the basic level scene term and *Freq* the number of co-occurrences/N.

Adjectives					
Term	*Freq.*	*N*	*Term*	*Freq.*	*N*
Beach		n=2824	**Village**		n=12707
Sandy	0.26	553	Deserted	0.31	233
Deserted	0.10	233	Pretty	0.15	616
Eroded	0.09	288	Green	0.14	4993
Soft	0.08	181	Quiet	0.12	1005
Rocky	0.06	756	Lovely	0.10	919
Warm	0.06	241	Pleasant	0.10	781
Glacial	0.06	290	Beautiful	0.08	1177
Low	0.05	2866	Remote	0.07	510
Beautiful	0.05	1177	Unusual	0.07	1175
Lovely	0.02	919	Large	0.06	5273
Hill		n=16232	**Mountain**		n=1256
Steep	0.19	2131	Distant	0.03	810
Distant	0.18	810	Black	0.03	1893
Wooded	0.16	1053	Remote	0.02	510
Black	0.12	1893	Rocky	0.02	756
Rough	0.12	1367	Grassy	0.01	622
Grassy	0.11	622	Steep	0.01	2131
Round	0.11	1605	Natural	0.01	702
Big	0.10	1337	Dark	0.01	537
White	0.10	2642	Broad	0.01	655
Broad	0.09	655	Running	0.01	2480

5 Discussion

Experiment 1 looked at the relative occurrence of particular terms in relation to studies on geographic basic levels reported in the literature. The aim of this experiment was to investigate how comparable the basic level terms identified in the Geograph database were with those resulting from direct human subjects testing.

The results show a number of interesting differences between the Geograph data and the other work. One obvious difference is the presence of highly ranked anthropogenic features in the Geograph list. This is not surprising in the case of the Battig and Montague and Van Overschelde *et al.* lists since participants in these experiments were asked for 'natural earth formations'. However, Smith and Mark's experiments did elicit anthropogenic objects, but these were not found in their top ten. By contrast, such terms appear frequently in the Geograph result. In addition, those that are in the Smith and Mark's list tend to be on have a larger spatial extent (e.g. city) than those of the Geograph list (e.g. building, town and village).

A second observation is that certain concepts in the Geograph list appear to have been substituted. Hill appears to be used instead of mountain, sea rather than ocean, and loch rather than lake. On the one hand, these differences can be seen as relating to the variations in experimental methods. The literature studies asked participants to think of concepts in response to abstract questions whereas the Geograph data comes from people describing concepts that can be seen in a picture. For example, there are few actual examples of concepts such as canyons or volcanoes in Britain, which is likely why these were not frequently found. On the other hand, this substitution could suggest that the use of terms is not insensitive to geographical context, as found in previous empirical work [21]. In Britain the term hill appears to be much more widely used than mountain, and large bodies of open water which would be described as lakes in England occur more frequently in Scotland where the Gaelic appellation loch is almost exclusively used. Interestingly, lochs may be open to the sea and thus may describe a concept which overlaps with, but is not identical to lake, defined by WordNet as "a body of (usually fresh) water surrounded by land." In addition, whilst Britain is bordered by both oceans and seas, it would appear that the term sea is more conventionally used.

Table 4 compared the relative rankings of the most frequently used 'natural earth formations' of Battig and Montague. For better consistency in terminology, the terms and rankings for the question "a kind of geographic feature" were used in the case of Smith and Mark. It is interesting to note that the term rankings of Van Overschelde *et al.* and Smith and Mark using students from the US correlate significantly at the $p<0.05$ level, but that neither of these lists correlate significantly with either those of Battig and Montague or derived from Geograph. Again, this might suggest that the conceptualisations are sensitive to locale. Of note is that the substitutions (e.g. sea for ocean) increased correlations though not sufficiently to be significant. One possible suggestion is therefore that the difference is not only one of terminology but also of environmental familiarity (for example British landscapes consist less of lakes and mountains and more of hills and valleys). The Geograph ranking was significantly correlated with that of Battig and Montague, though only where no terminology substitutions were applied. However caution is required to avoid over interpreting these results since the significance of the Spearman's rank correlation is very sensitive when comparing so few categories.

Experiment 2 aimed to investigate the ability of the Geograph dataset to provide characteristic sets of concepts describing semantics for particular examples of basic level places. It used word lists of elements (parts) and adjectives in relation to two basic levels (village and hill) identified in the previous experiment and two (mountain and beach) suggested by Tversky and Hemenway. The selection of these levels was to both consider places that were typical to the area covered by the Geograph dataset and ones that would allow comparison with Tversky and Hemenway's results in terms of contents. In addition, it was intended to examine the ability of the technique to provide sets of terms that are both sensible and distinct enough to discriminate between levels even, in the case of hill and mountain, when they could be seen as somewhat synonymous.

It is interesting to note (Table 5 and 6) that there are strong differences in the terminology used for each of the basic levels. In terms of elements, beach was identified with coastal landforms and materials like rock, sand and sea. This compares well with the results of Tversky and Hemenway whose list contained water and sand. The elements of village were instead very functional and architectural and typical of what might be expected in typical British villages e.g. cottages, a village green, a shop, a pub and a church. There are some similarities between hill and mountain, particularly in the use of perceptual terms such as top, summit, and ridge. However, differences can also be seen both in terms of the language used for similar concepts, for example path and forest were found for mountains and track and trees for hills, and in terms of composition, for example valley and hill sides for hills and moorland and streams for mountains. Similarities for mountain were also found with the work of Tversky and Hemenway, for example animal, trail, and stream and; sheep, path, and stream. One interesting difference is where they found grass and tree, our results identified moorland and forest, which are more typical of the mountain landscapes found in Britain.

The results for adjectives were also quite distinct for each of the levels. Beaches, for example, tended to be described using process (eroded, glacial) and affective (soft, beautiful, warm, lovely) terms. Villages have some similarities with the use of affective terms, but tend to invoke more of a sense of an idyllic place (pretty, lovely, pleasant, beautiful), in addition there seems to be an emphasis on remoteness from other settlements (deserted, quiet, remote). The adjectives for hill and mountain are in many ways similar to the description of their elements describing perceptual characteristics and land cover. To an extent these are shared (steep, distant, broad, black, grassy) but there are also differences. Hills are described as being big, round and wooded whereas mountains are described in terms that are more desolate such as remote, natural, rocky and, dark.

Overall the results are very promising. The sets of concepts found for the basic levels fit well with the authors' conceptions of the types of places they describe, at least in a British context. They are also diverse enough to allow discrimination between levels. For mountain and hill there are a number of commonalities (which is expected for basic levels, see [30] p.123). However these tend to reflect the perceptual characteristics of the landforms which are inherently similar. Other characteristics such as land cover, infrastructure and remoteness were much more differentiated suggesting that the two concepts cannot be seen as synonymous. Nonetheless, a degree of caution is required as, as can be seen in Tables 5 and 6, the occurrence values (n) for mountain concepts are generally an order of magnitude lower than for hill. This reflects the scarcity of mountains within the dataset in general and makes the level sensitive to idiosyncratic descriptions.

The experiments presented thus far are preliminary work and have a number of important limitations, related to the dataset used, the set of assumptions and, the experimental methods themselves. Firstly, we assume that the Geograph dataset will help us to build a concept ontology of how people describe *place*. The dataset itself is very large (346270 images taken by a total of 3659 individuals) which we hope gives some universality to the results. However, as demonstrated in Experiment 1 the results (for example the use of the term hill instead of mountain) appear sensitive to locale, and the participants in the Geograph experiment are likely to be a self-selecting group (they have access to the internet and digital cameras and an interest in documenting the geography of the UK), which implies in turn that the results may be biased. Secondly, the terms found are limited to those we searched for, and although we based our search terms on theory we cannot rule out synonyms having much higher frequencies. Third, we did not treat toponyms in a special way – we assume that if hill is mentioned the picture itself and thus the place is a hill – in other words that a toponym is a instantiation of a type of place. More importantly, terms which are found in toponyms (e.g. "Black Hill") are not being used to describe a location and thus our co-occurrence experiments should be controlled for such effects. However, in most cases it appears unlikely that the adjectives found are likely to be parts of toponyms. In general we did not control for either word sense (i.e. does running refer to an activity or describe a property of a stream) or context (i.e. is the term really being used in conjunction with the basic level in question). Nonetheless, we believe that the results presented demonstrate the potential richness of new sources of data as possible methods of exploring the creation of, for example, landscape concept ontologies, which have until now been mostly limited to human subject experiments with correspondingly much smaller data volumes.

6   Conclusions and Further Work

The research reported here has argued that as geographic information technologies expand into new areas and new aspects of life they need to be supported by more extensive ideas of geography, better reflecting how people experience and think about the world in everyday life. Here, theory on *place* has been proposed to provide support in exploring the problem of image description and retrieval.

The work has demonstrated a novel approach to eliciting concepts of *place* as they are expressed in natural language through the analysis of a large database of images and their captions from a collaborative web project (Geograph). This resource has be shown to compare favourably to approaches in the literature that have been based on human subject testing, though caveats have been acknowledged in relation to sources of bias. The use of online resources has the advantage that it can help to obviate the need for costly and time-consuming participant experiments.

The potential of the resource to develop an ontology of *place* concepts for use in indexing and search of images has also been explored. This was based on the idea of basic level scene types and characteristic terms (elements and adjectives) that commonly co-occur with these. In particular, it was found that adjectival terms (less studied previously) are important descriptors and discriminators of places.

These experiments will be expanded in future work to populate a more comprehensive concept ontology, considering both a wider variety of basic levels and a number of other place related aspects (e.g. activities). The influence of toponyms and ambiguity on our results also be examined, as well as the issue of how to detect such concepts from analysis of spatial and other data.

Acknowledgements. We would like to gratefully acknowledge contributors to Geograph British Isles, see http://www.geograph.org.uk/credits/2007-02-24, whose work is made available under the following Creative Commons Attribution-ShareAlike 2.5 Licence (http://creativecommons.org/licenses/by-sa/2.5/). The research reported in this paper is part of the project *TRIPOD* supported by the European Commission under contract 045335.

References

1. von Ahn, L, Dabbish, L.: Labelling Images with a Computer Game. In: CHI 2004, Vienna, Austria (April 24-29, 2004)
2. Armitage, L.H., Enser, P.G.B.: Analysis of user need in image archives. Journal of Information Science 23(4), 287–299 (1997)
3. Balasubramanian, N., Diekema, A.R., Goodrum, A.A: Analysis of User Image descriptions and Automatic Image Indexing Vocabularies: An Exploratory Study. In: International Workshop on Multidisciplinary Image, Video, and Audio Retrieval and Mining, Sherbrooke, Quebec, Canada (October 25-26, 2004)
4. Battig, W.F., Montague, W.E.: Category norms for verbal items in 56 categories: a replication and extension of the Connecticut Norms. J. of Experimental Psychology 80(2), 1–46 (1969)
5. Bjarnestam, A.: Text-based Hierarchical Image Classification and Retrieval of Stock Photography. In: The Challenge of Image Retrieval Conference, Newcastle, UK (1999)
6. Canter, D.: The Psychology of Place. Architectural Press, London (1977)
7. Couclelis, H.: Location, place, region, and space. In: Abler, R.F., Marcus, M.G., Olson, J.M. (eds.) Geography's Inner Worlds, Rutgers University Press, pp. 215–233 (1992)
8. Craik, K.H.: Appraising the Objectivity of Landscape Dimensions. In: Krutilla, J.V. (ed.) Natural Environments: Studies in Theoretical and Applied Analysis, Resources for the Future, Baltimore, pp. 292–346 (1971)
9. Curry, M.R.: Toward a Geography of a World Without Maps: Lessons from Ptolemy and Postal Codes. Annals of the Association of American Geographers 95(3), 680–691 (2005)
10. Eakins, J.P., Graham, M.E.: Content-based image retrieval, JISC Technology Applications Programme Report 39 (October 1999)
11. Edwardes, A.J.: Re-placing Location: Geographic Perspectives in Location Based Services, Ph.D Thesis, University of Zurich (2007)
12. Edwards, J.A, Templeton, A.: The structure of perceived qualities of situations. European J. of Social Psychology 35, 705–723 (2005)
13. Egenhofer, M.J., Mark, D.M.: Naive geography. In: Kuhn, W., Frank, A.U. (eds.) COSIT 1995. LNCS, vol. 988, Springer, Heidelberg (1995)
14. Fisher, P., Unwin, D.: Re-presenting Geographical Information Systems. In: Unwin, D.J., Fisher, P. (eds.) Re-presenting GIS, pp. 1–17. Wiley & Sons, London (2005)
15. Goodrum, A.A.: Image Information Retrieval: An Overview of Current Research. Informing Science 3(2), 64–67 (2000)
16. Gustafson, P.: Meanings of place: everyday experience and the theoretical conceptualizations. J. of Environmental Psychology 21, 5–16 (2001)
17. Harrison, S., Dourish, P.: Re-place-ing space: The roles of place and space in collaborative systems. In: CSCW 1996. Proc. Computer Supported Cooperative Work, Boston, MA, pp. 67–76. ACM Press, New York (1996)
18. Jordan, T., Raubal, M., Gartrell, B., Egenhofer, M.: An affordance-based model of place in GIS. In: SDH 1998. Proc. of 8th Intl. Symp. on Spatial Data Handling, International Geographic Union, Vancouver, Canada, pp. 98–109 (1998)

19. Kasmar, J.V.: The Development of a Usable Lexicon of Environmental Descriptors. Environment and Behavior 2, 153–169 (1970)
20. Llobera, M.: The nature of everyday experience: examples on the study of visual space. In: Unwin, D.J., Fisher, P. (eds.) Re-presenting GIS, pp. 171–195. Wiley & Sons, London (2005)
21. Mark, D.M, Turk, A.G.: Landscape Categories in Yindjibarndi: Ontology, Environment, and Language. In: Kuhn, W., Worboys, M.F., Timpf, S. (eds.) COSIT 2003. LNCS, vol. 2825, pp. 28–45. Springer, Heidelberg (2003)
22. Markey, K.: Interindexer Consistency Tests: A Literature Review and Report of a Test of Consistency in Indexing Visual Materials. Library and Information Science Research 6, 155–177 (1984)
23. Marsh, E.E., White, M.D.: A taxonomy of relationships between images and text. J. of Documentation 59, 647–672 (2003)
24. Relph, E.: Place and Placelessness, Pion, London (1976)
25. Rosch, E.: Principles of categorization. In: Rosch, E., Lloyd, B.B. (eds.) Cognition and Categorization, Erlbaum, Hillsdale, NJ (1978)
26. Shatford, S.: Analyzing the subject of a picture: a theoretical approach. Cataloguing and Classification Quarterly 6, 39–62 (1986)
27. Smeulders, A.W.M., Worring, M., Santini, S., Gupta, A., Jain, R.: Content-Based Image Retrieval at the End of the Early Years. IEEE Transactions on Pattern Analysis And Machine Intelligence 22(12), 1349–1380 (2000)
28. Smith, B., Mark, D.M.: Geographical categories: an ontological investigation. Int. J. Geographical Information Science 15(7), 59–612 (2001)
29. Tuan, Y.-F.: Space and Place: The Perspective of Experience, U of Minnesota Press (1977)
30. Tversky, B., Hemenway, K.: Categories of Environmental Scenes. Cognitive Psychology 15, 121–149 (1983)
31. Van Overschelde, J.P, Rawson, K.A, Dunlosky, J.: Category norms: An updated and expanded version of the Battig and Montague (1969) norms. J. of Memory and Language 50, 289–335 (2004)
32. Vogel, J.B., Schiele, B.: A Semantic Typicality Measure for Natural Scene Categorization. In: Rasmussen, C.E., Bülthoff, H.H., Schölkopf, B., Giese, M.A. (eds.) Pattern Recognition. LNCS, vol. 3175, Springer, Heidelberg (2004)

Towards the Geo-spatial Querying of the Semantic Web with ONTOAST

Alina Dia Miron, Jérôme Gensel, Marlène Villanova-Oliver, and Hervé Martin

Laboratoire d'Informatique de Grenoble,
681 Rue de la Passerelle BP 72,
38402 Saint Martin d'Hères Cedex France
{Alina-Dia.Miron, Jerome.Gensel, Marlene.Villanova-Oliver,
Herve.Martin}@imag.fr

Abstract. One of the challenges raised by the construction of the semantic Web lies in the analysis and management of complex relationships (thematic, spatial and temporal) connecting several resources. The automatic discovery of such relations will improve the current capabilities of existing search engines. Spatial information plays an important role in the resources available on the Web, thus, integrating spatial criteria into queries addressed to search engines would increase the power of expression of the formulation and will improve the search result. However, ontology languages of the semantic Web, OWL in particular, still do not have the expected specific characteristics for a well adapted representation and exploitation of spatial data. We present here ONTOAST, a spatial ontology modeling and reasoning system. ONTOAST manages qualitative spatial relations which can be used to express spatial queries. It also handles the inference of new qualitative relations, thus increasing the spatial-based search capabilities.

Keywords: Geo-spatial ontology, qualitative spatial relations, spatial annotation, spatial query, OWL, AROM, ONTOAST.

1 Introduction

Most of Web documents contain spatial references, in the form of addresses or place names, which are, in the majority of cases, ignored or hardly used by search engines. The task of information retrieval could use localization information in order to refine or to better specify queries. For instance, when searching a person, whose name is known (e.g.. "François Martin"), a regular syntactic search engine usually returns several results, which do not all refer to the same person. A solution to refine the answers is to use information concerning the spatial localization of the searched person (his home town, his work address…). For instance, we can enhance the initial query ("François Martin") by adding the residence town: "Grenoble". Existing search engines demand exact word matching and thus can exclusively retrieve pages that contain "François Martin" and/or "Grenoble". Resources containing the zip code 38000, or the alternative metaphorical name "capital of the Alpes", etc. are ignored.

J.M. Ware and G.E. Taylor (Eds.): W2GIS 2007, LNCS 4857, pp. 121–136, 2007.

Moreover, the presence of the two elements on the same web page does not necessarily mean that François Martin lives at Grenoble. For enhancing query expressivity, we are interesting in analyzing spatial relations expressing knowledge like: François Martin works in the *south of* Grenoble, *near* the University Campus...

Exploring geo-spatial information, either expressed as data or meta-data, requires the use of some dedicated representation and reasoning formalisms, adapted to this kind of information and compatible with the semantic Web technologies [1]. The semantic Web was imagined as an evolution of the current one towards a gigantic and distributed knowledgebase integrating semantic models linked to resources via annotations. Those semantic models, also known as ontologies [2], are a mean, for human users (via different search engines) but also for software agents, to easily exploit and retrieve information. But in order to reach the goal of *"omnipresent semantic definitions"* an extra effort is required from users, for they need to specify the meaning of resources they define and use. Ideally, each Web page should contain, beside its informal data (pictures, text, links, etc.), a formal semantic description of its content and meta-data about its creator, its localization...

Proposed by the W3C to support the semantic Web approach, OWL [3] has become the standard language for representing ontologies. Streaming from researches in the domains of Description Logics, Conceptual Graphs and Frame languages, OWL offers a great modeling expressivity as well as a wide variety of reasoning engines (Pellet, RacerPro, Fact...). Nevertheless, regarding the representation of spatial or/and temporal data, an expressive language like OWL shows to be too generic and not very well intended for modeling this kind of information. This drawback as well as the lack of spatial inference engines has motivated our work of adapting the structure of AROM [4], an Object-based Knowledge Representation System, for better responding to the growing demand of geo-spatial data representation and reasoning. So far, AROM comes with two extensions i) AROM-ST [5] which supports the management of space and time in knowledgebases with applications in the field of geomatics and ii) AROM-ONTO [6], a version of AROM compatible with OWL-DL in terms of power of representation. AROM-ONTO offers the modeling structures necessary to define ontologies *"à la OWL"* and thus opens AROM to the semantic Web annotations.

Our first objective is to integrate these two extensions into a new module: ONTOAST (for ONTOlogies in Arom-ST). The main advantage of ONTOAST lies not only in its capacity to model and reason about spatial features of ontological concepts but also in its compatibility with the standard ontology language OWL. In order to complete this extension, we propose to integrate into ONTOAST a set of qualitative spatial relations, in order to increase the flexibility and the expressivity of the language and to allow some spatial reasoning even when precise numeric data is unavailable. For this purpose, AROM Algebraic Modeling Language (AML), extended here with *topology*, *orientation* and *proximity operators*, is used to construct complex spatial relations. Finally, exploiting the compatibility between AROM and OWL, ONTOAST supports the definition of geo-spatial ontologies for the semantic Web, as well as the formulation and resolution of qualitative spatial queries over these ontologies. For this purpose, ONTOAST exploits the reasoning mechanisms offered by AROM (instance classification, value definition through AML equations ...).

This paper is organized as follows. Section 2 details the context of our work. Section 3 presents the object-based representation system AROM, its two extensions AROM-ST and AROM-ONTO as well as their coupling which results in the ONTOAST system. Section 4 describes the three types of qualitative spatial relations integrated into ONTOAST: topology, direction and distance. Section 5 concludes by giving the future directions of our work.

2 Context

Two challenges raised by the semantic Web, and hopefully the geo-spatial semantic Web are, on the one hand, the handling of flexible and imprecise queries (and thus data), and on the other hand, the definition of algorithms performing an effective and exhaustive analysis of complex relations (thematic, spatial or temporal) between several resources. Solutions which support flexible (close to natural language) query formulation, in particular those coming from the field of qualitative reasoning, can be adapted to address the information on the Web. Researches in the *qualitative* domain [7] improve the flexibility by adapting inferences to the human way of thinking and allowing deductions even when precise numerical data is missing. Furthermore, Geographical Information Systems (GIS) build on database management techniques, offer long-established, efficient solutions for handling numeric as well as qualitative spatial relations between geographical objects. Qualitative relations are rarely pre-calculated and stored in databases, but are rather determined by geometric calculations launched on the fly. For instance, it is highly unlikely that a geospatial database will explicitly store all topological relations between one country and all the cities in the world. Instead it is plausible that, using the topological operators based on geometric calculations, the inclusion constraint is checked at the very moment where this kind of question is raised.

The problem of qualitative data handling can be easily transposed from the GIS domain to the semantic Web, since the later can be seen as a gigantic database containing geo-spatial information associated with resources. It would then be valuable to adapt the semantic Web representation formalisms and languages to suit the particularities of spatial information. Additionally, it is necessary to define new inference engines capable of exploiting the spatial data by deducing new knowledge on demand.

At this time, with the standard ontology language OWL, it is possible to create expressive ontologies which model complex domains and to reason with facts and axioms contained within these ontologies by using external inference engines. Nevertheless, the typing system used by OWL is rather limitative and does not provide yet the required characteristics for a suitable representation and exploitation of spatial data. As a matter of fact, it only offers a set of predefined types proposed by XML Schema which cannot be easily extended. This makes the modeling of spatial data a difficult task. The authors of [8] have shown that if the space representation is not a fundamental limitation of OWL, it is still far from being intuitive. In this direction, Dolbear and Hart [9] have studied the conversion of an ORACLE spatial database into an OWL ontology. The spatiality of concepts (in our example the geometric contour or the point localization) was modeled by sets of coordinates which mitigate the lack of dedicated primitive types in OWL (see Fig. 1).

Fig. 1. The geometric description of a department and a city using the OWL language

Still, even if possible to reproduce in OWL, spatial primitives like points, polygons or areas would be exclusively used for object descriptions. We can only attach these descriptions to the resources they annotate (the city of Grenoble or the 38th French department). For example, if one knows the geographical coordinates (the contour) of France and those of the city of Grenoble, one can associate them respectively with the resources which represent France and the city of Grenoble. Inferring that Grenoble is a French city requires a spatial reasoner capable of deducing new spatial relations from existing knowledge, but, up to now, the existing inference engines do not offer such geo-spatial inference capabilities. As a result, for modeling the spatiality of resources, one can exclusively exploit the explicit descriptions embedded in the referred ontologies. Or it is physically impossible to pre-calculate and store all existing spatial relations between all spatial objects, especially in a giant open environment like the Web.

These qualitative spatial relations could be inferred using mathematical operators like those defined by GIS. As an illustration, let us suppose that the coordinates of France and the *inclusion* relation with Grenoble are known. A topological operator would be able to infer the fact that Grenoble is located in Europe, in the North of the Mediterranean Sea, etc.

3   From AROM to ONTOAST

AROM [4] is a generic tool for knowledge modeling and exploitation, in the family of Object oriented Knowledge Representation Systems (OKRS). Knowledge representation in AROM relies on two concepts of equal importance: *classes* and *associations*. AROM *classes* describe a set of *objects* sharing common properties and constraints. Object properties are modeled by *variables* while *constraints* are expressed by algebraic equations binding variables of different classes of the current knowledgebase. One of the originalities which differentiate AROM from other OKRS, lies in the explicit representation of relations between *classes* by mean of *associations*. AROM uses *associations* similarly to those found in UML, for organizing links between objects with common structure and semantics. These links, described by means of *roles*, can also possess properties modeled by association *variables* and *constraints*. Classes and associations are hierarchically organized by the specialization relation. Another interesting feature of AROM is an algebraic modeling language (AML) which can be used for writing operational knowledge in a declarative manner. This language allows one to specify the value of a *variable* using numerical and symbolic equations between the various elements composing the knowledgebase. It can also be used to apply integrity *constraints* for classes or associations, and to query the knowledgebase using a predefined formalism (AROM Query). The AROM AML relies on a set of predefined operators for AROM types: basic *arithmetic operators (addition, subtraction, power, whole part, etc)*, *comparison operators (superior, inferior, different, etc)* or *trigonometrical operators (sin, cosin, tangent, etc)*. AROM also integrates most of the inference mechanisms found in object knowledge representation systems such as: default value, inheritance, procedural attachment, filtering and instance classification.

Another advantage when using AROM is the presence of an extensible type system [10]. On top of this type system we have built a new module AROM-ST [5] which partially integrates the GML data model proposed by OpenGIS [11]. More precisely, AROM-ST, in its current version, deals with a set of simple geometrical types: *Point, Polyline, Polygon, Line* and *LinearRing*, as well as a set of complex types: *Multipoint* (set of points), *MultiLine* and *MultiArea*. In order to allow an adapted handling of spatial attributes we have also extended the AML's core with a set of spatial operators [5], namely *topological operators* expressed by binary predicates testing the relative position of two objects in space (*disjoint, touches, overlaps, inAdjacent, within/contains, crosses* and *equals*), *set operators* allowing the manipulation of space as a set of regions (*union, intersection, symmetricalDifference and difference*) and *measurement operators (dimension* and *distance)*. Modeling and exploiting spatial knowledgebases with AROM-ST both become easy and intuitive tasks.

In recent work [6], we have explored ways of adapting AROM to the semantic Web, by proposing a comparative study with OWL following three axes: representation, typing and inference. This comparison has shown an important deficit of AROM from the point of view of the representation expressivity. As a result, we have proposed an extension of the AROM meta-model aiming at reducing the representation gap between the two languages. The new meta-model, AROM-ONTO, proposes new representation structures (defined classes, object properties, property restrictions, etc) which bring AROM closer to OWL. Finally we have compared the typing systems and the inference mechanisms offered by the two languages and establish important complementarities. For instance, description logics reasoners built or adapted for OWL ontologies handle concept level inferences (i.e. concept classification, consistency checking, etc.) while AROM offers instance level inferences (i.e. AML equations, procedural attachment, instance classification, etc.).

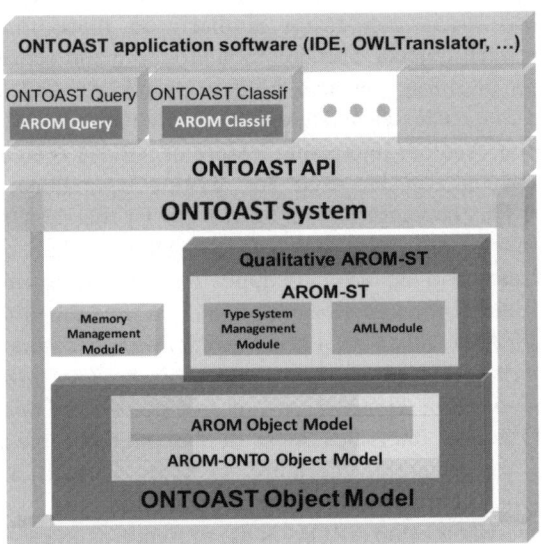

Fig. 2. The modular organization of the ONTOAST

ONTOAST (for ONTOlogies in Arom-ST) is a version of AROM which includes both the spatial types and operators managed by AROM-ST, and the core extension proposed by AROM-ONTO to support the compatibility with OWL (see Fig. 2). ONTOAST is a spatial ontology modeling and management system creating a bridge between the GIS domain and the semantic Web. Furthermore, ONTOAST integrates a set of predefined qualitative spatial relations used to complete the available data on modeled objects as well as to allow a more flexible query formulation. Those qualitative relations can be automatically inferred from existing knowledge when they are needed, or explicitly defined by users.

4   Use Case

For illustrating our approach, we consider the simple case of annotating a Web page which describes a set of spatial concepts (Fig. 3), using Protégé [12] and the iAnnotate plug-in [13]. The Web page is dedicated to French speaking countries. Annotations are defined with respect to, on the one hand, a chosen ontology (the window in the middle) which contains the definitions of concepts we refer to (Department, Lake, Country, etc) and, on the other hand, the classes which correspond to the geometrical types (point, polygon, etc.).

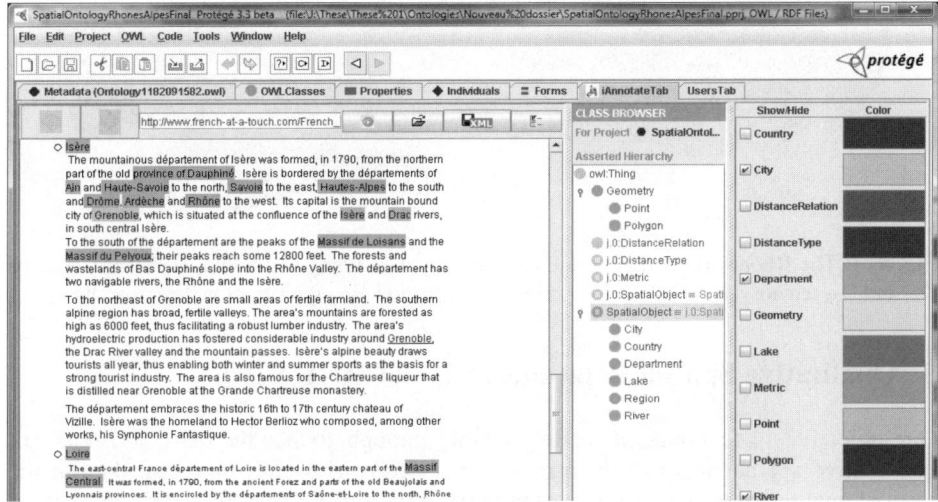

Fig. 3. Example of a practical case of annotating with spatial information a web page dedicated to French regions and departments

By directly selecting the text to be annotated on the web page (left tab), we can create instances of concepts defined by the reference ontology (center tab). Those instances are underlined with the color corresponding to the concepts they materialize. For example, all the departments are underlined in light green. Annotated instances are automatically added to the ontology, which can be further enhanced by defining the data type and the object properties that characterize the new resources. We can, for example, attach geometries to each object or specify the fact that the French department named the Drôme shares a relation *SouthOf* with the French department called the Rhône.

The resulting ontology can be easily converted to ONTOAST using an OWL-AROM translator based on XSLT rules. An illustration of the result is given in Fig.4. At the instance level, the nine spatial objects (the 8 departments and the lake) are associated with their exact geometries (contours modeled by polygons). Let us consider the existence of a ninth department, the Alpes de Hautes Provence (D_{04}), for which geometrical data are missing, but which is related to the model by an explicitly stated spatial relation it shares with the department D_{38}. The following section shows queries which are solved using the spatial reasoning capabilities offered by ONTOAST.

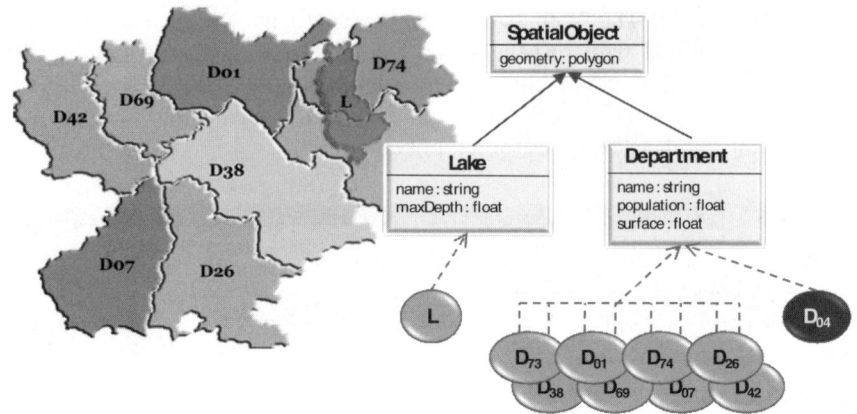

Fig. 4. a) The Rhône-Alpes region, the departments it contains and a lake L situated both on D_{74} and D_{73} territory. b) ONTOAST spatial ontology modeling the situation described by a).

5   Qualitative Spatial Relations in ONTOAST

Our objective is to obtain a model flexible enough to handle the coexistence of quantitative spatial data in the form of exact geometries and imprecise data in the form of qualitative spatial relations. These two kinds of information must complement each other and offer advanced means of reasoning. At this time, we take into account three categories of qualitative relations: *topology*, *orientation* and *distance*. They can result either from an explicit user declaration, or be automatically inferred from existing knowledge. Therefore, when a user request is submitted, the answer will be built: i) using the explicit knowledge, if the required relation is already stored in the base, ii) by carrying out qualitative inferences based on the composition of explicitly stated qualitative relations between objects, iii) by deducing qualitative relations using some numerical estimation and computation methods and iv) by applying qualitative reasoning both on explicit spatial relations and on deduced ones. Sections 5.1, 5.2 and 5.3 present examples for these four cases. Moreover, the main reasoning mechanisms offered by AROM, AML equations and instance classification, will be extended to take into account the spatial AML operators and the predefined qualitative spatial relations. ONTOAST will thus provide spatially adapted inferences for Semantic Web ontological knowledge.

In order to integrate the qualitative spatial relations into ONTOAST, we have created a special type of association: AromQSA (for Arom Qualitative Spatial Association), which generically represents the three categories of relations that we propose (see Fig. 5). A spatial qualitative association is formally defined as an association linking spatial objects, even when their exact geometries are not available.

Topological relations are inspired by the RCC8 calculus [15] and define the possible spatial configurations between two regions. *Distance associations* model ternary relations which refer, beside the described objects, a reference object. We also consider three types of *proximity relations* which respectively describe the fact that an object A is *Closer*, *Farther* or approximately (with a predefined error tolerance) at the same distance (*Equidistant*) from an object B, than the reference object C. Binary associations (specializations of *Direction*, Fig. 5) model the nine cardinal positions (N, S, E, W, SE, SW, NE, NW) between two spatial objects A and B. The following sections illustrate the use and behavior of these three associations.

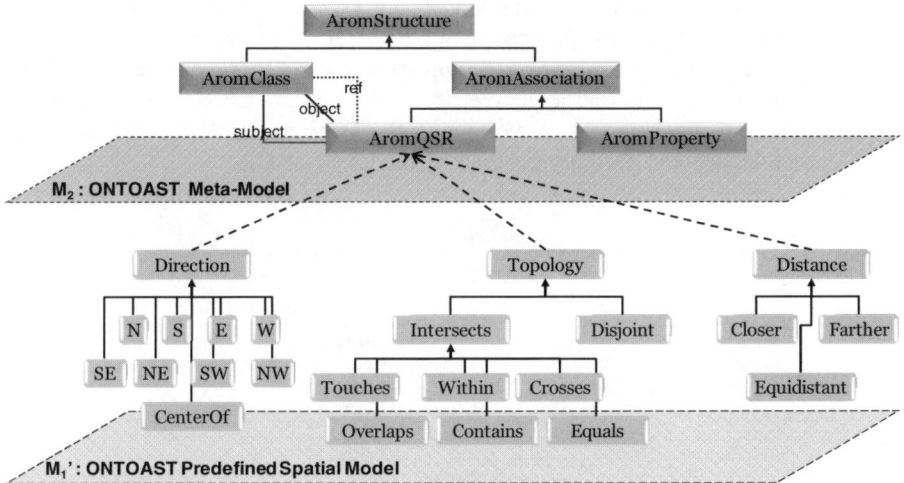

Fig. 5. The predefined qualitative spatial relations added to ONTOAST model and their definition in terms of meta-model structures

5.1 Topological Relations

The *topological* relations we have considered are inspired by the RCC-8, one variant of the Region Connection Calculus (RCC) [15] [16]. RCC-8 considers regions as being represented by sets of points and defines their intersection using the C(x,y) primitive. A similar calculus was developed by Egenhofer [17], who considered regions as being associated with three sets of points: *the interior*, *the exterior*, and *the boundary*. He models spatial relations between two regions in terms of 3x3 matrixes, called the 9-intersection, and finally identifies the same base relations as the RCC-8 calculus: *Disjoint, Intersects, Touches, Within, Crosses, Overlaps, Contains* and *Equals*. Except *Contains*, *Within* and *Crosses*, the other relations are symmetrical. Furthermore, the *Within* and *Contains* relations model opposite situations, same as *Disjoint* and *Intersects*. They all translate in terms of association the spatial operators discussed in section 3. Our idea is to use AML spatial operators on geometries attached to objects in order to check different qualitative configurations. Depending on the results, the system automatically constructs these relations.

In order to illustrate our approach let us go back to the example showed by Fig.4. At the ontology level, we insert a specialization (*Covers*) of the topological relation *Overlaps* between the lake and the departments (Fig. 6). This relation refines the intension of the predefined association by adding two attributes *accessibility* and *protectedArea* which characterize the area common to the connected objects. The fact that *Covers* is a specialization of *Overlaps* is symbolized by a graphic pictogram.

Let us assume that the objective is to answer the following questions: i) *what topological relations stand between the lake L and the departments D_{73} and D_{42}?* and ii*) which are the departments partially covered by the lake L?* When confronted to similar queries, the system first checks whether the required relation exists or not in the knowledgebase. In the case of question i), the ontology explicitly contains the tuple *covers1*, linking the lake L to the department D_{73}, so the reasoning stops there because the answer is found. On the other hand, between L and D_{42} there is no explicit topological relation. In this case, the query will be answered by successively applying the AML topology operators on the geometries of the two objects (ie: D_{42} *disjoint L?*, D_{42} *touches L ?*, D_{42} *overlaps L?*...). Following the same principle the system infers that the second department partially covered by the lake L is D_{74}.

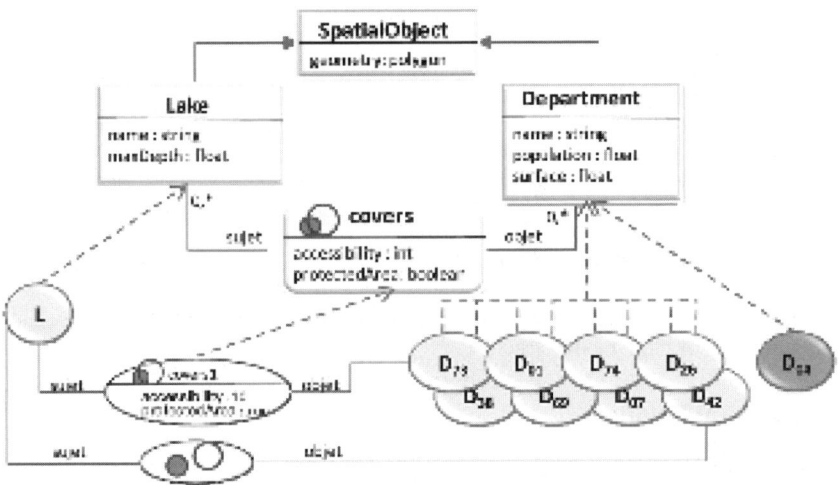

Fig. 6. The ONTOAST spatial ontology modeling topological relations

Concerning the reasoning process, in the absence of geometrical data, we adopt the composition table introduced by [18]. This table was formally derived using the properties characterizing the topology relations (for example the transitivity) and indicates, for each pair of existing topological relations, the relation obtained when composing the two.

The consistency of the ontology is checked by the system, when confronted with a user statement which explicitly introduces a new spatial relation. In order to manage the possible conflicts, we consider that geometric reasoning are always correct, while for explicitly stated data, the confidence associated with data source is taken into account. The system will not accept two contradictory declarations. For example, disposing of geometries for D_{42} and D_{38}, the system will reject a user declaration which states that a *Crosses* relation lies between the two. The reason is the existing conflict with the relation *Disjoint* between D_{42} and D_{38}, resulting from numerical computation.

5.2 Distance Relations

When talking about distances, the first element to identify is the used metric. Are we interested by the Euclidian distance, the shortest distance in a road network, or by the time it takes to travel (by train, by car, by plain…), etc.? Another question is that of defining the chosen extent for the spatial objects. Are we comparing points or surfaces or both? Our general purpose is to express and reason about regions. The main problem we have to face is the translation between numeric data (geographical frontiers) and qualitative data (qualitative distances). How to define the distance between two areas? Is it the *minimum* distance between the *frontiers points*, the *average* distance between *all the points* of the considered surface, the distance between *administrative or gravity centers*, etc. We propose to consider as predefined distances the ones presented above but, as a way of generalizing our approach, we intend to give the opportunity to the user to define its own distance through an AML equation.

Numeric distances are mapped onto quantitative distances using fuzzy sets [19] or mutually exclusive distance intervals of increasing ratio [20]. For simplicity reasons, in this first version of ONTOAST, we have chosen to implement the second alternative, in spite of its reduced expressivity power. For example, let us consider three cities A, B and C, situated respectively at 24, 26 and 155 miles from a reference point P. On a scale composed by two symbols: *near* (maximum 25 miles away from P) and *far* (more than 25 miles away from P), B and C will be assigned to the same category *far*. The fact that B is *almost near P*, almost at the same distance as A, but considered in the same category of distance as C embodies a loss of precision.

There are two categories of qualitative distances: the *absolute* ones and the *relative* ones. *Absolute distances* are binary relations between two spatial objects: *the subject* and *the object*. They are obtained by dividing the space into several sectors, according to the considered metric and the desired granularity. Each sector is centered on *the object* and is associated with a qualitative symbol (*i.e. very close, close, far, very far*, etc).

Next, the distance computed by numeric methods is positioned on this scale in order to obtain the corresponding absolute qualitative distance. For instance, using the Euclidian distance and the city of Grenoble as reference *object*, we obtain the situation illustrated by Fig. 7 a). We can easily observe that, according to this classification, there are four towns *far* from Grenoble: Vif, Vizille, Autrans and La Terrasse and only one town *very close*: St. Martin d'Heres. If we consider *the shortest road* distance with respect to Vif, we obtain the situation described by Fig.7 b), which says that Autrans and Vizille are both *close* to Vif, even if Vizille is, geographically speaking, *very close*. The complexity of this method lies in the difficulty of taking into account the *context of definition* (or the scale). For instance, the *close* relation may have different semantics when used to model some distance between buildings or when used to model distance between countries.

Relative distance takes into account the context by considering, besides the two connected objects, a *reference* entity. Using the distance d between *the object* and *the reference*, space is divided in three sectors *Closer* (containing all objects located at a distance d_i from *the reference*, $d_i<d$), *Equidistant* (containing all objects located at a distance d_j from *the reference*, with $d \le d_j \le d+\alpha$, where α is a predefined precision)

and *Farther* (containing all objects located at a distance d_k from *the reference*, with $d_k > d + \alpha$) (Fig.7c). *Subject* positions are modeled according to those three sectors. For instance we can deduce, on an Euclidian space, that "Vizille is further from Vif than Grenoble is" (Fig.7c).). For the moment, ONTOAST takes into account only this kind of simple relative distances.

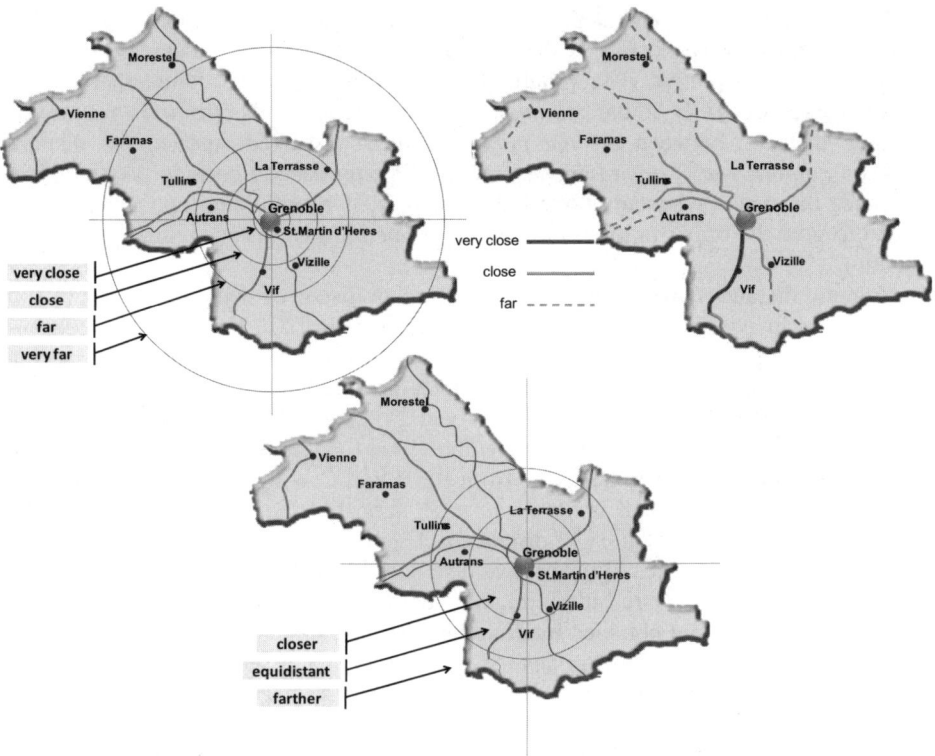

Fig.7. Distance relations between cities belonging to the French department of Isère. a) *absolute relation* based on the Euclidian distance b) relative distances based on traveling time c) relative distances based on the Euclidian distance.

Reasoning facilities offered by ONTOAST for qualitative distance relations are based on predefined composition tables. These inferences sometimes lead to significant difficulties. For instance, [21] shows the case of a sequence of aligned points $p_1, p_2, ..., p_n$ such that each point p_i is in a relation *close* with his predecessor, p_{i-1}. The question is from which j so that $j > i$ can we say that p_j is *far* from p_i? Moreover, when combining distances it is often necessary to take into account additional information like the orientation of the considered points or the properties of the considered metric. For instance, if B is far from A and C is far from B, then C can either be very far from A if the three points are aligned with B between A and C, or close to A if the angle between AB and BC is small. Furthermore, distances can behave in various "non mathematical" ways [14]. For instance, when considering the distance in terms of time needed to travel, an uphill journey will take longer than the return downhill journey. In this case, the relation is clearly non symmetrical.

5.3 Direction Relations

Using natural language, orientation of spatial entities is expressed in terms of qualitative categories such as: "on the right of the building", "in front of the house", "in the southwest of Paris", "in the north of France", etc. If the *context of definition* is a limited space so that it can be entirely perceived by the human eye without any movement, the most frequently used qualitative categories are: "in front of", "behind", "on the left", "on the right" with respect to the observer position or a chosen reference object. Larger spaces like geographic continents, which cannot be properly apprehended by moving and which require an intermediary representation (like maps) are usually characterized in terms of cardinal directions (N, S, E and W). Obvious mappings can be defined between these two sets, since they convey the same relations in different contexts.

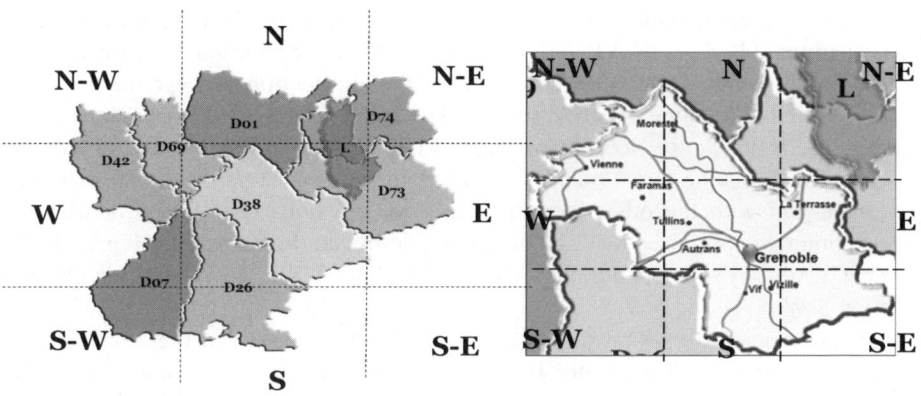

Fig. 8. Orientation relations between regions: a) *the subject* (D_{07}) crosses several directions (West, South-West and South) compared to the reference object (D_{38}); b) the *reference* object (D_{38}) has intrinsic directions

In the literature, cardinal directions are considered as ternary relations involving a *subject*, a *reference object* and a *frame of reference*. [22] suggests two ways of modeling orientation. The first method partitions the space into nine cone-shape areas of acceptance, having the *reference object* as origin. The idea is to consider objects represented by points that are assigned to different classes of direction (north, southeast, east, northwest, etc.) depending on their position relatively to the considered cones. However, things get a little more complicated when talking about extended objects represented by polygons, not only because they can cross several direction sectors [18], but also because they have intrinsic directions (e.i. the north of the country). ONTOAST represents and reasons about cardinal relations between regions, following the approach presented by Goyal and Egenhofer [23].

	N	NE	E	SE	S	SW	W	NW	C
N	N	N	NE	e	c	w	NW	NW	N
NE	NE	NE	NE	e	e	c	n	N	NE
E	NE	NE	E	SE	SE	s	c	N	E
SE	e	e	SE	SE	SE	s	s	c	SE
S	c	e	SE	SE	S	SW	SW	w	S
SW	w	c	s	s	SW	SW	SW	w	SW
W	NW	N	c	s	SW	SW	W	NW	W
NW	NW	n	n	c	W	w	NW	NW	NW
C	N	NE	E	SE	S	SW	W	NW	C

Fig. 9. Cardinal directions composition table as proposed by [22]

Using the *footprint* (minimum bounding box) of the *reference* region (D_{38}), we partition the space as shown in Fig.8.a). For each sector of the resulting grid, it is possible to infer whether *the subject* is contained in the sector and if so, to which extent. Numeric computations and performed for this purpose, using nine specialized AML operators (*N, S, E, W, NE, NW, SE, SW, Center*), one for each direction and one for the central sector. These operators calculate the intersection between the geometry of the *subject* and the nine direction sectors. The result is expressed as percentage. For instance, D_{07} is 3% *South*, 46% *South-West* and 39 % *West* of D_{38}. Finally, as a result, the system chooses the direction corresponding to the biggest percentage if there is a substantial gap with the other identified direction relations (*i.e.* 45%) or uses the composition table proposed by [22] for the two dominant directions (see Fig.9). In the example given by composing the relations W and S-W, we obtain the general direction S-W.

Locating a spatial region with respect to the known regions in the model, in the absence of its exact geometry, involves interesting reasoning. For instance, assuming our spatial ontology contains a unique relation "*South from*" linking the departments D_{04} to D_{38} (Fig.6), the system will easily infer that D_{04} also shares a *South of* relation with D_{01} and additionally that it is *farther* from D_{01} as D_{38} is. Still, not all such queries get a definitive answer. If we consider the existence of a department D_{10}, situated *West* from D_{73}, it is impossible to infer its relative position to D_{38} without using additional information (*i.e.* its geometry, the distance from D_{73}, etc.). And even if the systems establishes D_{10} and D_{38} as being on the same direction compared to D_{73}, it is difficult to know whether D_{10} is *West* from D_{38} or *between* D_{38} and D_{73}.

Answering queries that consider both inclusion and direction relations, like "Which cities are located *South* of the department D_{38}?", is done by the analysis shown in Fig.8b). The considered space is reduced to the minimum bounding box that includes the surface of the department, and is split into 9 equal areas. Further, the system applies the same reasoning as for usual extended objects, considering the central sector as new *reference*. For instance, we can argue that Grenoble and Autrans are situated in the *center of* the French department Isere and that Vif and Vizille are cities located in the *south* of this department.

6   Conclusions and Perspectives

This paper presents ONTOAST, a spatial ontology modeling and management system. It manages both quantitative data and qualitative spatial relations which can be used, on the one hand, to define spatial object, attributes and relations, and, on the other hand, to queries the modeled knowledge. ONTOAST is the result of combining two prior AROM extensions: AROM-ST which handles spatial types and operators and AROM-ONTO which ensures the compatibility with OWL. Moreover, in order to allow the formulation of spatial queries close to natural language, ONTOAST integrates three types of qualitative relations: *topological*, *orientation* and *distance*. It also handles the inference of new qualitative relations, increasing thus the spatial-based search capabilities.

Our researches are now diverted towards new analysis and querying algorithms. In particular, we intend to provide ONTOAST with new capabilities such as semantic analytics [24] that will allow the discovery of implicit intrinsic connections between objects contained by ontologies.

References

1. Berners-Lee, T., Hendler, J.A., Lassila, O.: The Semantic Web. Scientific American 284(5)002E, 34–43 (2001)
2. Gruber, T.R.: Towards Principles for the Design of Ontologies Used for Knowledge Sharing. In: Formal Ontology in Conceptual Analysis and Knowledge Representation, Kluwer Academic Publishers, Dordrecht (1993)
3. McGuinness, D., van Harmelen, F.: OWL Web Ontology Language -Overview (2005), http://www.w3.org/TR/2004/REC-owl-features-20040210/
4. Page, M., Gensel, J., Capponi, C., Bruley, C., Genoud, P., Ziébelin, D., Bardou, D., Dupierris, V.: A New Approach in Object-Based Knowledge Representation: the AROM System. In: Monostori, L., Váncza, J., Ali, M. (eds.) IEA/AIE 2001. LNCS (LNAI), vol. 2070, pp. 113–118. Springer, Heidelberg (2001)
5. Moisuc, B., Davoine, P.-A., Gensel, J., Martin, H.: Design of Spatio-Temporal Information Systems for Natural Risk Management with an Object-Based Knowledge Representation Approach. Geomatica 59(4) (2005)
6. Miron, A., Capponi, C., Gensel, J., Villanova-Oliver, M., Ziébelin, D., Genoud, P.: Rapprocher AROM de OWL... Langages et Modéles á Objets. Toulouse, France (in French) (2007)
7. Cohn, A., Hazarika, S.M.: Qualitative Spatial Representation and Reasoning: An Overview. In: Fundamenta Informaticae, vol. 46, pp. 1–29. IOS Press, Amsterdam (2001)
8. Katz, Y., Grau, B.C.: Representing Qualitative Spatial Information in OWL-DL. In: Proceedings of OWL: Experiences and Directions (2006)
9. Dolbear, C., Hart, G.: So what's so special about spatial? Ordnance Survey of Great Britain (2006)
10. Capponi, C.: Type extensibility of a knowledge representation system with powersets. In: Raś, Z.W., Skowron, A. (eds.) ISMIS 1997. LNCS, vol. 1325, pp. 338–347. Springer, Heidelberg (1997)
11. OpenGIS home page: OpenGIS Consortium, OpenGIS Reference Model (2004), http://www.opengis.org/

12. Protégé: La Création d'Ontologies Web Sémantique avec Protégé-2000 (2000), http://www.emse.fr/~beaune/websem/WS_Protege-2000.pdf
13. iAnnotate Tab documentation page: http://www.dbmi.columbia.edu/~cop7001/iAnnotateTab/iannotate_doc.htm
14. Cohn, A.: Qualitative Spatial Representations. In: IJCAI-99 Workshop Adaptive Spatial Representations of Dynamic Environments (1999)
15. Randell, D.A., Cui, Z., Cohn, A.G.: A spatial logic based on regions and connection. In: Proc. 3rd Int. Conf. on Knowledge Representation and Reasoning, pp. 165–176. Morgan Kaufmann, San Mateo (1992)
16. Randell, M., Cohn, A.G.: Modeling topological and metrical properties of physical processes. In: Proceeding 1st International Conference on the Principles of Knowledge Representation and Reasoning, Los Altos, pp. 55–66 (1989)
17. Egenhofer, M.: Point-set Topological Spatial Relations. International Journal of Geographical Information Systems, 161–174 (1991)
18. Egenhofer, M., Rodriguez, M.A., Blaser, A.: Query Pre-processing of Topological Constraints: Comparing a Composition-Based with Neighborhood-Based Approach. In: Hadzilacos, T., Manolopoulos, Y., Roddick, J.F., Theodoridis, Y. (eds.) SSTD 2003. LNCS, vol. 2750, pp. 362–379. Springer, Heidelberg (2003)
19. Dutta, S.: Qualitative Spatial Reasoning: A Semi-Quantitative Approach Using Fuzzy Logic. In: Buchmann, A.P., Smith, T.R., Wang, Y.-F., Günther, O. (eds.) SSD 1989. LNCS, vol. 409, pp. 345–364. Springer, Heidelberg (1990)
20. Hong, J.: Qualitative Distance and Direction Reasoning in Geographic Space. PhD.Thesis, University of Main (1994)
21. Renz, J.: Qualitative Spatial Reasoning with Topological Information. LNCS (LNAI), vol. 2293, pp. 31–40. Springer, Heidelberg (2002)
22. Frank, A.U.: Qualitative Spatial Reasoning: Cardinal Directions as an Example. International Journal of Geographical Information Science 10(3), 269–290 (1996)
23. Goyal, R.K., Egenhofer, M.: Cardinal Directions between Extended Spatial Objects. IEEE Transactions on Knowledge and Data Engineering (2001)
24. Sheth, A., Arpinar, I.B., Kashyap, V.: Relationships at the Heart of Semantic Web: Modeling, Discovering, and Exploiting Complex Semantic Relationships. Technical Report, LSDIS Lab. Computer Science University of Georgia. Athens, GA 30622 (2002)

Modeling Environmental Process Using Semantic Geospatial Web Service

Shanzhen Yi[1] and Bo Huang[2]

[1] Centre of Information Engineering and Simulation,
Huazhong University of Science and Technology, Wuhan, China
yishanzhen@tsinghua.org.cn
[2] Dept of Geography and Resource Management
The Chinese University of Hong Kong, Shatin, NT, Hong Kong
bohuang@cuhk.edu.hk

Abstract. Environmental problems have been modeled and analyzed using GIS for long time, and they are often described by a set of interacting models from multiple disciplines. In this paper a composite process for environmental problem with multiple models is analyzed. A system dynamics modeling method, STELLA, is introduced to illustrate how to model the composite process with cause-effect relationships. Compared with other environmental modeling methods, environmental modeling supported by semantic geospatial Web service has the advantages of achieving interoperability and allowing real-time and online updating of the deployed models. A prototype system for Modeling Environmental Process using Semantic Geospatial Web Service, i.e. MEPSGWS, has been designed, which includes process ontology, rules for cause-effect relations, GML-based parameter type, Web service description language, and interoperable protocol. As an application of MEPSGWS, a watershed composite process is chosen to illustrate how to model a composite environmental process with semantic geospatial Web service.

Keywords: Semantic Web service, Environmental system, Process ontology, Spatial modeling and simulation.

1 Introduction

Integration of GIS and environmental modeling has been extensively studied. While the integration could use either the tightly coupled or loosely coupled approach, there are still challenges on real-time updating environmental models with the latest new findings and modeling cause-effect relationships (i.e. connection between activities or events) between models in a composite environmental process. A composite environmental process may consist of many environmental models from different disciplines in the modular form, hence environmental and ecological system modeling should be capable of handling compound process and models [1, 2].

The importance of Web services has been recognized and widely accepted in industry and academic research community. As an extension of Web service, Semantic Web Service provides powerful expressiveness of service descriptions and

J.M. Ware and G.E. Taylor (Eds.): W2GIS 2007, LNCS 4857, pp. 137–147, 2007.

formal modularization of service layers, thereby offering an opportunity for the integration of multiple environmental models in a distributed GIS environment with a new way. This paper attempts such a service method to model complex environmental process.

Many approaches for integrating environmental models have been proposed, such as the Modular Modeling System (MMS) [3], the Interactive Component Modeling System (ICMS) [2, 4], and the Spatial Modeling Environment (SME) [5]. Among them, MMS is a Unix-based tightly coupled system. ICMS is a PC-based software product, developed to facilitate the rapid development and delivery of catchment science for managers. SME is a distributed object-oriented simulation environment which incorporates the set of code modules that actually perform the spatial simulation. These approaches have resulted in the development of concept of environmental modeling frameworks which bring together suites or libraries of modules, but fall short of providing the methods for composite process modeling with cause-effect relations and real-time online updating of models deployed in distributed computing environment.

The composite process modeling of environmental systems, for example, watershed environment process, presents the characteristics of hierarchical, multi-level cause-effect relationships. In the process of watershed development, a series of impact on ecological and environmental system have emerged [6]. The impacts include land use change, degradation of water resources and water quality, non point pollution, and damage to ecological system. The impact pathway is complex, and the modeling approach of the composite process is an interactive integration of multi-discipline models.

STELLA is a modeling method for complex system with compound models [7, 8]. It has a number of advantages for watershed composite process modeling, but its support of modularity is very limited [9]. There are no formal mechanisms that could put individual STELLA models together and provide their integration in the Web and Internet environment. Hence, it may not benefit from the advantages of semantic Web service, such as interoperability, open environment, coordination of cross-depart-mental processes, and real-time online updating of models.

In this paper we provide a framework, i.e. Modeling Environmental Process using Semantic Geospatial Web Service (MEPSGWS), which could integrate real-time online updating of STELLA-based composite process models and cause-effect relations. In the remainder of the paper, section 2 describes the composite process of environmental systems with cause-effect relationships using STELLA modeling method, and section 3 presents the method of modeling a composite process using Semantic Web Service. Section 4 illustrates the modeling of composite process with MEPSGWS. Finally, section 5 concludes the paper.

2   Composite Process of Environmental System

Composite process of environmental system refers to that the environmental processes are composed of multiple events and sub-processes with complex cause-effect relationships. Watershed is a typical environmental system that captures water in any form to a common water body. Watershed processes are considered as a composite process which includes physical, ecological and social processes.

The processes of watershed development and the caused change of ecological environment are complicated with multiple factors, such as land use/land cover change, water quality and soil loss. Watershed processes and environmental impact pathways are modeled by cause-effect relationships [6]. In a composite process, event and process are similar concepts. One event may be another sub-process, one person's process is another's event [10]. In cause-effect relationships, the event that represents cause and the event that represents effect are also defined as processes in our approach.

2.1 Cause-Effect Linkages of Composite Process

The importance to address cumulative environmental impacts in Environmental Impact Statements (EIS) of large development projects is increasingly underlined. However, cumulative impacts are generated through complex impact pathways, involving multiple root causes and lower and higher order effects, interlinked by cause-effect relationships [6]. The term 'cause-effect relationship' here refers to the connection between an event and its consequence, e.g. an initial activity (root cause) leading to a first-order effect, or lower order effect leading to a higher order effect. The causes and receptors may be defined at different spatial scales. The cause-effect relationships are illustrated in Fig. 1.

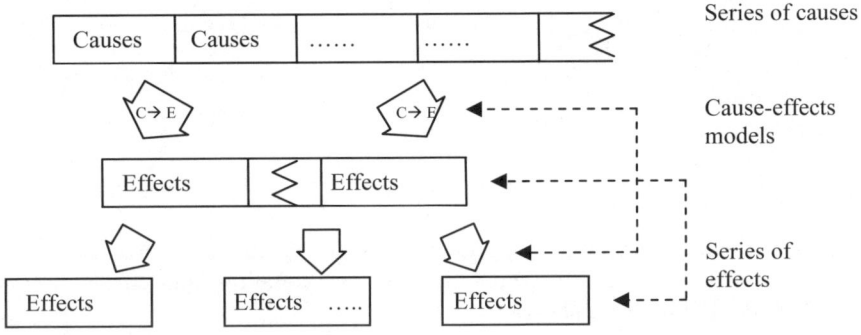

Fig. 1. Cause-effect relationships

2.2 STELLA Method for Modeling Composite Process and Cause-Effect Relationships

In order to illustrate how to model the composite process of environmental system with cause-effect relationships by environmental modeling methods, the STELLA method is introduced [7, 8]. STELLA is a graphical programming language and software, which is used for system dynamics modeling. STELLA delivers models by means of some of the icon-based systems. The components of stocks, flows, converters and connectors in STELLA are represented with the following icons (Fig. 2):

Fig. 2. The icons of components

Connectors indicate the cause-effect relationship between diagram components. *Stocks* are used to represent system components that can accumulate material over time. *Flows* represent components whose values are measured as rates. *Converters* represent constants, variables, functions, or time series.

A simplified rain and soil loss model is represented by STELLA as shown in Fig. 3 [18]:

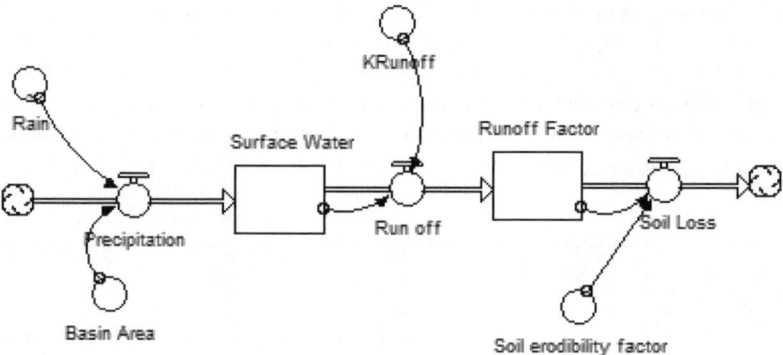

Fig. 3. A STELLA model for composite process of rain and soil loss

STELLA provides method to compose mathematic equations of application models with related constants, variables and functions. The equations and constants for the STELLA model in Fig. 3 are shown below:

Soil loss model is described by the following statements:
```
INFLOWS:
Run_off=Surface_Water*KRunoff {units of m³ of water}
OUTFLOWS:
Soil_Loss=Runoff_Factor*Soil_erodibility_factor {units of m³}
Surface_Water(t)=Surface_Water(t-dt)+(Precipitation-Run_off)*dt
INIT Surface_Water=0.0

Runoff_Factor(t)=Runoff_Factor(t-dt)+(Run_off-Soil_Loss)*dt
INIT Runoff_Factor=0.0
```

Precipitation and runoff models are described by the following statements:
```
INFLOWS:
Precipitation=Basin_Area*(Rain*0.1) {gives units of m³/hr}
OUTFLOWS:
Run_off=Surface_Water*KRunoff {units of m³ of water}
```

```
Basin_Area=8000000 {units of m²}
KRunoff=0.4
Soil_erodibility_factor=0.01
```

Rain model is described by the following statements:
```
Rain = GRAPH(TIME) (hr, mm)
(0.00, 0.00), (1.00, 0.1), (2.00, 0.45), (3.00, 1.00), (4.00,
2.15), (5.00, 2.60), (6.00, 1.45), (7.00, 1.20), (8.00, 0.55),
(9.00, 0.2), (10.0, 0.00)
```

STELLA-based modeling method has a lot of advantages, but it is limited in supporting modularity [9]. As modeling method based on Semantic Web Service has the advantages of interoperability, open environment, real time online updating of models, and coordination of cross-departmental processes, following sections explores the modeling method of composite process using Semantic Geospatial Web Service.

3 Modeling Composite Process with Semantic Geospatial Web Service

Instead of STELLA that should represent all system processes in a special and closed modular environment, Semantic Web Service provides a framework to integrate models in an open environment. Semantic Web Service defines the ontology of process modules and interfaces by Web Service Description Language (WSDL) [17], which integrates models in an interoperable environment. Particular implementations of modules are flexible, and a wide variety of model components are to be made available through web service protocols such as SOAP (simple object access protocol)[17]. The input and output parameters of these models are usually encoded by Extensible Markup Language (XML). In our approach, Geographic Markup Language (GML) [13] is employed to represent geographic and environmental data parameters. The architecture of Semantic Geospatial Web Service for composite process of environmental system has been designed, as shown in Fig. 4.

The upper part of the architecture is used to describe models and their parameters in a composite process, including components of ontology language for Geospatial Web service, rule language for cause-effect relationships and geographic data type for parameters. The lower part includes description language for Web service and interoperable protocols, such as WSDL and SOAP.

The OWL-services language (OWL-S) (OWL stands for Web Ontology language) [11, 12, 14] and SWRL (Semantic Web Rule Language) [15, 16] provide ontology to describe a composite process, which are used in the upper part of the proposed architecture.

A composite process in OWL-S is a kind of processes composed of simple or atomic processes. With the control construct and rule language, such as IF-THEN-ELSE rules [19], the cause-effect relationships in environmental processes could be described for modeling and simulation analysis.

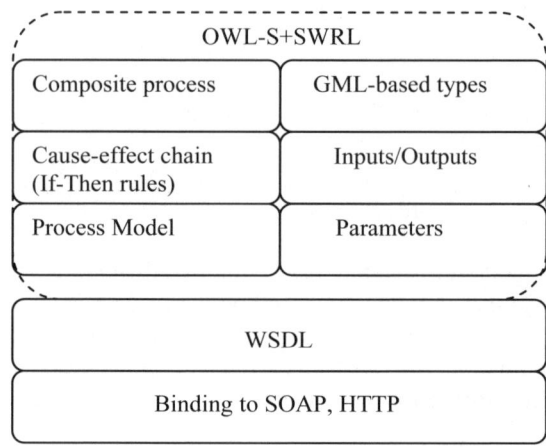

Fig. 4. The architecture of MEPSGWS

The parameters for models include geospatial features in geographical process analysis. GML is an XML based language to represent geospatial and environmental data for integration and interoperation in the Web environment, such as Web mapping. GML-based data types are employed to represent parameters of inputs and outputs in the Semantic Web Service of composite environment process.

The interface of a Web service and interactions with other services are described by the Web Services Description Language. WSDL is used to compose XML document which describes technical details of a Web service. The protocol and data format specifications for a particular service port type constitute a reusable binding. A port has been defined in WSDL by associating a network address with a reusable binding, and a collection of ports defines a service.

The lower layers are based on general network protocols, such as HTTP and SOAP, which is a standardized enveloping mechanism using XML for communicating document-centric messages and remote calls for procedure and model.

3.1 Process Ontology in OWL-S

The OWL-S includes a set of ontologies for describing the properties and capabilities of Web services. The OWL-S is designed to support effective automation of various Web Services related activities, such as service discovery, composition, execution, and monitoring. In this research, we use OWL-S to describe composite process of environmental systems with cause-effect relationships and geospatial parameters.

The OWL-S process ontology provides a vocabulary for describing the composition of Web services. This ontology uses atomic process, simple process and composite process to describe Web service behavior—i.e., primitive and complex actions should have preconditions and effects. In our approach, we focus on the composite process, which is a collection of processes. Each process has the properties of parameter, precondition (represented by expression), input and output.

The decomposition of composite process can be specified by using control constructs such as *Sequence*, *Choice* and *If-Then-Else*. *Sequence* gives a list of control constructs to be done in order. *Choice* calls for the execution of a single control construct from a given bag of control constructs. The *If-Then-Else* class is a control construct that has properties of ifCondition, then and else. Its semantics is intended as "Test If-condition; if True do Then, if False do Else."

3.2 Modeling Parameters with GML

In the composite process of OWL-S, the data type of parameters of a process should be specified. When modeling a geographic environmental process, the parameters usually include geospatial data types. GML has been an adopted specification of the OpenGIS Consortium (OGC) [13], and is rapidly emerging as a industry standard for the encoding, transport and storage of all forms of geographic information. Hence, it is used in our approach to represent input and output parameters.

Every parameter in the process has a type, specified using a URI. A GML type parameter is defined with OWL-S as following XML code:

```
<owl:DatatypeProperty rdf:ID="parameterType">
  <rdfs:domain rdf:resource="#Parameter"/>
  <rdfs:range  rdf:resource="&gml; anyURI"/>
</owl:DatatypeProperty>
```

Input and output of a process are subclasses of parameters. The parameter type is specified by GML, such as multipolygon, defined as:

```
<owl:Class rdf:ID="Input">
   <rdfs:subClassOf rdf:resource="#Parameter"/>
</owl:Class>

<owl:Class rdf:ID="Parameter">
   <rdfs:subClassOf>
      <owl:Restriction>
        <owl:onProperty rdf:resource="#parameterType" />
     <rdf:datatype="&gml;#multipolygon"/>
      </owl:Restriction>
   </rdfs:subClassOf>
</owl:Class>
```

4   Illustration of MEPSGWS for Composite Process Modeling

The ontologies of watershed composite process are required which are used to develop conceptually sound models, to effectively represent cause-effect relationships, to enhance interoperability between models developed in different domains, and to provide the opportunity for the real-time online updating of models. To accomplish these goals, we need to express these processes not only in terms of their properties but also in terms of their relationships with other processes. Geographic and environmental data types also needed to define the input and output of process and model. A collection of such process descriptions using OWL-S could then form the foundation of composite process that will be used in a simulation framework.

The MEPSWS provides a modeling environment for composite process. A geospatial Web service flow for STELLA model of rain and soil loss process presented in Fig. 3 is shown in Fig. 5.

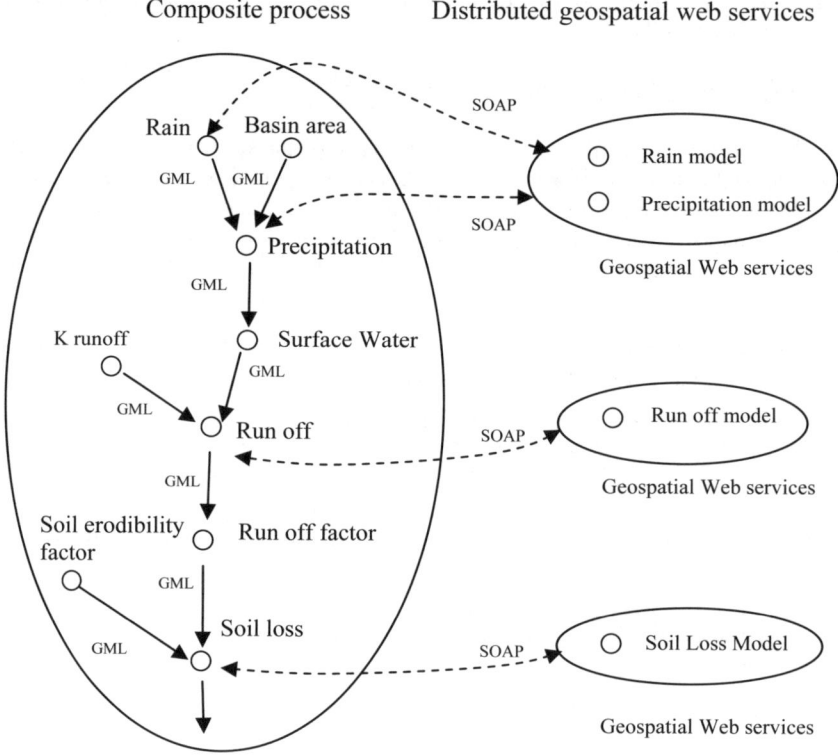

Fig. 5. Composite process and geospatial Web services

The definitions of composite process and atomic process in a watershed system are defined by terms of OWL-S, such as rain model, precipitation model, run off model and soil loss model. A fragment of the definition of composite and atomic processes is shown below:

```
<owl:CompositeProcess rdf:ID=•Watershed process•>
  <process:Composed of>
   <process:Sequence>
    <process:Components rdf:parse Type=•Collection•>
     <process:AtomicProcess rdf:about=•#Rain model•/>
     <process:AtomicProcess rdf:about=•#Precipitation model•/>
     <process:AtomicProcess rdf:about=•#Runoff Model•/>
     <process:AtomicProcess rdf:about=•#Soil Loss Model•/>
    </process:Components>
   </process:Sequence>
  </process:Composed of>
</owl:CompositeProcess>
```

```
<process:AtomicProcess rdf: ID=•Rain model• />
     <process:hasInput   rdf:resource=•#Initial-content•/>
     <process:hasInput   rdf:resource=•#Source-term•/>
     <process:hasOutput rdf:resource=•#Rain•/>
</process:AtomicProcess>

<process:AtomicProcess rdf:ID=•Precipitation model•>
     <process:hasPrecondition rdf:resource=•#expr1•/>
     <process:hasInput   rdf:resource=•#Rain•/>
     <process:hasInput   rdf:resource=•#Basin area•/>
     <process:hasOutput rdf:resource=•#Precipitation•/>
 </process:AtomicProcess>
```

In order to model the cause-effect relationships in the composite process, the cause processes are modeled as precondition of the effect processes. OWL-S provides If-Then-Else control construct for the condition expression. For example, the precondition of *Precipitation model* represented by *expr1* includes the processes of *Rain* and constants *Basin area*. When the condition, represented as rdf resource *RainPossible* and *BasinAreaPossible*, is satisfied, then the expression of precondition is true. The condition expression is shown as following:

```
<owl:Expression: rdf:ID=•expr1•>
  <process:IfThenElse>
     <process:IfConditon  >
        <expr:AND >
            <expr:condition rdf:resource=•#RainPossible•/>
            <expr:condition rdf:resource=•#BasinAreaPossible •/>
            </expr:AND>
      </process:IfConditon>
      <process:Then>
          <expr1:condition rdf:ID= •value•> true </expr1>
      </process:Then>
      <process:Else>
         <expr1:condition rdf:ID= •value•> false </expr1>
     </process:Else>
  </process:IfThenElse>
</owl:Expression>
```

The definition of possible condition is represented by rdf resource in the form of following expression. In the condition expression, one of the conditions *rainPossible* is represented as *Rainmodel* is already defined (i.e. Rainmodel != UNDEF) :

```
<expr:condition rdf:ID=•RainPossible•/>
   <expr:expressionBody>
       #Rainmodel != UNDEF
   </expr:expressionBody>
</expr:condition>
```

The parameter types of input and output are based on the data type defined by GML specification. For example, the input parameters in precipitation model are represented by Graph-Time.gml.

```
<process:Input rdf:ID=•Rain• >
   <process: parameterType rdf:resource=•Graph-Time.gml•/>
</process:Input>
```

The protocol of SOAP offers interoperability across a wide variety of platforms. It enables interoperability by providing a generalized specification for invoking methods on objects and components using HTTP calls and XML data formats. WSDL define the parameter and type as messages. An OWL-S atomic process usually corresponds to a WSDL *operation*. The operation binding is also defined for connecting to actual services supplied by the provider's server, such as the SOAP server. For example, the rain model was developed as Web service using Java class *RainModel*. A fragment of description of services defined by WSDL is shown below:

```
<wsdl:service name="RainModel" >
 <wsdl:port name="ComputeRain" binding="tns:RainModel-Binding">
     <wsdl:soap:address
      location="http://localhost:8080/soap/servlet/rpcrouter"/ >
 </wsdl:port>
</wsdl:service>
```

In our example, the service class *RainModel* was deployed in the Apache SOAP server using SOAP Service Manager. The rain model could be updated by an expert in real-time and online in the server site with new findings. A simplified rain model has been discovered in [20].

5   Conclusions

In the modeling of watershed complex environmental process, cause-effect relationships have been considered because of the multi-level characteristics of the process. A compound model for simplified watershed rain and soil loss process has been described using the STELLA method to illustrate the cause-effect relationships in an environmental process.

The method and architecture of modeling environmental process using semantic geospatial Web service has been designed. MEPSGWS provides a compound modeling method for environmental process in an open and interoperable environment, which has resolved the problem of real-time and online updating of models and the modeling of cause-effect relationships in the composite process.

To illustrate the feasibility of proposed approach, MEPSGWS has been applied to model the composite process of watershed rain and soil loss. The composite process is defined by OWL-S, GML, WSDL and SOAP, and a service for the rain model has been deployed using the SOAP server. Our future work includes composing services using a middleware for composite process simulation, which will make the deployment and the call of the services more flexible.

References

1. Fotheringham, A.S.: GIS-based Spatial Modelling: A Step Forward or a Step Backwards. In: Fotheringham, A.S., Wegener, M., Masser, I., Salge, F. (eds.) Spatial Models and GIS, New Potential and New Models, pp. 21–30. Taylor & Francis, Abington (2000)
2. Rizzoli, A.E., Davis, J.R., Abel, D.J.: Model and data integration and re-use in environmental decision support systems. Decision Support Systems. 24, 127–144 (1998)

3. Leavesley, G.H., Markstrom, S.L., Brewer, M.S., Viger, R.J.: The modular modeling system (MMS). In: The physical process modeling component of a database-centred decision support system for water and power management. Water, Air and Soil Pollution, vol. 90, pp. 303–311. Springer, Netherlands (1996)
4. Interactive Component Modelling System – ICMS, http://www.clw.csiro.au/ products/icms/index.html
5. Maxwell, T., Costanza, R.: A language for modular spatio-temporal simulation. Ecological Modelling 103, 105–114 (1997)
6. Brismar, A.: Attention to impact pathways in EISs of large dam projects. Environmental Impact Assessment Review 24, 59–87 (2004)
7. Elshorbagy, A., Ormsbee, L.: Object-oriented modeling approach to surface water quality management. Environmental Modeling & Software 21, 689–698 (2006)
8. Shi, T., Gill, R.: Developing effective policies for the sustainable development of ecological agriculture in China: the case study of Jinshan County with a systems dynamics model. Ecological Economics 53, 223–246 (2005)
9. Voinov, A., Fitz, C., Bounmans, R., Costanza, R.: Modular ecosystem modeling. Environmental Modeling & Software 19, 285–304 (2004)
10. Worboys, M.: Event-oriented approaches to geographic phenomena. International Journal of Geographical Information Science 19, 1–29 (2005)
11. The OWL Services Coalition, OWL-S: Semantic Markup for Web Services, W3C Member Submission (November 22, 2004), http://www.w3.org/Submission/2004/SUBM-OWL-S-20041122/
12. Maluszyński, J.: On integrating rules into the Semantic Web. Electronic Notes in Theoretical Computer Science 86, 1–11 (2003)
13. Cox, S., Daisey, P., Lake, R., Portele, C., Whiteside, A.: Geography Markup Language Version 3.0 OGC Document, Number 02-023r4 (2002), http://www.opengeospatial.org/ specs/
14. OWL-S: Semantic Markup for Web Services, W3C Member Submission (November 22, 2004), http://www.w3.org/Submission/OWL-S/
15. Horrocks, I., Patel-Schneider, P.F., Bechhofer, S., Tsarkov, D.: OWL rules: A proposal and prototype implementation. Journal of Web Semantics 3, 23–40 (2005)
16. Mei, J., Boley, H.: Interpreting SWRL Rules in RDF Graphs. Electronic Notes in Theoretical Computer Science 151, 53–69 (2006)
17. Christensen, E., Curbera, F., Meredith, G., Weerawarana, S.: Web Services Description Language (WSDL) 1.1 (2001), http://www.w3.org/TR/2001/NOTE-wsdl-20010315
18. Maxwell, T., Voinov, A., Costanza, R.: Spatial Simulation Using the SME. In: Costanza, R., Voinov, A. (eds.) Landscape Simulation Modeling, A Spatially Explicit, Dynamic Approach, pp. 21–42. Springer, New York (2004)
19. Yi, S.Z., Xing, C.X., Liu, Q.L., Zhang, Y., Zhou, L.Z.: Semantic and interoperable WebGIS. In: The 2nd IEEE International Conference on Web Information Systems Engineering, Kyoto, Japan, pp. 42–47 (December 3, 2001)
20. Docine, L., Andrieu, H., Creutin, J.D.: Evaluation of a simplified dynamical rainfall forecasting model from rain events simulated using a meteorological model. Phys. Chem. Earth (B) 24, 883–997 (1999)

Historic Queries in Geosensor Networks

Stephan Winter[1], Sören Dupke[1,2], Lin Jie Guan[1], and Carlos Vieira[1,3]

[1] Department of Geomatics, The University of Melbourne, Victoria 3010, Australia
[2] Institute of Geoinformatics, University of Münster, Germany
[3] Departamento de Engenharia Civil, Universidade Federal de Vicosa, Brasil
winter@unimelb.edu.au

Abstract. This paper addresses—to our knowledge, for the first time—the problem of querying a geosensor network—a sensor network of mobile, location-aware nodes—for historical data. It compares different network architectures and querying strategies with respect to their performance in reconstructing events or processes that happened in the past, trying to support the hypothesis that reconstruction is possible within the limited capacities of a geosensor network only. In a concrete case study, these queries are studied in a simulated peer-to-peer ride sharing system.

1 Introduction

Mining for historic data is relevant to study complex systems. Just querying the current states and plans of self-organizing, mobile and autonomous agents may never show the higher patterns of behavior emerging from their activities. The running example in this paper will be an ad-hoc peer-to-peer system providing shared rides for participating pedestrians (clients) and vehicles (hosts) in urban traffic [1,2,3]. In this system, agents negotiate shared rides with local peers, move, and regularly update their travel arrangements according to new opportunities emerging in their current neighborhood. This means, a client's complete trip can only be queried in hindsight, after completion.

Considering recent trends in ubiquitous computing and ambient intelligence, complex systems in geographic scales and environments will be observed, managed and analyzed by intelligent computing platforms that are sensor-enabled, mobile or embedded, and wirelessly connected to peers or web-based services. This connection of the intelligent computing platforms can be based on different architectures. Location-based services, as one example, connect the mobile device of an individual agent in a complex system directly with a central service via mobile communication operators. Intelligent transportation systems, as another example, typically have a hierarchical structure of localized, regional and centralized sensor analysis and traffic management operations. Finally, peer-to-peer services such as the mentioned shared ride system, or any other geosensor network, work by local collaboration and have by design no need for a centralized service.

This paper addresses the problem of querying the mobile intelligent computing platforms for historical information. While this problem is trivial for data collected in a central or even in a federated database, it is not trivial for data circulating in geosensor networks. In geosensor networks, agents can, for relatively cheap costs, temporarily

J.M. Ware and G.E. Taylor (Eds.): W2GIS 2007, LNCS 4857, pp. 148–161, 2007.

store their communication messages, producing by that way local memories. But these memories have interesting properties. Most notably, the geosensor network node is mobile, i.e., its location at the time of a query is no longer related to the location of the queried event. Secondly, the node is volatile, i.e., it cannot be accessed all the time. And thirdly, the node has limited battery power, which limits communication, and limited storage capacity. The latter means that local memories decay quickly as newer information has to overwrite older one. At the same time it is highly undesirable to create a central registry of all sent and received messages in the network, due to the accumulating communication costs, a general bandwidth problem, and a communication bottleneck for agents closer to the central registry.

Research questions in this context are manifold. For a geosensor network of intelligent mobile computing platforms, is it possible to abstain at all from a central collection of data and still query historical data? What are the costs of such a solution, in terms of communication effort and completeness? And if other communication architectures are integrated: what are the additional costs for serving these (hierarchic or central) structures with data? Do the costs balance the advantage of a central data set?

This paper compares different network architectures and querying strategies with respect to their performance in reconstructing historic events or processes, trying to support the hypothesis that reconstruction is possible within the limited capacities of a geosensor network only. These queries are studied along the case study of the mentioned peer-to-peer ride sharing system. Thus, the hypothesis reads exactly: The reconstruction of shared rides from local memories can be achieved with small local storage overhead and low communication overhead.

Architectures and their performances are investigated in a simulation of the ride sharing system. The simulation consists of two phases: the travel phase of a single client to generate evidence of a trip in the memories of multiple agents, and then the attempt of an investigator to reconstruct the client's trip. Observing both phases in the different communication architectures, one can assess and compare the communication effort as well as the investigator's success rate.

This paper presents in Section 2 background knowledge of the problem and approach, introducing the required aspects of geosensor networks and intelligent transportation systems. Section 3 explores the evidence for past events within messages exchanged in the network. Section 4 discusses then ways of storing these messages for different types of geosensor networks. Then Section 5 presents the simulation of a reconstruction of past shared rides, and Section 6 demonstrates the behavior of this simulation. Section 7 summarizes the findings and gives an outlook.

2   Background

2.1   Mobile Geosensor Networks

Mobile sensor networks are specializations of wireless sensor networks [4,5,6] with mobile network nodes. If the mobile nodes are equipped with positioning sensors, they are also called geosensor networks [7,8,9].

Critical resources in mobile sensor networks are energy (on-board batteries) and communication bandwidth. Hence, nodes in mobile sensor networks communicate via short-range wireless technologies in synchronized short communication windows. Messages are received from neighbors in radio range, or from more remote nodes by multi-hop transmission relayed by intermediate nodes. The mobile sensor network topology is fragile and dynamic not only because the nodes move, but also because the nodes are volatile. They turn off and on for various reasons, in this context of the case study, for example, because pedestrians and vehicles do not participate in traffic all the time.

Communication in geosensor networks can serve three types of interest: information dissemination, information routing to a specific destination, and two-way negotiations.

1. Information dissemination can follow different strategies [10]:
   (a) Flooding strategy: An agent always informs all other agents within its communication range of information generated or received from other peers.
   (b) Epidemic strategy: An agent informs only the first n agents it encounters within its communication neighborhood after generation or reception of the information.
   (c) Proximity strategy: An agent informs other peers within its communication neighborhood only as long as the agent is within a certain threshold distance from the location the information refers to.
   None of these dissemination strategies are able to ensure information delivery between agents, either implicitly or explicitly. In the current context, an investigator disseminating a query can never be sure to reach the nodes with memories of interest.
2. Information routing reduces significantly the number of broadcasted messages compared to undirected dissemination [11,12]. However, in the present context, the location of nodes with interesting memories is unknown at the time of querying. Only the location of an immobile investigator may be used for geographical routing of query responses.
3. In contrast to the information dissemination problem, two-way point-to-point communication, as needed in the shared ride system [1], has to happen in one short communication windows assuming static connectivity within this window, due to the mobility of the nodes, and the corresponding fragility of the communication network. This fact limits the communication depth of negotiations to the immediate neighborhood of a node.

Applications of geosensor networks are based on the ad-hoc exchange of messages. Hence, for the reconstruction of historic events in geosensor networks these messages are in the focus of interest for any investigator. We will investigate different architectures and strategies to collect historic messages and reconstruct historic events.

2.2 Intelligent Transportation Systems as Geosensor Network Application

Future intelligent transportation systems will be built from autonomous mobile and immobile agents. Car manufacturers talk of the *autonomous vehicle*, referring to on-board nodes of vehicular ad-hoc networks, and including some immobile agents in the vehicle's environment for some applications[1]. Of course, intelligent transportation involves

---

[1] E.g., http://www.car-to-car.org, or http://www.internetits.org

other agents as well, such as people, goods, and non-automobile means of transport. Most of these agents' intelligent computing platforms would be battery-reliant. For the benefit of these agents, communication has to be limited and synchronized as in mobile sensor networks. We call this design *heterogeneous geosensor network* since it consists of nodes of different roles.

An intelligent transportation system can include immobile agents at the roadside (or along the lines of other means of transport). Nodes at the roadside are typically connected to regional or central databases, and are less restricted in terms of communication costs or protocols. We call this general design *hybrid geosensor networks*; elsewhere the specific network is also called vehicular ad-hoc network [13].

Agents in intelligent transportation systems have increasingly direct access to the Web, for example via 3G or wireless broadband, and hence, to other centralized services. We call this co-existence of a peer-to-peer network between mobile agents (a geosensor network) and direct Web access for each agent *mixed networks*.

3   Historic Queries in an Ad-Hoc Ride Sharing System

In this section the value of messages exchanged in a geosensor network during the course of their business is investigated. An *investigator* is a person or agent external to the geosensor network, who is interested in the reconstruction of specific past events in the network. For this purpose it is assumed that the investigator can communicate with the network and is accepted to have the authority to query the network.

As the running example, imagine police trying to reconstruct rides in the mentioned shared ride system [1,14,3]. Investigations may concern the identification of a host that was reported by a client, for instance. We will now study the information needs of an investigator, the evidence by messages, and the necessary analysis of the messages within this shared ride system.

3.1   Information Needs of an Investigator

A *shared ride* is the part of the trips of two agents that is jointly traveled. A client's trip can consist of multiple shared rides, of waiting times, and, if the client is mobile, of locomotion. A host's trip consists of locomotion only, during which the host vehicle can be unoccupied, or provides single, parallel or overlapping shared rides, depending on their passenger capacity. Any shared ride can be represented by a tuple $r = (c, h, t_s, t_e, \mathbf{s})$, where c is the client identifier, h the host identifier, t_s is the start time of the ride and t_e the end time of the ride, and $\mathbf{s}$ is the sequence of street network segments traveled together.

An investigator being interested in reconstructing rides first has to harvest all relevant exchanged messages, and then to reconstruct the realized shared ride(s) from this collected evidence.

3.2   Evidence for Events

The only evidence that exists about shared rides are the memories of the ride negotiations between clients and hosts. Since an investigator has to prepare for incomplete evidence due to the mobility and volatility of the network nodes, and also for unreliable evidence due to potentially manipulated memories by malicious agents, it is important to find redundant and independent evidence.

Negotiation messages fulfill this criterion. They were originally broadcasted by radio to the local public, i.e., within the communication range of the sender. Depending on the chosen dissemination strategy, messages can also be forwarded, i.e., re-broadcasted, over a few communication hops. A client broadcasts a *request* (e.g., by proximity dissemination), collects *offers* from local hosts (which are routed back), selects the best offer, and broadcasts a *booking* message (which is routed forward to the selected host). This messaging has generally witnesses.

The individual message types contain the following information:

- A request refers to the client's identity, the client's current position and the desired trip. Hence, a request specifies only $(c, \mathbf{s})$, but neither h nor t; ad-hoc planning always implies a desire for t_s ='as soon as possible'. With other words, requests are under-specified and lowly correlated with the rides made.
- In reply, hosts can offer a client any set of segments of their travel routes ahead. That means, an offer specifies a *potential* ride: $(c^p, h^p, t_s^p, t_e^p, \mathbf{s}^p)$. As potential rides, the investigator would need to know all offers and the client's cost criteria to reconstruct the selection of a client.
- In response to all offers, a client can book any one of them, or parts of them. Accordingly, a booking message specifies a *selected* ride: $(c^s, h^s, t_s^s, t_e^s, \mathbf{s}^s)$. Selected rides are plans; and plans can be changed in one of the next negotiation processes. This means that even booking messages are not a true image of the rides made. Only a travel plan for now, $t^s = now$, cannot be changed in the future; it will be realized at least for the initial segments $s_{1...i}$ until the plans are changed in one of the next negotiations. Let us call this subset of booking messages with $t^s = now$ the *travel messages*.

In short, evidence from the different message types would be partially contradicting and of mixed use. Hence, storing and querying all negotiation messages is inefficient. Considering the information value of a message type for an investigator, obviously a travel message ranks highest. Collecting only travel messages allows a complete and unambiguous reconstruction of a ride, which will be discussed in more detail in Section 3.3. And concentrating on travel messages only brings a radical reduction of memory load.

Storing only travel messages also lowers the communication load to retrieve all stored messages during an investigation. Travel messages are received and stored at least by the client and selected host. In general, also witnesses will be present. For a single shared ride, client c and host h will not change, but witnesses can change for each negotiation process. This means that witnesses may have only partial knowledge of a shared ride. However, this partial knowledge of witnesses provides the desired independent redundancy of the system. If either the client or the host manipulate their internal records after the shared ride, there may be other records in the network that will at least indicate inconsistencies with the manipulated record.

Agents can store only a limited amount of messages, and when their specified memory capacity is reached they start overwriting oldest messages by newer messages.

3.3 An Investigator's Analysis Strategies

Once the investigator has collected some travel messages, these messages can be recombined to trips. Given t, with $t_s \leq t \leq t_e$, and either c or h, the investigator is interested in the corresponding h or c, and in t_s, t_e, and s. The identity of the given agent c or h is constant over a ride. In fact, it is this property that defines start and end of a shared ride in travel messages:

- A shared ride starts when in a temporal sequence of a client's travel messages a new host h appears. This is the travel message sent immediately before the client got into the vehicle of a host.
- From here two options exist. Either the client got into the vehicle. Then the next travel message of this client, although not necessarily for the same host h, is for a new location because they had some ride with h. Or the client or host ignore the booking in the last moment. Then the next travel message of this client is for the same location, because no shared ride had taken place.
- A shared ride ends when (i) the shared ride had started before, and (ii) either a new shared ride starts, or the client has reached his destination.

In particular, the investigator cannot deduce s from the content of the first travel message of a shared ride. Since travel plans are regularly revised, even during a shared ride, s can change over time.

4   Exploring Historical Database Architectures

This section explores the precise nature of efficient data storage and querying strategies for reconstructing historic events in geosensor networks. The data of interest are messages that were exchanged in the geosensor network during their course of business, assuming that they contain all relevant information for the events that happened in the network.

Shenker *et al.* [15] identify three options to store messages in a geosensor network:

1. *external storage*, or *centralized warehouse approach*: The messages are sent to external storage where they can be further processed as needed.
2. *data-centric storage*: The messages are stored within the geosensor network at a node other than the agent node that generates the event.
3. *local storage*: Messages are stored locally, on board of the the agent node that broadcasts or receives the message.

These three options get a different flavor if one considers the three types of networks identified above for future intelligent transportation systems, or similar complex systems of autonomous agents in geographic space: the heterogeneous geosensor networks, the hybrid geosensor networks, and the mixed networks. Consider Table 1 for the following discussion.

External storage. In a *heterogeneous geosensor network* requires that every agent delivers each broadcasted or received message to an access point, since it might be needed

Table 1. Database and network architectures: reasonable combinations are marked by '+', and the ones investigated later are encircled

	heterogenous GSN	hybrid GSN	mixed network
external storage	−	⊕	⊖
data-centric storage	−	+	−
local storage	⊕	−	−

for any potential query. This burdens the network with extreme and unnecessary communication traffic, since the traffic cannot be related to a specific query (which is not posed at this stage). Moreover, the nodes near the access point become communication traffic hot spots and central points of failure, depleting of energy prematurely. External storage in *hybrid geosensor networks* is cheaper in terms of communication costs and increases demands on temporary local storage: immobile mediating agents in the environment that connect also to a central database are candidates to harvest information from passing-by mobile nodes. Communication costs consist then of the offloading of the mobile agents' local databases at mediating agents. Information loss can occur if local databases have to overwrite data before they can offload. Immobile mediating agents can also have a role as witnesses of negotiations by themselves. Collecting evidence by this way does not produce any communication overhead in the geosensor network. If mediating agents are placed sufficiently dense, even no other witnesses are required. External storage in *mixed networks* is trivial. The always-on connection to a Web service allows continuing registration of messages as they are sent or received, even concurrently for large numbers of agents. But the corresponding communication costs depend not only on the energy consumption of this communication channel; this channel is subject to fees.

Data-centric storage. In a *heterogeneous geosensor network* appears to work well over a range of conditions for a reasonably stationary and stable sensor network, however, it clearly fails in mobile and volatile geosensor networks. It works well though in *hybrid geosensor networks*, where the immobile mediating agents can also store messages locally instead of reporting to a central database. By this way, messages are stored close to the location of the events they are about. For many similarities with the external storage in hybrid geosensor networks, this case is not investigated individually. Data-centric storage is not applicable for *mixed networks*.

Local storage. Have the highest requirements on local storage capacity (or the risk of loosing information is highest), but do not incur any communication costs for storage. More importantly, the on-board storage allows messages to be filtered for a response to an investigator. If, for example, an investigator comes with narrow and infrequent queries, local storage may be more efficient than external storage. Local storage can be realized in all types of discussed networks.

The three competing database designs and network architectures require different querying strategies. Three cases are studied in more detail (Table 1): the costs of accessing locally stored data in heterogeneous geosensor networks, the costs of external storage in a hybrid geosensor network, and external storage in a mixed network.

5 Design and Implementation of the Simulation

In this section a discrete event simulation is designed to compare the three different approaches in the communication behavior and network design. The comparison will concentrate on the degree of completeness of reconstructions of shared rides, and the required communication effort.

5.1 The Simulation Parameters

The simulation happens in a regular street network of an 11×11 grid. In previous work we have shown that a grid is sufficient to approximate real street networks for the simulation of the shared ride system, see [2,?] for the evidence. In this grid network all agents are moving with the same speed: one street segment per time unit. After each time unit they are located at street junctions and negotiate. Radio range is globally set to one grid segment. The negotiation process is always initiated by a client broadcasting a request. The offer and booking messages are send in a directed way and therefore they are broadcasted at most three times. In the simulation client c seeks to find a ride from the junction $x_{3,5}$ to junction $x_{8,5}$. Their negotiations happen within a range of three hops. All these parameters are fix for the following simulations. Varying them would linearly influence the results, hence, we report only results for these parameters.

In the simulation all host agents are active for a constant number of time units, traveling random routes, and then become off-line immediately after they finish their trips. They remain off-line for a constant number of time units before they enter traffic again. This cycle is shifted for all hosts such that the number of hosts in traffic is kept constant. The off-line duration was arbitrarily chosen for these experiments.

An investigator is present at the central junction of the street network, which is also a fix parameter. The investigator's location is thus chosen as close to the historical event as possible. Since all agents involved in the trip have moved since the trip took place, this does not mean anything. The investigator can be placed anywhere and should get similar results. The choice in the simulation was made for design reasons of the experiment: the simulation world is a limited space, and only if the investigator is near to the center, its communication paths are somehow isometric.

The investigation is observed for its *communication effort* and its *completeness* of the reconstruction of the rides of a single client c between his start and destination location. Completeness can be represented as the percentage of reconstruction of the client's route, expressed by the ratio between the (correctly) collected street segments divided by the total number of segments traveled by the client. The communication effort is measured by the number of broadcasts in the investigation process.

5.2 Stages of the Simulation

The simulation is divided into two different stages: the first stage is used to create the local databases of agents about a travel history of client c, and the second one is used to reconstruct the rides of the client, observing the success rate and communication effort of the reconstruction.

5.3   The Implemented Alternative Investigation Strategies

Local storage in a heterogeneous geosensor network. The investigator's request is broadcasted using the flooding strategy with a radio range of one segment per hop. Communicating the request is integrated within the heterogeneous geosensor nodes' communication paradigm: in synchronized time windows, agents receive messages from near agents, and they process the messages received at the previous communication window (forwarding or responding). Between these communication windows the agents continue moving, and they can move out of range from other nodes, which can lead to situations where a request is not received by all agents.

When agents receive an investigator's request, they process the message and query their local database to find matching records. If they find matches, agents will respond to the investigator, their response being forwarded through the network. Each request is only responded once by an agent, even if received multiple times.

The implemented process reaches a steady state (in finite environments), but the investigator has no chance to recognize when this steady state has been reached. For this reason a conservative heuristics is applied in the simulation, derived from the size of the grid network. Each query terminates automatically after 200 time units.

External storage in a hybrid geosensor network. In this approach the investigator's request is supported by mediating stationary agents. Mediating agents act as witnesses and as offload stations for the mobile agents, as discussed above, and forward their collected evidence into a central database. The distribution of the mediating agents over the grid is regular at various densities.

External storage in a mixed network. This approach is implemented as an observer function of all the events in the simulation [16], but not further studied because of predictable outcomes: for the costs of sending in all travel messages through an always-on communication channel to a central service, all rides can be reconstructed completely.

6   Simulation and Test

The simulation is computed for the two remaining investigator strategies, and for a variation of the host densities in order to observe how the reconstruction strategy improves for various densities. For each setup 1000 experiments are conducted; the presented results show average values.

For both strategies, the investigator performs a query: "Along which segments $\{s_k\}$ did client c travel by shared rides from $x_{3,5}$ (the client's location at t_0) to $x_{8,5}$ (the client's destination, reached at the end of simulation phase one, at t_d), and with whom did c travel?".

Figure 1 compares the results of the investigation in the heterogeneous geosensor network (only mobile agents, left group) with the results in the hybrid geosensor network (righthand four groups). Both investigations are made in networks with host densities of 0.1, 0.4, 0.7, and 1.0 host per junction. In the experiments with hybrid networks, mediating agents are only witnesses; they do not serve as offload points (i.e., the evidence collected by mobile agents is just ignored). The mediating agents are placed in

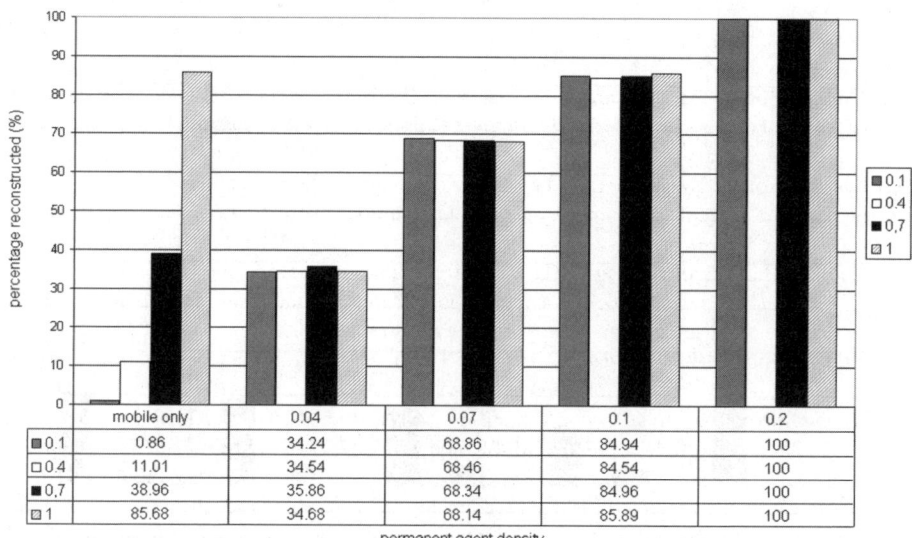

Fig. 1. Percentage of reconstruction of the trip of client c, with an off-line duration of 24 time units. The mobile agents do not transfer their memory to the mediating agents.

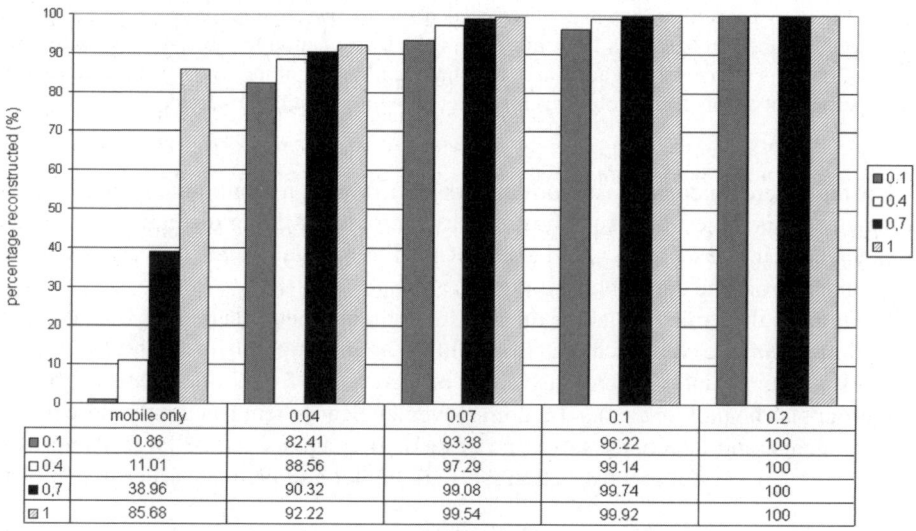

Fig. 2. Percentage of reconstruction of the trip of client c, with an off-line duration of 24 time units. The mobile agents transfer their memory to the permanent agents, when they are in communication range.

regular order in the grid network. Their density varies from 0.04 to 0.2. There is no communication overhead for this investigation.

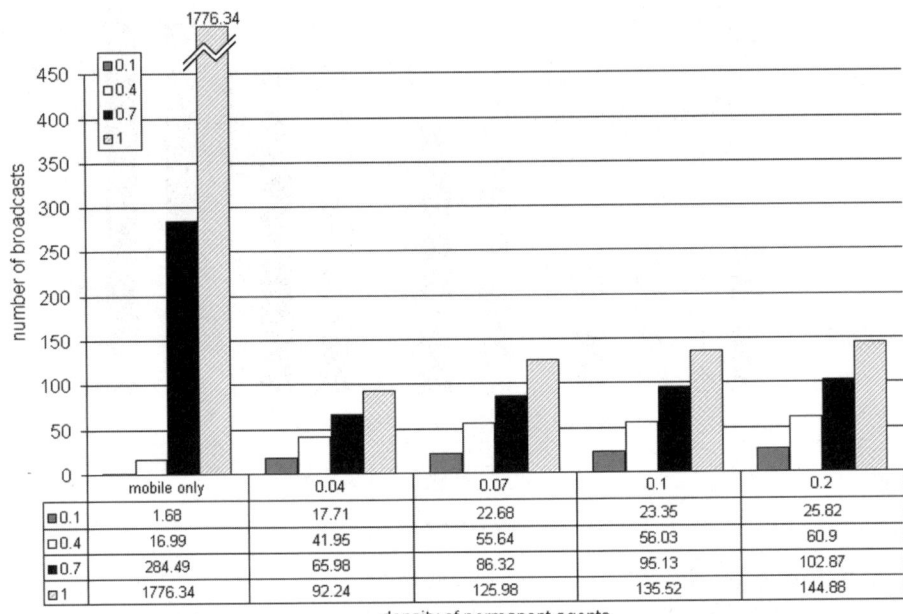

The chart shows number of broadcasts vs density of permanent agents:

	mobile only	0.04	0.07	0.1	0.2
▣0.1	1.68	17.71	22.68	23.35	25.82
▢0.4	16.99	41.95	55.64	56.03	60.9
▪0.7	284.49	65.98	86.32	95.13	102.87
▢1	1776.34	92.24	125.98	135.52	144.88

Fig. 3. The number of messages that where needed for the investigation process. In the mobile scenario these are all broadcasts from the start of the investigation to the end of collecting the answers. In the permanent agent approach we count the numbers of messages that are transferred to the permanent agents from mobile agents, before the investigation is started.

For the heterogeneous geosensor network (left group), a monotonic increase of the completeness is noticeable as the host density increases. Adding mediating agents, the completeness increases with the density of mediating agents, but this appears to be independent from the current host density.

In Figure 2 the same results for the investigation in a heterogeneous geosensor network (left group) are compared with the results of an investigation in a hybrid geosensor network where mediating agents also act as offload points for mobile agents (righthand four groups). The mobile agents' random movement leads them frequently close to mediating agents, and hence, the coverage of mediating agents can be largely improved by this way. Successful reconstruction rates are higher and plateau earlier in complete reconstruction.

Figure 3, finally, shows the communication overhead for the different settings in the previous experiment (Figure 2). In the heterogeneous geosensor network (left) all broadcasts are counted from the start of the investigation to the end of collecting the responses. In the hybrid geosensor network (the next four groups) the messages are counted that are broadcasted by mobile agents when offloading their memories in the range of mediating agents.

The number of broadcasted messages by the mobile agents levels with higher densities of mediating agents since then all witnessed messages are offloaded. The latter observation assumes that mobile memories are sufficiently dimensioned. In the experiment they witnessed only one client. With more clients, and more ongoing negotiations around, witnesses may loose some messages before they meet the next mediating agent. This happens already for the lower mediating agent densities in the present experiment, leading to less reported witnessed travel messages. With a lower mediating agent density also the risk grows that some messages are lost because of host volatility.

The comparison of both investigation processes shows that the hybrid geosensor network strategy (with offloading) is much cheaper in terms of the communication effort even for a single client and specific rides. While the number of messages in the heterogeneous geosensor network investigation process increases exponentially with more agents in the network, it only increases linearly for the permanent agent approach (until it plateaus). The linear increment means that the number of messages that have to be transferred per witnessing agent stays constant. So the communication effort is balanced among the agents in the network.

The pro-active character of the hybrid geosensor network approach means that an investigation in this case can be conducted instantly, while the heterogeneous geosensor network approach takes a while to collect the travel messages from the mobile agents. Also, the database of the investigator is all-embracing and serves for any ad-hoc request or further analysis.

7   Conclusions

This paper investigates the possibilities to reconstruct historic events in geosensor networks. For their deployment, geosensor networks do not require any real-time registration of events, and hence, any external interest in historic in-network events has to be assessed by its own communication overhead and storage capacity requirements with the mobile agents.

The three investigated solutions—local storage in heterogeneous geosensor networks (1), external storage in hybrid geosensor networks with witnessed evidence from mediating agents only (2), or with evidence from all agents (3)—show very different behavior in this respect. All rely on access to previously exchanged messages by a central investigator. Messages are witnessed, hence stored redundantly, and therefore robust against malicious manipulation. In the example of the shared ride system, it was sufficient to use travel messages only, a subset of all exchanged messages. The relative low number of travel messages compared to all messages in a negotiation process supports the second postulate of the hypothesis.

The investigation in a heterogeneous geosensor network, as the first solution, is to a large extent dependent on the density of the mobile agents. For low densities, communication overhead would be acceptable, but reconstruction rates are low. The reason is that the network is not sufficiently connected to transport requests and responses. For high densities, communication overhead explodes. Hence, this solution does not support the hypothesis.

The second solution, an investigation by evidence witnessed by mediating agents only, produces no communication overhead in the heterogeneous geosensor network, and realizes full tracking of all rides in near-real time if the density of mediating agents is large enough to witness all negotiations. Lower densities of mediating agents lead to incomplete reconstructions of some trips. However, the hypothesis is fully supported.

The third solution, with mediating agents witnessing and serving as offload points for mobile agents' memories when they pass by, shows a behavior somewhere in between. It has the largest reconstruction rates, because the lack of witnessing mediating agents with their lower density can be balanced by the mobile agents' memories. The additional communication overhead in the heterogeneous geosensor network is linear with the number of and lengths of rides made. In cases where infrastructure costs for mediating agents have to be balanced with communication costs in the mobile agents, this solution is a viable alternative, also supporting the hypothesis.

These outcomes can serve as a starting point to design efficient storage policies for evidence of events in geosensor networks. Self-controlling applications are required for developing trust of users or the public, for reputation mechanisms for individual participants, and for liability and safety measures. In return, storage should not destroy the advantages of distributed peer-to-peer systems, and should also keep an eye on privacy.

Future research can further optimize both strategies, the heterogeneous geosensor network approach as well as the hybrid geosensor network approach. In hybrid geosensor networks, for example, a witnessing mobile agent could check internally whether a received travel message was broadcasted within the range of a mediating agent, and then skip this message from storing. Furthermore, mediating agents could act in the name of the investigator. They could, for example, disseminate locally an investigator's request, and collect responses. Dissemination could be controlled by proximity, and responses could again be geographically routed.

A further area of research is the vulnerability of the investigation in the presence of malicious agents. Different scenarios and strategies to mislead the investigator can be developed and investigated in further simulations. From these results one can expect additional measures or strategies to protect the system from misuse.

Acknowledgments

The authors would like to thank Yun Hui Wu for her contributions to the simulation system. Carlos Vieira gratefully acknowledges funding by the Brazilian Government (CNPq grant 201060/2004-3).

References

1. Winter, S., Nittel, S.: Ad-Hoc Shared-Ride Trip Planning by Mobile Geosensor Networks. International Journal of Geographical Information Science 20(8), 899–916 (2006)
2. Raubal, M., Winter, S., Tessmann, S., Gaisbauer, C.: Time geography for ad-hoc shared-ride trip planning in mobile geosensor networks. International Journal of Photogrammetry and Remote Sensing (Special Issue on From Sensors to Systems: Advances in Distributed Geoinformatics) (2007) (Paper accepted in December 2006)
3. Wu, Y.H., Guan, L.J., Winter, S.: Peer-to-peer shared ride systems. In: Nittel, S., Labrinidis, A., Stefanidis, A. (eds.) GeoSensor Networks 2006. LNCS, Springer, Heidelberg (2007) (Paper accepted in December 2006)

4. Akyildiz, l.F., Su, W.J., Sankarasubramaniam, Y., Cayirci, E.: Wireless Sensor Networks: A Survey. Computer Networks 38(4), 393–422 (2002)
5. Zhao, F., Guibas, L.J.: Wireless Sensor Networks. Elsevier, Amsterdam (2004)
6. Gandham, S., Musunnuri, R., Rentala, P., Saxena, U.: A Survey on Self-Organizing Wireless Sensor Network. In: Zurawski, R. (ed.) The Industrial Information Technology Handbook, pp. 1–16. CRC Press, London (2005)
7. Nittel, S., Stefanidis, A., Cruz, I., Egenhofer, M., Goldin, D., Howard, A., Labrinidis, A., Madden, S., Voisard, A., Worboys, M.F.: Report from the first workshop on geo sensor networks. SIGMOD Record 33(1), 141–144 (2004)
8. Stefanidis, A., Nittel, S.: GeoSensor Networks. CRC Press, Boca Raton, FL (2005)
9. Nittel, S., Labrinidis, A., Stefanidis, A. (eds.): Advances in Geosensor Networks. LNCS. Springer, Heidelberg (2007)
10. Nittel, S., Duckham, M., Kulik, L.: Information Dissemination in Mobile Ad-hoc Geosensor Networks. In: Egenhofer, M.J., Freksa, C., Miller, H.J. (eds.) GIScience 2004. LNCS, vol. 3234, pp. 206–222. Springer, Heidelberg (2004)
11. Intanagonwiwat, C., Govindan, R., Estrin, D.: Directed Diffusion: A Scalable and Robust Communication Paradigm for Sensor Networks. In: Pickholtz, R., Das, S.K., Caceres, R., Garcia-Luna-Aceves, J.J. (eds.) MobiCOM 2000. Sixth Annual International Conference on Mobile Computing and Networking, pp. 56–67. ACM Press, Boston (2000)
12. Gerla, M., Chen, L.J., Lee, Y.Z., Zhou, B., Chen, J., Yang, G., Das, S.: Dealing with Node Mobility in Ad Hoc Wireless Network. In: Bernardo, M., Bogliolo, A. (eds.) SFM-Moby 2005. LNCS, vol. 3465, pp. 69–106. Springer, Heidelberg (2005)
13. Eichler, S., Schroth, C., Eberspächer, J.: Car-to-car communication. In: Proceedings of the VDE-Kongress - Innovations for Europe, Aachen, VDE, vol. 1, pp. 35–40 (2006)
14. Winter, S., Raubal, M.: Time Geography for Ad-Hoc Shared-Ride Trip Planning. In: Aberer, K., Hara, T., Joshi, A. (eds.) MDM 2006. 7th International Conference on Mobile Data Management, pp. 6–6. IEEE Computer Society, Nara, Japan (2006)
15. Shenker, S., Ratsanamy, S., Karp, B., Govindan, R., Estrin, D.: Data-Centric Storage in Sensornets. ACM SIGCOMM Computer Communication Review 33(1), 137–142 (2003)
16. Guan, L.J.: Trip Quality in Peer-to-Peer Shared Ride Systems. Master thesis, The University of Melbourne (2007)

Wireless Positioning Techniques – A Developers Update

S. Rooney, K. Gardiner, and J.D. Carswell

Digital Media Centre, Dublin Institute of Technology, Aungier St., Ireland
{seamus.a.rooney,keith.gardiner,jcarswell}@dit.ie

Abstract. This paper describes the current efforts to develop an open source, privacy sensitive, location determination software component for mobile devices. Currently in mobile computing, the ability of a mobile device to determine its own location is becoming increasingly desirable as the usefulness of such a feature enhances many commercial applications. There have been numerous attempts to achieve this from both the network positioning perspective and also from the wireless beacon angle not to mention the integration of GPS into mobile devices. There are two important aspects to consider when using such a system which are privacy and cost. This paper describes the development of a software component that is sensitive to these issues. The ICiNG Location Client (ILC) is based on some pioneering work carried out by the Place Lab Project at Intel. (Hightower et al., 2006) The ILC advances this research to make it available on mobile devices and attempts to integrate GSM, WiFi, Bluetooth and GPS positioning into one positioning module. An outline of the ILC's design is given and some of the obstacles encountered during its development are described.

Keywords: Location based services, Location prediction, WiFi positioning.

1 Introduction

The ICiNG Location Client (ILC) is a part of an overall project called ICiNG. The ICiNG project is about researching a multi-modal, multi-access concept of e-Government (DIT, 2006). It develops the notion of a 'Thin-Skinned City' that is sensitive to both the citizen and the environment through the use of mobile devices, universal access gateways, social software and environmental sensors. Intelligent infrastructure enables a Public Administration Services layer and a Communities Layer.

The ICiNG project will set up test-beds in high-profile European locations such as Dublin, Barcelona and Helsinki to act as 'City Laboratories' for researching, evaluating and demonstrating technologies and services using intelligence in the environment.

In the ICiNG project, the fundamental requirement for this type of service interaction is location. The ILC software component is about researching and developing an open source standalone location determination application that will be able to use a hybrid of different technologies to determine its location. For example, the ILC in any given environment will be able to return a location providing there is

J.M. Ware and G.E. Taylor (Eds.): W2GIS 2007, LNCS 4857, pp. 162–174, 2007.

at least one of the beacon technologies available and these include GPS, WiFi, GSM and Bluetooth. The ILC has being developed using a modularized component architecture so that each scanning component can and does run independently of the other technologies being available. This approach also has the advantage of being able to make any changes that need to be made with as little disruption as possible. This makes it easier to update specific modules and even add or remove modules without affecting the overall performance of the ILC. The ILC will not communicate with any external applications that are not installed on the device. Instead, it uses a Record Management Store (RMS) to store a limited amount of information and only at the users consent. The ILC is by design an open source, network independent, location determination mobile application that can utilise GPS, WiFi, GSM and Bluetooth information or any combination of them, to calculate location. (Kilfeather et al., 2007)The paper is organised as follows: Section 2 gives a brief summary of the related work carried out by the Place Lab research group. Section 3 gives a brief description of the ICiNG test beds, privacy considerations, location considerations and introduces some specific terminology. It then gives a more detailed description of the ILC architecture and its various modules and finishes by highlighting various development issues encountered so far. Finally, section 4 gives the conclusions for the project to date and comments on the expectations for the ILC's location determination methodology.

2   Related Work

2.1   Place Lab

Place Lab was an Intel research project that ended in 2006. The goal of Place Lab was to try and exploit freely available radio signals (GSM, WiFi, and Bluetooth) for location determination of a user's HHD (Hand Held Device). They did this by building a database of the known transmitting locations (beacons) of these signals, and then used the *assumed distances* (e.g. based on signal strength) from the user HHD to these radio beacon locations in a trilateration[1] procedure to determine the user's location. (Without initial knowledge of the direction to each of the beacons, a triangulation procedure using angles is not possible.) However, at the time, certain technologies were not yet advanced enough to take Place Lab fully mobile as some of the signal spotters worked on mobile phones (i.e. Bluetooth) while others needed a laptop (i.e. WiFi).

In the case of the ILC, the technologies that prevented Place Lab from bringing this technology fully mobile have advanced to the stage where mobile devices now offer integrated WiFi, GPS and Bluetooth capabilities. This advancement of these

---

[1] Trilateration is a method of determining the relative positions of objects using the geometry of triangles in a similar fashion as triangulation. Unlike triangulation, which uses angle measurements (together with at least one known distance) to calculate the subject's location, trilateration uses the known locations of two or more reference points, and the measured distance between the subject and each reference point. To accurately and uniquely determine the relative location of a point on a 2D plane using trilateration alone, generally at least 3 reference points are needed.

technologies is the basis for the ILC's design and like Place Lab, the ILC is designed to determine the position of a mobile device using passive monitoring that gives the user control over when their location is disclosed, laying the foundation for privacy-observant location-based applications. (LaMarca et al., 2005)

2.2 Location Calculations

The ILC will implement and test several different location calculation algorithms to determine the best possible location. After initial testing we will look at improving upon one or more of these algorithms while continuously testing for the most accurate results. There may be a performance sacrifice on a mobile device to achieve increased accuracy or there may be improvements to algorithms that could deem one algorithm more suitable to a mobile device than another. Only after some careful testing will it be possible to make these assumptions. A description of the algorithms that we will implement and test initially is given in the following.

2.2.1 Centroid Location Determination

This is one of the simplest algorithms that will be used by the ILC. It involves taking into account the locations of all known beacons in the area and then positioning the user mathematically at the centre of them. This approach ignores many things that could improve the location determination including beacon signal strength, confidence in the beacons location and environmental issues (i.e. tall buildings, hills, buses, etc.). (Hightower et al., 2006)

2.2.2 Weighted Centroid Location Determination

This is very similar to the standard Centroid algorithm but it takes into account other values in the RMS database when calculating a location. For example, the signal strength of each beacon can be taken into account to further determine if the ILC is nearer one beacon or another. Also, as previously stated, there will be a test bed used to test the ILC that will have fixed beacons where the exact position of each beacon is known. Initially, we will only be using these beacons but later during testing, we will carry out some Wardriving in and around the test bed area and add unknown beacons to the RMS database that have estimated locations. In this case, the beacons we know the exact locations of will get a greater weight when calculating location, than beacons that are detected using Wardriving.

2.2.3 Assisted ILC

Much like assisted GPS, assisted ILC will use an Assistance Server but on the phone. For example, if a mobile device was within signal distance of a known Bluetooth beacon and then moves out of range of that Bluetooth beacon this information would not be discarded straight away. Assisted ILC will use this information to help correct the location determination for a period of time. As time passes the weight of the correction will also decrease. Also depending on the type of beacon being used for the correction, there will be a weight attached to the expected accuracy it is expected to correct for.

Initially we have decided to implement these three algorithms and test them out in our test bed environment to determine which one offers the best overall performance in computational speed, resource efficiency and energy consumption and at the moment we are currently only testing the first algorithm "Centroid" with the first prototype ILC. There are other techniques that were considered, e.g., Particle Filters (Hightower and Borriello, 2004) and Fingerprinting, which are used in RADAR (Bahl and Padmanabhan, 2000) but initially it was decided that these might be too costly in terms of performance on a mobile device. Another possible technique is Gaussian Processes for Signal Strength-Based Location Estimation (Ferris et al., 2006) which seems more optimal for mobile devices, but initially we aim to test the first three Location Determination Algorithms deciding later if a different approach is required.

3 ICiNG

3.1 ICiNG Test Bed

The general approach of the Dublin test bed is to bring existing wireless and wired infrastructure belonging to the Dublin Institute of Technology and Dublin City Council together within an experimental on-street wireless network (Figure 1). Access to the new wireless network will, in the first instance, be open, free and supported by the Dublin ICiNG project team.

The ICiNG street signs will be the WiFi/Bluetooth Access points and the ILC will use its internal logic to determine a best guess location from the known positions of the street signs and other radio signals in the area.

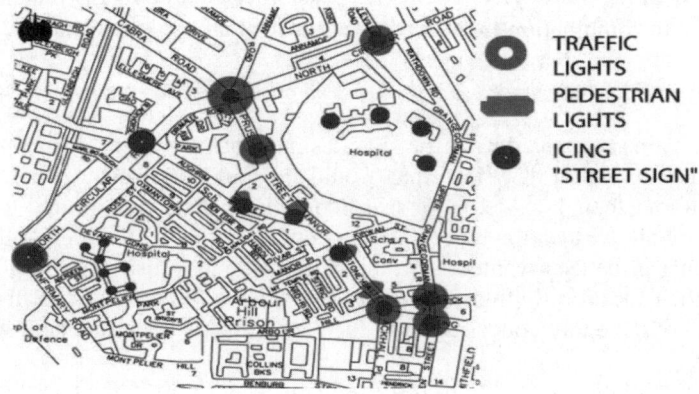

Fig. 1. Grangegorman Map: The Dublin Test-bed for the ICiNG System with the wireless access points highlighted

3.2 Specific Terminology

The *Icing Mobile Client (IMC)* refers to the complete set of application components on the mobile device. The IMC is comprised of:

- The *ICING Location Client (ILC)* whose purpose is to calculate and make available the device location to the MDA based on GPS, Bluetooth, and WiFi Beacon information.
- A number of *Mobile Device Applications (MDA)*. An example of an MDA is an accessibility application that enables users of the ICiNG system to report accessibility issues to the City Council using a Jabber client[2] extension.

3.3 Location Considerations

There are many Location Based Services (LBS) identified in the ICiNG project. These range from providing a location tracking sensor network to retrieving metadata based on a mobile device's location. While these services are heterogeneous in nature they all require a method of determining the location of a particular device or sensor. There are many existing systems available to provide this location information, some using cell services provided by mobile telecoms providers or others using satellite technology such as GPS.

However, as discussed previously each of these technologies and services have inherent advantages and disadvantages. Some services operate well in urban areas and in areas of high cellular radio density while others perform well where line of sight to satellites in the GPS system is established. Also, beyond the purely technical or technological considerations to be taken into account in location determination are issues of privacy and safety which location technologies raise. (Vossiek et al., 2003)

Fortunately, the issue of deciding which of these technologies to use is being somewhat mooted by the increasing trend of mobile devices to incorporate multiple access technologies in the same platform. The availability of GSM, WiFi, Bluetooth and GPS on the same device offers the possibility of intelligently using all these technologies in combination to improve location availability and accuracy.

3.4 Privacy and Cost

One of the main concerns during the ILC design phase was that of user privacy. We did not want to design a system that would or could be used to track a citizen's location without their knowledge or prior consent. Instead, we wanted to develop a system where all the location determination could be done on the mobile device itself and then, only if the user wanted, it would be possible to inform the rest of the ICiNG system of their location. In this way, the user will have full control over their location and would not have any concerns about their movements being tracked without their knowledge.

For users that allow the ILC to disclose their location to ICiNG services, we also wanted a system that would address any other privacy concerns they might have. Of these issues, we identified the following to be particularly important; location information retention, location information use, and location information disclosure.

---

[2] An open, secure, ad-free alternative to consumer IM services like AIM, ICQ, MSN, and Yahoo. Jabber is a set of streaming XML protocols and technologies that enable any two entities on a network to exchange messages, presence, and other structured information in close to real time.

For ICiNG, it was decided at an early stage that location information retention, if it needed to happen at all, would be only for a short task-specific time frame depending on whether the user was partaking in specific studies or if they had signed up to a service that required their movements to be tracked. For example, a parent could register their child's mobile phone to such a service to monitor their child's whereabouts.   Another issue that needed to be addressed was that of disclosing movements of users to 3rd parties, which the ILC never directly does.

Although we have addressed privacy as a consideration, the very nature of the ILC is that for the location information to be useful, a 3rd party application needs access to it - and the ILC has no control over what 3rd party applications do with this information. However, 3rd party applications do need to access the record store that holds the devices location and this can and will be restricted to verified applications. Unfortunately we are not able to enforce our policies onto other applications and acknowledge that providing such a facility does require the 3rd party application to address privacy in their own design. Although we have no control over what happens the information once it is used by another application a system such as Casper (Mokbel et al., 2007) should be considered if there is any communication with a ermote server to provide a location based service.

The ILC was designed to be a monetarily zero cost solution to location determination. The only cost we see for the user is the time cost to install the application and the bandwidth cost to receive freely available radio signals. There is also a CPU cost of running the application and after testing the initial prototype, we will endeavor to reduce the overall processing cost of the application using more efficient algorithms, etc.

So far we have discussed some of the different positioning techniques available, and some of their advantages and disadvantages. We identified concerns that most people would have about their privacy being infringed upon and what the possible solutions to these issues are. We also noted, due to technology limitations of the day, what Place Lab was not able to do in bringing a fully mobile location based system to fruition, and how ICiNG would take the next step and extend their work by designing and developing such a system. The next part of this paper gives a more detailed overview of the ICiNG system, focusing on the ILC.

3.5   ILC – ICiNG Location Client

For the ILC, we use a combination of technologies to develop a location determination system that integrates the best features of all technologies available. By using all these technologies together, any disadvantage that each individual technology has diminishes. The ILC is designed as provider-network independent, privacy sensitive and zero cost (in terms of network resource usage) software component that allows mobile devices to determine location by a "best guess" methodology.   The prototype ILC is designed to run on a Series 60 (3rd Edition) mobile phone running the Symbian operating system (version 9.x), although other platforms and operating systems could be accommodated with relatively minor changes. Figure 2 shows how this architecture works together.

There is a set of defined rules that the ILC uses when searching for beacons in the RMS Database. It is dictated by the degree of accuracy that should be expected depending on the type of beacon. Although all beacons in the RMS Database are read, it first looks for Bluetooth beacons and if any are found it will discard the any other beacons used for determining the location; hence the degree of accuracy should be <10 meters. Next, if there are no Bluetooth beacons, it looks for a GPS reading. If there is no GPS reading then it looks for WiFi beacons and if there are no WiFi beacons, it looks for GSM beacons. As the technologies used changes, so will the degree of accuracy the tracker is providing. Even though beacons with a smaller weight for accuracy are disregarded in the location determination returned to the MDA, this beacon data does not get totally discarded. If the more accurate beacons become unavailable and the ILC switches to the less accurate beacons in its Database for triangulating position, then the less accurate location gets a correction applied based on its proximity to the last known more accurate beacons. (Rooney et al., 2007)

The above rule set and the respective weights that will be applied to each technology is still being refined and will only become clear after prototype testing with different configurations, a phase that has just commenced. Finer points like the expected accuracy from each technology, how the constant use of each technology effects the battery life on a mobile device and hence how many times each signal should be scanned to get the maximum battery life/accuracy ratio will all be tested and reconfigured in time. The different characteristics each of these technologies brings will determine the most efficient and accurate rule set for combining these technologies to work seamlessly together. Currently, we are only testing using the first basic algorithm and plan on doing extensive testing with it. Once this is completed, and we have data to work with, we will move onto implementing and testing the next two algorithms.

3.5.1 Java Bluetooth Spotter

The Bluetooth spotter polls the Bluetooth Terminal hardware and scans for any Bluetooth devices in range, and any devices found in range are returned to the tracker module.

In the case of the Bluetooth spotter a list of the currently detected devices is returned with some exceptions. Take the case of other mobile phones in the vicinity, for location determination this data is inadequate because mobile devices are not stationary and have no permanent location associated with them. Therefore, any Bluetooth beacons that are classed as mobile are filtered out of the list before being returned to the tracker module.

3.5.2 Java GPS Spotter

The GPS Spotter communicates with a GPS receiver to obtain Lat/Long coordinates and returns these coordinates to the tracker module. In the initial implementation of the ILC an external GPS receiver is used. The GPS spotter is used to improve the accuracy of the ILC when used outdoors as the density of WiFi access point's naturally decreases.

3.5.3 Java and C++ GSM Spotter

The Java GSM spotter sends requests to the C++ GSM spotter to retrieve the current GSM tower information and returns it to the tracker module.

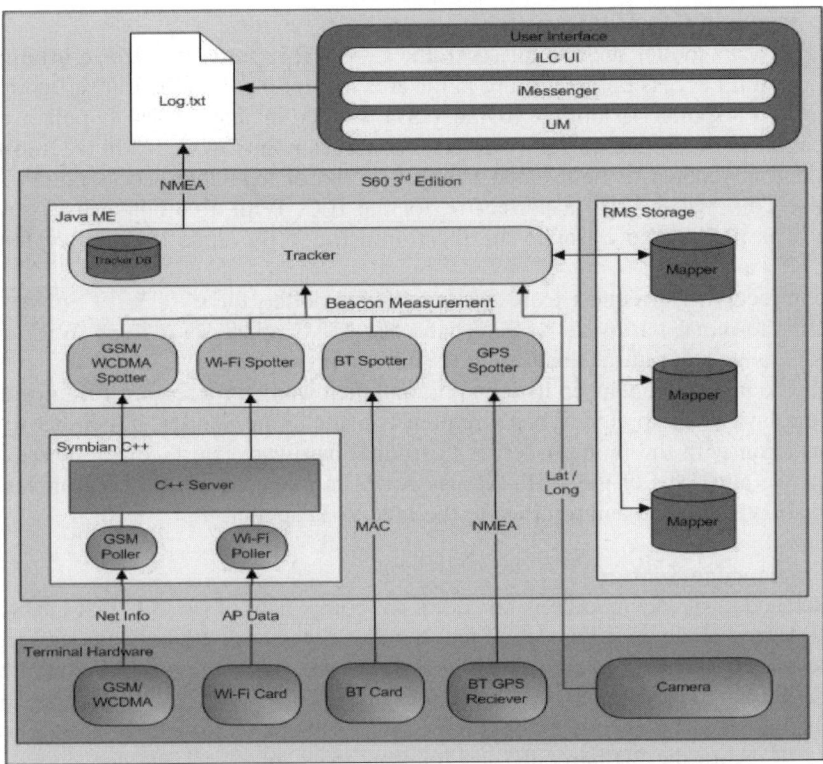

Fig. 2. The architecture of the Icing Location Client (ILC)

In this case, the GSM spotter only returns one set of information at any one time as there is a restriction inherent in the Symbian S60 architecture that prevents any application from obtaining a full list of the currently available masts. This restriction somewhat affects the accuracy of the ILC when in a remote area where the availability of WiFi and Bluetooth is virtually zero as trilateration cannot occur without at least three detectable beacons. Although we are considering different approaches to get around this, none have so far been implemented or tested. For example, a short list of GSM towers that have been scanned in the last 30 seconds and then performing a trilateration of the recorded towers while giving greater weight to the cell towers that were scanned nearer the current time is one such solution.

Upon receiving a request from the Java GSM Spotter, the C++ GSM spotter polls the GSM terminal hardware to determine the cell information about the currently active cell tower and returns this data back to the GSM spotter.

It does this by first opening up a listening port that listens for beacon requests from the Java GSM spotter. Upon receiving a request it polls the mobile device terminal hardware for network information. When the data is returned, the information is collated and sent through the open socket. A flag is then set to notify the server to wait until another request arrives from the Java GSM Spotter.

3.5.4 Java and C++ WiFi Spotter

The Java WiFi spotter sends requests to the C++ WiFi spotter to retrieve information about the WiFi access points present in the area and returns it to the tracker module.

The WiFi spotter is similar to the GSM spotter in that it has to poll a native application to obtain the currently detectable access points in the vicinity. It also has the added advantage of being able to return a list of available access points in the vicinity. This makes WiFi a perfect fit for the ILC. With a complete list of access points it is possible to estimate the location using WiFi alone if there are three or more AP's available.

Upon receiving a request from the Java WiFi spotter, the C++ WiFi Spotter polls the WiFi terminal hardware to determine the MAC addresses of any WiFi access points in range and returns this data to the Java WiFi spotter.

This module first opens up its own port and then waits for a request for information from the Java WiFi spotter. When a request is made to the Spotter, it then has to open a connection with the monitor server (Terminal hardware interface). It subsequently obtains the attributes of the WiFi Access points in range. When this is complete, the information is collated and returned to the Java WiFi spotter.

3.5.5 Semacode Spotter

The Semacode spotter module is an additional component of the ILC that enables the user to take a photo of a Semacode tag. Semacode tags are 2D barcode stickers that can be placed at virtually any location, for example, on lampposts. It is possible to encode the barcode with any type of information usually a URL. In the case of the ILC the barcodes are encoded with the location of the barcode. The Semacode spotter is used to record the 2D barcodes and translate them into Lat/Long coordinates. The spotter then returns these accurate coordinates to the tracker module. If the Semacode tags are positioned correctly using highly accurate GPS, the location obtained from using the tags is the most accurate data available to the ILC. Therefore, when a Semacode tag is decoded and a location obtained from it, this position is deemed to be highly accurate and all other estimates are overridden.

3.5.6 Tracker Module

The Tracker module is responsible for collecting beacon information from the four Java spotters and organizing the beacon information into a local tracker database. Using this database of beacons the tracker uses trilateration to determine the current position of the mobile device. It is at this point in the process that other positioning algorithms can be used to try and estimate a more accurate position, some of which will be discussed in section 3.5. The process used by the tracker is as follows,

1. When the ILC starts, the tracker module implements four spotters for Bluetooth, WiFi, GSM and GPS.
2. Upon validation that each hardware component is present, the tracker module polls the four Java spotters for any beacons in range of the phone.
3. The Java spotters then return information about any beacons in range.
4. The tracker module then stores this information in the tracker database.
5. The tracker module then checks to see what beacons are in the tracker database.

6. Depending on the beacons available, the tracker chooses which beacons to use and then compares them against known beacons in the RMS Mapper Database.
7. If the current beacons are unknown, then the tracker module will revisit the tracker database to select the next viable beacons to check against the RMS Mapper Database.
8. Once the tracker module's beacons have been checked against the RMS Mapper Database, it then attempts to trilaterate the best possible position based on the current beacon information.
9. When the position has been determined, it is inserted into the RMS Location Database with the current date and time. This information is then only available to other applications that have been verified as legitimate users of the ILCs location calculation. This verification is done when the application is being packaged.

3.5.7 Tracker Database
The tracker database is a list of the beacons that are currently in range of the mobile device. Each element in the list is compared against the known beacon database, or Mapper database, in order to determine their locations and subsequently trilaterate the devices position.

3.5.8 RMS Mapper Database
The RMS Mapper database is a database of known beacons and their associated locations. This database is used by the Tracker module to lookup beacon locations which are then used to trilaterate the current position of the mobile device. Initially this database only contains data about known beacons and has been manually recorded and is very accurate. In subsequent versions of the ILC it will be possible to use databases created directly from Wardriving and that can be downloaded from websites like [www.wigle.net]. In this scenario, the ILC will be capable of connecting to an online database of known access points and download beacon information for the current location. Also, a more effective possibility will be to use Bluetooth to push local beacon data to mobile devices when they enter certain Bluetooth areas.

3.6 Development Issues

Development for mobile devices and in our experience Symbian C++ and J2ME does not come without its own problems, few of which are documented and only through community forums can a solution be found in most cases. As we have found very little documentation available we decided a section on the pitfalls we have encountered could be helpful to the readers of this paper.

3.6.1 Computer Hardware
One of the main issues we encountered was that the development environment would not work correctly on different types of hardware, mainly on laptops. There is no recommend hardware besides the standard information you get in relation to CPU speed, memory size and HDD size. We encountered numerous intermittent problems while trying to develop on different laptops with different specifications and numerous clean installs of the operating system. In the end, when we switched to a

desktop 95% of the intermittent problems we were having were solved. There is also some issue with the integration of the different software packages you need to install. For example, each phone comes with the Nokia PC suite which can be installed and uninstalled with no issues. But if you install the Carbide j SDK and the J2ME Series 60 SDK which is used to develop java applications for Series 60 phones, some problems occur. For Example, if you need to uninstall the Nokia PC suite, the uninstaller also removes various windows drivers.

3.6.2   Security Certificates

Developing on the new Symbian Series 60 3rd edition platform has the addition of a new security level to deal with. While developing and creating a new application, the SDK automatically generates what is called a UID for each different application type. When you attempt to test your application on an actual device there are some restrictions. Even though it runs fine in the emulator, it will not install on the device itself. The solution is to go to www.symbiansigned.com, download an application called the DevCertRequest and request a development certificate that includes the capabilities you want for your application. This application will generate a public key for the desired application. Using this key return to the Symbian signed website and upload the key to receive the certificate. On receipt of the certificate, recompile the application with the certificate and this will generate a new version of your application which can then be installed on the device for testing.

3.6.3   Development Environment

Setting up the environment for the Carbide C++ requires that Active Perl 5.6.1.631 is installed. If any other version except this exact one is used, then the developer will not be able to compile applications.

3.6.4   GSM Cell Tower Access

When the initial concept for trilaterating position using GSM cell towers was conceived it was thought that it would be possible to scan for all cell towers in the area and trilaterate using the nearest 3-6 towers. However, this is not the case and during the development it was discovered that it is only possible to get the cell tower information of the currently connected cell tower. Further research has revealed that it would have been possible to do this if we were developing under the Windows CE platform but the Symbian S60 operating system has put in place restrictions that prevent a list of cell towers from being obtained.

3.6.5   WiFi MAC Restriction

When we first developed our WiFi module we discovered that while we are able to get the SSID and the signal strength of the WiFi access point, we are unable to get the MAC address of it. This is because Symbian had also put in place a restriction that prevents access to some of the attributes. This problem has since been resolved and a patch was released that allows the developer to access this data, though rather crudely.

3.7 Future Considerations

ZigBee (IEEE 802.15.4)

ZigBee is the name of a new high level communication protocol using small, low power digital radios based on the IEEE 802.15.4 standard. It has being designed for devices that require a low data rate, a long battery life and is intended to be a cheaper alternative than other likewise technologies such as Bluetooth but not a direct competitor. It was decided that this is still an emerging technology and until there are enough real world devices that use this, we would not develop a module for it. However, although there is no current module for this wireless technology, due to the inherent open ended design of the ILC,, it would not require any major changes to the ILC as only an extra module would be needed to add this functionality. Such is the case with any future technology that becomes popular with mobile devices.

4 Conclusions

The main purpose of the ILC is to provide privacy sensitive, low cost and open source positioning to mobile devices. The initial use for this technology is to provide position estimates to a suite of e-Government applications developed as part of the ICiNG project. A prototype ILC has been developed that integrates GSM, WiFi, Bluetooth and GPS into two Symbian S60 applications that scan for wireless beacons in the area and based on a known database of beacons, provides an estimated or best guess location. In doing so, the ILC is designed in such a way that it attempts to deal with some of the issues relating to mobile positioning, in particular that of privacy and cost. Using the ILC does not incur any network costs in relation to everyday usage and access to the devices position is left to the discretion of the user.

The initial ILC prototype uses some simple trilateration algorithms to determine its position based on the currently visible beacons. The first prototype has been implemented but no test results are available at the time of publication. The next step in this research is to investigate some other algorithms with the aim of increasing the accuracy of the calculated location, the performance issues in terms of CPU load and battery life, using all or a combination of the technologies mentioned and by testing them in environments that have different types of wireless infrastructure, in particular the performance difference associated with using a sectorised WiFi infrastructure. Testing will confirm how each of the trilateration algorithms perform on mobile devices and if there are any performance issues, and if so, where do these issues arise and what modifications are needed to improve on them. Again, when we have concrete test results, we will publish them and further enhance the efficiency and usability of this location based system.

References

Bahl, P., Padmanabhan, V.N.: RADAR: An In-Building RF-based User Location and Tracking System. In: IEEE Infocom 2000, Tel-Aviv, Israel (2000)

Ferris, B., Hahnel, D., Fox, D.: Gaussian Processes for Signal Strength-Based Location Estimation, Robotics: Science and Systems, Philadelphia, Pennsylvania (2006)

Hightower, J., Borriello, G.: Particle filters for location estimation in ubiquitous computing: A Case Study. In: Davies, N., Mynatt, E.D., Siio, I. (eds.) UbiComp 2004. LNCS, vol. 3205, pp. 88–106. Springer, Heidelberg (2004)

Hightower, J., LaMarca, A., Smith, I.: Practical Lessons from Place Lab. IEEE Pervasive Computing 5(3), 32–39 (2006)

Kilfeather, E., Carswell, J., Gardiner, K., Rooney, S.: Urban Location Based Services using Mobile Clients: The ICiNG Approach, GISRUK, Maynooth, Ireland (2007)

LaMarca, A., et al.: Place Lab: Device Positioning Using Radio Beacons in the Wild. In: Gellersen, H.-W., Want, R., Schmidt, A. (eds.) PERVASIVE 2005. LNCS, vol. 3468, Springer, Heidelberg (2005)

Mokbel, M.F., Chow, C.-Y., Aref, W.G.: The New Casper: A Privacy-Aware Location-Based Database Server. In: International Conference on Data Engineering, Istanbul, Turkey (2007)

Rooney, S., Gardiner, K., Carswell, J.: An Open Source Approach To Wireless Positioning Techniques, Mobile Mapping Technology, Padua, Italy (2007)

Vossiek, M., et al.: Wireless local positioning. Microwave Magazine 4(4), 77–86 (2003)

Framy – Visualizing Spatial Query Results on Mobile Interfaces

Luca Paolino, Monica Sebillo, Genoveffa Tortora, and Giuliana Vitiello

Dipartimento di Matematica e Informatica, Università di Salerno
Via Ponte don Melillo,
84084 Fisciano(SA), Italy
{lpaolino,msebillo,tortora,gvitiello}@unisa.it

Abstract. The widespread use of mobile devices in basic map navigation tasks has recently attracted researchers on usability problems arising from the reduced visualization area and the limited interaction modes allowed by small keypads. In this paper we analyse the most common approaches suggested in the literature and propose a new visualization technique, named *FRAMY*, which is based on one of them and exploits a novel interaction metaphor for picture frames to provide hints about off-screen objects. It was conceived to cover a wider range of spatial data visualization tasks, which may simultameously involve different geographic layers.

Keywords: Mobile devices, geographical information systems, spatial visualization methods, spatial functions.

1 Introduction

People use maps in a number of tasks, including finding the nearest relevant location, such as a gas station, or for hand-optimizing a route. Using a map, users can easily compare alternative locations, such as the selection of restaurants. Users can see how far away a restaurant is from his/her current location, and whether it lies close to other locations the user considers visiting. Advantages deriving from maps are uncountable, they provide people with the ability to show and analyze the world as they perceive it, namely by verifying topological, directional and metrical relationships. However, a complete and exhaustive evaluation of a region is possible just in case users are able to:

1) have a global or wide enough view of the map portion they are taking into account and simultaneously
2) have a vision of the map where features can be well distinguished from each other.

   In other cases, comparisons as well as evaluations of spatial relationships may result very difficult because users have either no sufficient information or confused information.

J.M. Ware and G.E. Taylor (Eds.): W2GIS 2007, LNCS 4857, pp. 175–186, 2007.
© Springer-Verlag Berlin Heidelberg 2007

This is the case for maps visualized on mobile devices, where viewing a wide portion of the map conflicts with showing features in a well defined and separate fashion, due to the small dimensions of screens.

As an example, if we perform a search for the closest hotel from our current location, the result may fall either inside or outside the screen. In the first case, we do not have problems to reach the hotel because we can distinguish it on the map. Vice versa, in the second case, we do not know which direction we have to follow to arrive at the target and we have to gropingly advance.

Visualizing the entire map is not the right solution. As a matter of fact, objects resulting from a query task might appear improperly overlapped, if a reduced map scale is adopted. As an example, if we are visualizing a layer showing the locations of farmhouses located in a given region and we want to see all of them together, we might want to represent the entire region on the screen, hence using a reduced scale for the map. In that case, there is a high probability that, below a certain proximity threshold, farmhouses will be overlapped and will then be hard to distinguish.

In the literature, several techniques have been proposed to display large information spaces on desktop screens. An interesting paper by Chittaro classifies the different approaches adopted so far to visualize maps and images on small screens [2]. The author argues that the techniques traditionally adopted to combine panning and zooming with either compression (the so-called *overview-detail* approach) or distortion operations (known as *focus-detail* approach) may still present severe limitations when applied to mobile devices, due to the very small screen dimensions. A promising approach, specifically conceived for visualizing distance and direction of off-screen locations, consists in providing visual references to off-screen areas, by augmenting the detail view with interactive visual references to the context. Such an approach has been successfully adopted in systems like Halo to enable users to complete map-based route planning tasks [1]. This suggested us that a similar idea could be adopted and extended for performing more complex GIS activities on a mobile device, which also involve spatial queries on different map layers.

Thus, in order to provide an appropriate tradeoff between the zoom level needed to visualize the required features on a map and the amount of information which can be provided through a mobile application, we propose a new visualization technique, named *FRAMY*, which exploits an interaction metaphor for picture frames, to provide hints about off-screen objects, resulting from a given query. The idea is to visualize a frame along the border of the device screen. The frame is divided into several semi-transparent colored portions, corresponding to different sectors of the off-screen space. For a given query, the color intensity of each frame portion is proportional to the number of resulting objects located in the corresponding map sector. Thus, the frame may indicate both the distance and direction of specific Point Of Interests (POIs), as in Halo, but it may also represent the amount of POIs located towards a specific direction. In general, the frame portion color intensity represents a summary of data located besides the screen border. Moreover, the number of frame portions can be interactively increased so as to refine the query results, indicating, e.g., the exact direction where certain objects can be found. If the query involves several layers, nested frames are visualized along the borders, each one corresponding to a different layer, with a different color.

FRAMY is currently being experimented on a prototype of a mobile GIS application, called *MapGIS*, designed to perform typical GIS operations and queries, and combines typical mobile interaction controls with the proposed visualization technique.

The paper is organized as follows. In Section 2 we give a brief overview of the common approaches followed to visualize spatial data on mobile devices. In Section 3 we describe the MapGIS application and explain how the proposed visualization technique supports the corresponding functionalities. In Section 4 we illustrate three typical scenarios where *MapGIS* can be employed, which show the effectiveness of the *Framy* technique. Some final remarks conclude the paper.

2 State of the Art

One of the simplest ways to analyze the information depicted in maps shown on mobile devices is the classic *Pan & Zoom* method.

Maps may be seen at different levels of detail by applying the *Zoom* operation while the visualized region of interest may be changed applying the *Pan* operation. Typical styles of interaction concerning with the former are the selection of the detail level by means of specific buttons (one for each level), by sliders, by menus or by managing the wheel of the mouse. The *Pan* operation is usually implemented by dragging the mouse over the map or by scrolling panels.

An approach combining Pan and Zoom together was presented in [5]. At the start of an action, two concentric circles are placed so that the location of the action on the display is at their centre. As the user drags the pointing device, a direction line is drawn between the starting position and its current location, indicating the direction of work. If the pointer remains within the smaller circle, no scaling of the information space occurs, only scrolling. As the pointer moves further away from the starting position, the scroll rate increases. When the pointer moves beyond the inner circle, both scaling and scrolling operations take place.

Several papers have been also presented concerning with the technique named "Jump & Refine"[4][9][10]. This technique segments the map into several sub-segments, each of which is mapped to a key on the number keypad of the smartphone. By pressing one of the buttons, the region represented by that button is automatically puts in evidence on the screen. At this point the user may either pan the visualized region in order to reach the required zone or recursively apply the segmentation for augmenting the detail level.

Another interesting approach has been presented in [1]. The software implement-ting this methodology focuses the attention on discovering points of interest located outside the visualized region. As a matter of fact, each point of interest is the center of a circle whose ray raises or decreases as we get close or far from that point. In such a way, we may verify anytime the proximity of the point. A variation of this approach is CityLights [7]. Here, compact graphical representations such as points, lines or arcs which are placed along the borders of a window to provide awareness about off-screen objects located in their direction.

In the past, other interesting techniques for visualizing data on small devices have been also presented. Generally, they may be divided in two main categories, Overview&Detail and Focus&Context [6].

In order to have the awareness of a particular region and the context in which it stay, Overview&Detail techniques present simultaneously two maps where one refers to a limited/detailed region and cover the entire screen while the second one is the whole scaled map located just in a corner. Optionally, the detailed region is also highlighted by a rectangle on the whole map. Panning the first map changes the rectangle position on the second one. Even though today this approach is largely used, it was firstly described in [2]. The *perspective overview-display* is another example of a Overview&Detail technique. Here, two completely distinct areas are used to display the overview (top) and detail (bottom) of an image. To save valuable screen space while displaying the overview, the image is prospectively warped.

The Focus&Context technique is based on the reasonable assumption that the attention of users is focused on the center of the screen. FishEye View[8] is an example of this technique. Here, the center of the map is clearly displayed in the center of the screen while distortion augments starting with less distortion near the focus to strong distortion near the borders of the available display size. Based on the FishEye approach, three possible implementations of the distortion function are provided in [6], uniform context scaling, belt-based context scaling, non-uniform context scaling.

Finally, an approach merging together Overview&Detail and Focus&Context is presented in [6]. Two different images are used. These images are created by a transformation function. To display the context, the whole image in the logical coordinate system is scaled to the available display size. The focus is created as during panning, but using an equal or smaller size than the available display size. Afterwards the images are combined by the transformation function by applying a fixed or time-variant α-blending.

3 The *Framy* Visualization Technique Supporting a Mobile GIS Application

In this section we describe the *MapGIS* application, namely a tool which allows mobile device users to perform typical GIS operations and queries whose results are depicted according to the *Framy* visualization methodology.

MapGIS is based on a typical GIS architecture realized in a J2ME platform for mobile applications. It allows users to gather information about localities on the considered map by querying the data pre-loaded and stored in a structured repository of the device. It manages the browsing of the map by using a specific pointer that is moved all round the screen through the directional keys on the device.

Two kind of functionalities are currently available on *MapGIS*, that is:

- basic spatial operations, and
- advanced selection operations.

Panning and zooming are examples of basic operations. They allow users to easily explore maps everywhere and at different scales. So, in order to accomplish such tasks as easily and accessible as possible and guarantee a good level of usability, rapid controls have been provided through the device keyboard.

As for the advanced selection operations, they can be used to process both geographic and descriptive data in order to answer specific user's requests about the real world data. As an example, in *MapGIS* a search functionality is provided on descriptive data by pointing out specific localities of the map. This functionality represents the generally called operation "Spatial Selection". It allows users to reference every point located on the map by means of the browsing pointer and returns the descriptive data which are strictly connected to the pointed locality.

Besides the usual route planning functionalities, which also rely on some optimization algorithms, *MapGIS* provides an advanced search functionality, which allows to localize points of interest of specific categories (theaters, parks, museums, cinemas, etc) by specifying some descriptive information set up by users.

Once the user performs the spatial requests, results are visualized on the map by applying the *Framy* visualization method which helps users to discover results located off-screen.

The user may choose to divide the map into any number *n* which is a power of 2. Starting from the center of the screen a fictive circle is drawn and the map is divided into *n* sectors of equal width ($360°/n$). Figure 1 gives an idea of this subdivision using $n = 8$, that is we divide the screen in 8 sectors. We use the notation Ui – invisible- to represent the parts within a sector which lie outside the screen, whereas Vi – visible-represents the inside screen parts.

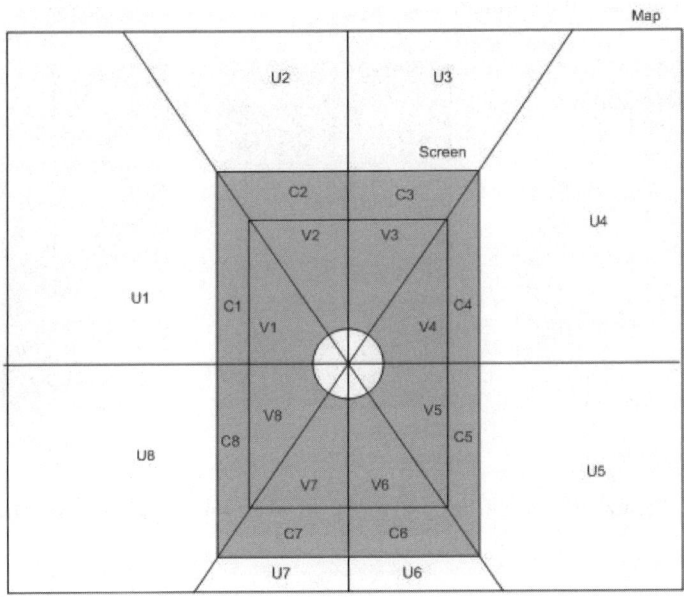

Fig. 1. The (off-) screen subdivision by *Framy*

The idea underlying this approach is to inscribe a semi-transparent area (*Cornice*) inside the screen. This area is partitioned in $C1, ..., Cn$ portions, each identified by the intersection between the *Cornice* and the *Vi* parts.

Once the user has posed a query, each Ci is colored with a different intensity on the basis of an aggregate function *f* applied on the resulting set associated with *Ui*.

Formally, we state that the color intensity is:

$$\text{Color(Ci)} = \text{f(Ui) for each } i \in \{1, ..., n\} \tag{1}$$

where *f* is any aggregate function which calculates a numeric value starting from a set of spatial data.

As an example, let us suppose that *f* is the *count* operation applied on the Hotel layer where features are selected only if they have four stars, namely:

```
SELECT count(geometry)
FROM Hotel
WHERE category = 4;
```

Then, the higher the number of selected features is within *Ui*, the more intense the color *(Ci)* is.

Analogously, if *f* computes the shortest distance from the current position then the Ci with the highest color intensity will indicate the *Ui* containing the closest feature.

With respect to other kind of approaches, *Framy* is not only addressed to visualize a results of a specific function (i.e. the proximity of features), but may be customized according to chosen aggregate function. As a matter of fact, the user may even choose the layer onto which it applies and the granularity level of the subdivision, that is how many parts compose the cornice.

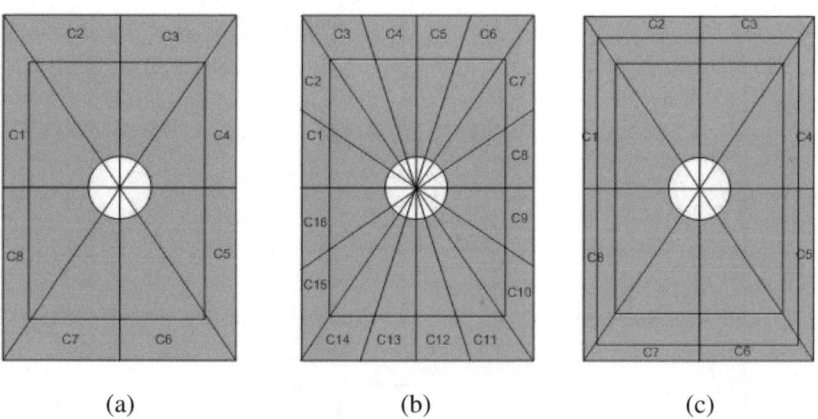

(a)   (b)   (c)

Fig. 2. (a) a Framy subdivision using 8 parts, (b) one using 16 parts, and (c) a concentric subdivision

For this reason, as an example, the user may apply the shortest distance function on both the Hotel layer and the Parking area layer in order to locate the Ci with the closest hotel and the Cj with the closest parking area, respectively. Moreover, in order to identify the direction to reach a particular target, the user may increase the number of subdivisions so that the Ci identifying the correct direction is as smaller and more precise as possible (see Fig. 2 (a) and 2 (b)).

Another important characteristic of *Framy* is the possibility to create more cornices and visually compare them. As a matter of fact, as shown in Figure 2(c) cornices may be nested in order to compare color intensity inside a sector. This is to say, the user may visualize the closest hotels on a cornice and the closest parking areas on a nested cornice in order to discover which hotel/parking pair is best located with respect to the current position. This may support user to choose the best place to reside by taking into account the hotel proximity and the number of points of interest in a map.

Another way to use *Framy* is by checking the color variation as the center of the map changes. As an instance, if the cornice is used to locate the closest hotel for each Ci, we can verify whether we are approaching to a particular hotel by controlling the intensity of the color in the chosen direction. This is to say, the hotel is approaching by a panning operation when the color gets more intense.

Defining Color Intensity

Let us present the technique we adopted for assigning colors to cornice. It is based on the Hue, Saturation, Value (HSV) model that we briefly recall in the following.

The HSV model, also known as HSB (Hue, Saturation, Brightness), defines a color space in terms of three components, namely:

- *Hue*, the color type (such as red, blue, or yellow). It ranges from 0 to 360 in most applications. Each value corresponds to one color. As an example: 0 is red, 45 is a shade of orange and 55 is a shade of yellow (see Fig. 3 for the complete color scale).

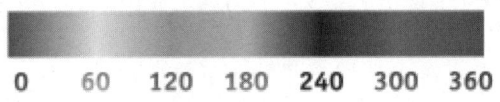

0   60   120   180   240   300   360

Fig. 3. The hue scale

- *Saturation*, the intensity of the color. It ranges from 0 to 100%. 0 means no color, that is a shade of grey between black and white. 100 means intense color. Sometimes it is called the "purity" by analogy to the colorimetric quantities excitation purity and colorimetric purity
- *Value*, the brightness of the color. It ranges from 0 to 100%. 0 is always black. Depending on the saturation, 100 may be white or a more or less saturated color.

At this point, let us now introduce the formulae we used to assign color intensities to the Cis . Let us suppose we are measuring the distribution of the n_i objects within the Ui sector and that we want to represent this value by using yellow tonalities. To reach this aim we set H to 60 (corresponding to the yellow color) as Figure 3 and Figure 4 depict, then we calculate the incremental value (step) that each object provides in term of tonality and finally, we multiply this value for the number of objects located in each Ui.

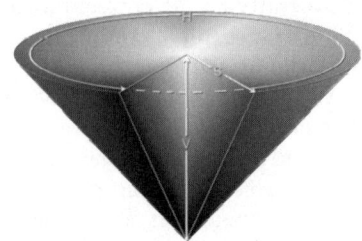

Fig. 4. The conical representation of the HSV model is well-suited to visualize the entire HSV color space in a single object

In order to calculate the step, we have to know the maximum and minimum number of objects that Ui could contain. As a matter of fact, the maximum one should correspond to the available most intense yellow, namely S=100, whereas the minimum one should correspond to absence of color, namely S=0. Then, the step will be:

$$Step=(MAX-MIN)/100 \qquad (2)$$

and the color for each Ci will be calculated by using the following triple:

$$Ci = (H_{Layer}, Si, V) \qquad (3)$$

where:

- H is assigned for each layer,
- $Si = step * n_i$
- $V = 100$

4   Exploiting *Framy* in Typical Spatial Query Scenarios

In this section three scenarios will be shown in order to exemplify three different modalities to use *Framy* as a visualization technique in *MapGIS*. In particular, in the first example we use *Framy* to indicate the quantity of objects belonging to the layer of interest. The second example shows how *Framy* may be used to reach a desired destination by using the intensity color variation. Finally, in the third scenario, we show how to use concentric cornices to compare different information.

Example 1: Framy as a visual synthesizing technique
The first scenario we describe shows the usage of *Framy* to support users in getting an idea about the distribution of the whole set of data over a map, disregarding of the actual visible map extension.

Figure 5 depicts the layers of the banks and streets in Sydney. The screen of a mobile phone is overlapped with them and the gray dotted lines indicate the chosen subdivision of the cornice, namely 8 portions. Finally, the center of the map is visualized as a cyan colored target.

It is possible to note that the color intensity of each Ci is proportional to the number of banks located in the corresponding off-screen sector. As an example, the C_is corresponding to the bottom-left side of the map are associated with a peripheral zone, then they result either transparent or colored with a very light yellow. On the other hand, the yellow color associated with the C_is on the opposite corner gets more intense as the corresponding sectors capture a greater quantity of banks.

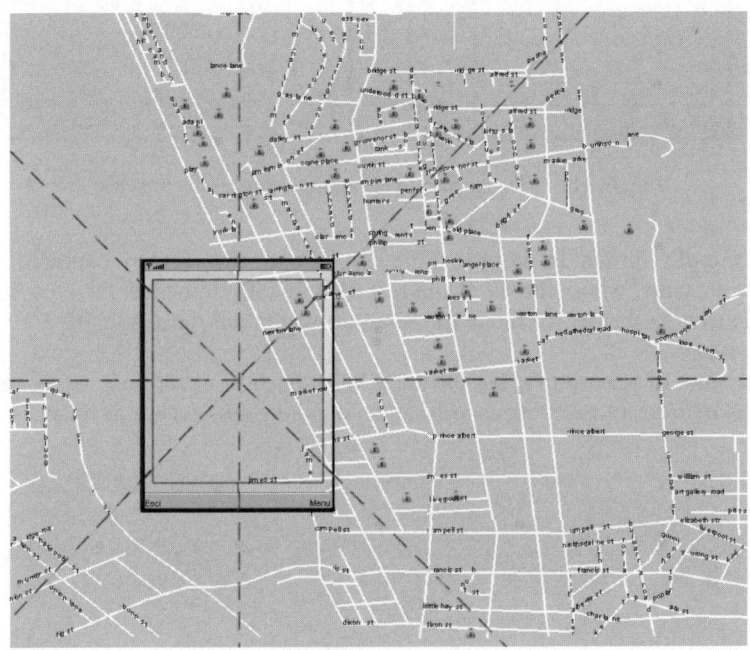

Fig. 5. Overlapping layers of interest with the screen of a mobile phone

Example 2: Locating a position by capturing color variations
By moving the device screen all around the map, the Ci colors are re-calculated and depicted at run-time in order to visualize the current situation around the map focus. This property provides users with the opportunity of understanding how the value of the aggregate function they have chosen changes.

This property is particularly interesting in many cases. As an example, let us suppose we are looking for the closest bank with respect to the map focus. In this case, the Ci which results more intensely colored indicates the sector containing the closest bank. Moreover, the color gets more intense as we move the map focus towards the request feature, thus indicating the right direction to it.

As previously said, this property may be further improved by dividing the screen in more portions. In such a way, choosing the right direction becomes easier because the most intense part is associated with a more narrow sector. At the limit, the number of divisions is so high that the exact direction may be indicated by a line.

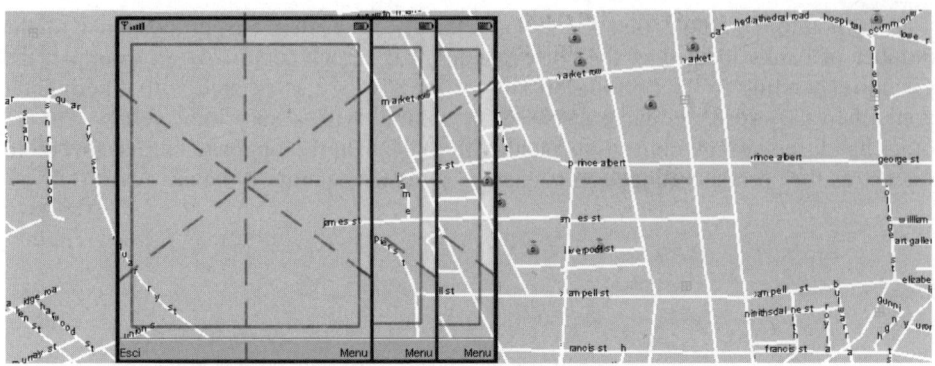

Fig. 6. The device screen taken in three successive instants

The example shown in Figure 6 depicts the device screen captured in three successive instants, where the Ci of interest dynamically indicates the closest bank. It is possible to note that as we move the screen towards the right side, the Ci color gets more intense.

For the sake of readability, we placed a large portion of the map overlapped with the screens in order to make it easier to understand the context we move in.

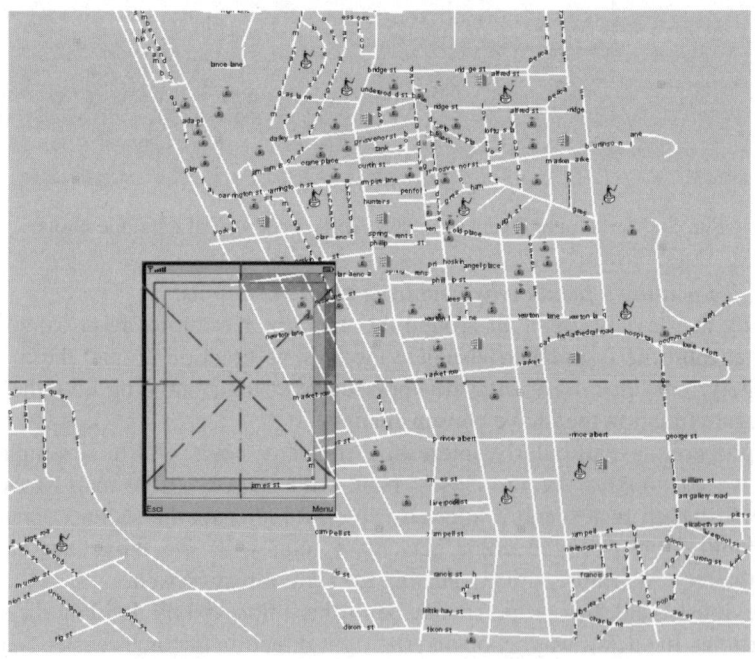

Fig. 7. A visual comparison by using Framy

Example 3: Framy as technique for visually comparing distributed information
In the last scenario, we show how to use *Framy* in order to compare values resulting from spatial aggregations.

In this example two layers showing hotels and monuments have been added to the map. Besides taking a look at the map, the request is to understand in which direction we should go in order to find a zone where it is easier to find simultaneously both a hotel and particular monuments.

In order to derive this information, we configured two concentric cornices – red for monuments and blue for hotels, respectively.

Given the data distribution, few Ci have been colored. In particular, colors are more intense on the right and up sides where the map shows numerous POIs. As a result, the top part better satisfies the initial requirement because the red and blue colors are simultaneously more intense with respect to other parts.

5 Conclusion and Future Work

In this paper, we defined a new method to visualize a large amount of geographic data on devices having small screens such as mobile phones, pda, etc.
Our solution adds a cornice inscribed inside the mobile screen which can be partitioned in agreement with data distribution on the map. Coloring such portions according to some spatial aggregate functions provides users with the opportunity to obtain a quantitative synthesis of features located off-screen, thus overcoming the inconvenient of a limited vision.

Some advantages come from this method. With respect to other kinds of approaches, this visual representation may be configured to work in different modalities. As a matter of fact, besides choosing the aggregating function on which the query is performed (e.g., count, distance, etc.), the user may choose the layer(s) on which the function is applied as well as the color representing the off-screen results. Moreover, the granularity level of the subdivision can be suitably adapted to the user's needs. Another important feature of this visual representation is the possibility to create more nested cornices and visually compare them. As a matter of fact, frames may be concentrically drawn in order to compare results of different aggregating functions.

In the future, we plan to accomplish some experiments to single out possible improvements into localizing information on mobile devices, also by comparing our approach with other specific methods. User experiments are also planned to verify the usability of this method in order to find out possible weakness.

An optimization of the color choice should be applied in order to avoid problems with different color intensities which cannot be distinguished.

References

1. Baudisch, P., Rosenholtz, R.H.: A Technique for Visualizing Off-Screen Locations. In: CHI 2003. ACM CHI Conference on Human Factors in Computing Systems, pp. 481–488. ACM Press, New York (2003)
2. Card, S.K., Mackinlay, J.D., Shneiderman, B.: Readings in information visualization - Using vision to think. Morgan Kaufman, San Francisco (1999)

3. Chittaro, L.: Visualizing information on mobile devices. IEEE Computer 39(3), 40–45 (2006)
4. Hachet, M., Pouderooux, J., Tyndiuk, F., Guitton, P.: Jump and Refine for Rapid Pointing. In: Mobile Phones in Proceedings of CHI 2007, pp. 167–170 (2007)
5. Jones, S., Jones, M., Marsden, G., Patel, D., Cockburn, A.: An evaluation of integrated zooming and scrolling on small screens. International Journal of Human-Computer Studies 63(3), 271–303 (2005)
6. Karstens, B., Rosenbaum, R., Schumann, H.: Information Presentation on Handheld Devices. In: Candance Deans, P. (ed.) E-Commerce and M-Commerce Technologies Ch. 2, Idea Group Publishing, USA (2004)
7. Mackinlay, J., Good, L., Zellweger, P., Stefik, M., Baudisch, P.: City Lights: Contextual Views in Minimal Space. In: CHI 2003. Proc. ACM CHI Conference on Human Factors in Computing Systems, pp. 838–839. ACM Press, New York (2003)
8. Rauschenbach, U., Jeschke, S., Schumann, H.: General rectangular fisheye views for 2D graphics. Computers & Graphics 25(4), 609–617 (2001)
9. Robbins, D.C., Cutrell, E., Sarin, R., Horvitz, E.: ZoneZoom: Map navigation for smartphones with recursive view segmentation. In: AVI 2004. Proc. International Conference on Advanced visual interfaces, pp. 231–234. ACM Press, New York (2004)
10. Rosenbaum, R.U., Schumann, H.: Grid-based interaction for effective image browsing on mobile devices. Proc. SPIE Int. Soc. Opt. Eng. 5684, 170–180 (2005)

PhotoMap – Automatic Spatiotemporal Annotation for Mobile Photos

Windson Viana[*], José Bringel Filho[**], Jérôme Gensel, Marlène Villanova Oliver, and Hervé Martin

Laboratoire d'Informatique de Grenoble, équipe STEAMER
681, rue de la Passerelle, 38402 Saint Martin d'Hères, France
{Windson.Viana-de-Carvalho,Jose.De-Ribamar-Martins-Bringel-Filho,
Jerome.Gensel, Marlene.Villanova-Oliver, Herve.Martin}@imag.fr

Abstract. The amount of photos in personal digital collections has grown rapidly making their management and retrieval a complicated task. In addition, existing tools performing manual spatial and content annotation are still time consuming for the users. To annotate automatically these photos using mobile devices is the emerging solution to decrease the lack of the description in photo files. In this context, this paper proposes a mobile and location-based system, called PhotoMap, for the automatic annotation and the spatiotemporal visualization of photo collections. PhotoMap uses the new technologies of the Semantic Web in order to infer information about the taken photos and to improve both the visualization and the retrieval. We also present a demonstration of the PhotoMap application during a tourist tour in the city of Grenoble.

Keywords: Semantic Web, spatial ontologies, location-based services, mobile devices, and multimedia annotation.

1 Introduction

The number of personal digital photos is increasing quickly, since taking new pictures gets simpler using digital cameras and the storage costs in these devices has decreased a lot. For this reason, searching for one particular photo among a huge number of pictures has become a boring and repetitive activity [9][15][18]. Besides time and geographic location of a photo shot provide critical information that can be applied for organizing photo collections and for increasing photo file descriptions [7][10]. Moreover, spatial and temporal information about a photo can be increased to improve shot context description. Web image search engines can exploit this contextual annotation instead of only querying filenames most of time. In addition, spatiotemporal annotation of a photo provides a support to conceive better browsing

---

[*] Supported by CAPES – Brasil.
[**] Supported by the Program Alban, the European Union Program of High Level Scholarships for Latin America, scholarship no. E06D104158BR.

J.M. Ware and G.E. Taylor (Eds.): W2GIS 2007, LNCS 4857, pp. 187–201, 2007.

and management image systems [10]. Several applications and research works such as Yahoo! Flickr[1], Google Panoramio[2], TripperMap[3] and WWMX [16] offer map-based interfaces for geotagging photos with location coordinates and textual descriptions. Despite the fact that digital cameras equipped with GPS receivers exist, these tools are not widely used in spite of. Hence, most of users have to manually geotagging their photos. The use of new mobile devices can change the photo annotation radically. These devices will have high-resolution built-in cameras and one can imagine that people will progressively change their traditional digital cameras by this new generation mobile terminals. In addition, these devices are progressively incorporating more sensors (e.g., GPS, RFID readers, and compass) that can provide a huge quantity of information about the *user context* when she/he uses her/his phone camera for taking a photo [17].

The interpretation and inference about the context data using Web Semantic technologies allows the generation of high quality information that can be employed for annotating photos automatically [9]. In this paper, we propose an automatic approach for annotating photos based on OWL-DL ontologies and mobile devices. We describe a vocabulary for representing captured and inferred context information. We present an ontology called ContextPhoto and a contextual photo annotation approach which improve the development of more intelligent personal image management tools. ContextPhoto provides concepts for representing spatial and temporal contexts of a photo (i.e., where, when) and SWRL rules for inferring the social context of a photo (i.e., who was nearby). In order to validate the context annotation process we propose, we have also designed and developed a mobile and Web location-based system. This new system, called PhotoMap, is an evolution of the related mobile annotation systems since it provides automatic annotation about the spatial, temporal and social contexts of a photo (i.e., where, when, and who was nearby). PhotoMap also offers a Web interface for spatial and temporal navigation in photo collections. The system exploits spatial Web 2.0 services for showing *where* a user takes her photos and the itinerary followed when taking them.

This paper is organized as follows: section 2 describes the existed image annotation approaches and illustrates a generic annotation process that can be used for designing of photo annotation systems; section 3 shows our ontology for photo annotation; section 4 gives an overview of the PhotoMap system describing its main components and processes; and, finally, we conclude in section 5 outlining underlining potential future works.

2 Image Annotation

In order to organize their personal photo collections, the majority of people modify the name of their image files for postponed search purposes. They organize their photos in file system directories. They usually label them with some spatial and/or temporal descriptions. For example, a photo's directory may contain a set of photos corresponding to a period (e.g., the winter of 2006), an event (e.g., a vacation trip), or

---

[1] http://www.flickr.com/
[2] http://www.panoramio.com/
[3] http://www.fickrmap.com/

a place (e.g., London photos) [7]. However, this simplistic organization does not make easy the image retrieval process when the photo number increases a lot.

The use of annotations can facilitate the management and the retrieval of photo collections. Photo annotation consists in associating metadata with an image. Annotation highlights the significant role of photos in restoring forgotten memories of visited places, party events, and people. Besides that, the use of annotations allows the development of more efficient personal image tools. For instance, *geotagging* photos allows the visualization of personal photos in map-based interfaces such as Flickr Map[4]. In addition, people can find images taken near a given location by entering a city name or latitude and longitude coordinates into geotagging-enabled image search engines. Despite of the obvious advantage of image metadata, manual process of annotation is a boring and time-consuming task. In addition, annotation tools that automatically generate content annotations are still not convenient for photo organization since they do not fill the so-called semantic gap problem [17][10]. We claim, as the authors in [10] and [15], that contextual metadata for automatic annotation is particularly useful to photos organization and retrieval, since, for instance, knowing the location and the time of a photo shot often describes the photo itself a lot even before a single pixel is shown.

Temporal information attached to a photo (i.e., the date and hour of the photo shot) is the most common contextual data acquired with traditional digital cameras. Image management tools as Adobe PhotoShop Album and [11] use this timestamp in order to organize photos in a chronological order. They can also generate temporal clusters in order to facilitate the image consultation task. Another important source of image metadata is the EXIF file (i.e., *EXchangeable Image File*). When one takes a digital photo, most of the cameras insert contextual data into the header of the image file. This image metadata describes information about the photo (e.g., name, size, and timestamp), the camera (e.g., model, maximum resolution), and the camera settings used to take the photograph (e.g., aperture, shutter speed, and focal length). EXIF also allows the description of geographic coordinates using GPS tags. Several research projects, such as **PhotoCompas** [10], **WWMX** [16], and Web applications (e.g., Panoramio, Flickr[4]) employ spatial and temporal metadata together in order to organize photo collections. They exploit the EXIF data for proposing photo navigation interfaces. For instance, **WWMX** is a Microsoft project whose goal is to browse large databases of images using maps. WWMX offers different possibilities of displaying images in a map and uses GIS databases in order to index the location information of the photos.

The major issues in these approaches are related to the lack of rich metadata associated with images. Most of the digital cameras only provide basic EXIF attributes. Besides that, the manual association of location tags with an image file is also a time-consuming task. Moreover, date/time and GPS coordinates are not the unique cues among all the context information that are useful to recall a photo [7] [15]. Our approach addresses these issues by combining mobile devices as a primary source of context metadata with spatial, temporal, and social reasoning. The number of built-in sensors for mobile devices will progressively increase such as GPS, digital compass, and high resolution cameras [17]. In addition, mobile devices are both

---

[4] http://www.flickr.com/map/

pervasive and personal. Mobile phones track the person and have clues about her current situation allowing the acquisition of contextual information. Spatial, temporal and social reasoning can be then applied to the initial contextual metadata in order to derivate high level context information. These enriched contextual metadata can be used for automatic image annotation when the user takes a photo with her camera phone. This method increases both the number and the quality of contextual information. The inferred context information enables the development of smart image management tools that offers organization, retrieval and browsing functions outperforming the capabilities proposed by today's geotagging photo applications.

The Fig. 1 shows an overview of an annotation process designed employing mobile camera phones. Unlike traditional digital cameras, mobile devices can execute external applications (e.g, Brew, J2ME applications) allowing the development of mobile photo annotation systems [1][10][9]. These applications can be designed for providing both manual content annotation and automatic context annotation. The manual annotation on the mobile device reduces the time lag problem that occurs in desktop annotation tools [1] since the user is still in the photo shot situation when she produces the photo metadata [10].

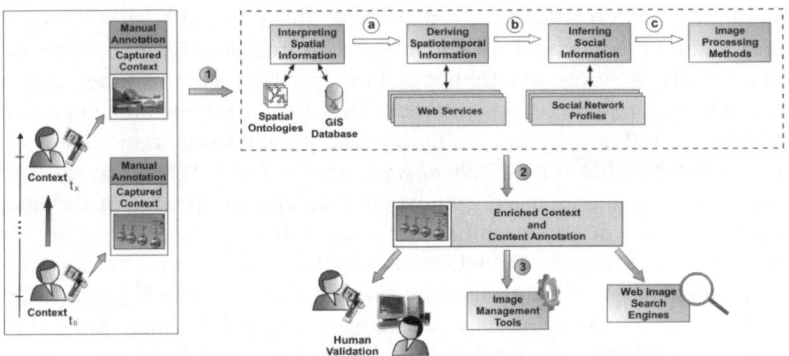

Fig. 1. Image annotation process designed with mobile camera phones

A photo annotation system can associate with the image file the manual annotation and the captured context by using an interoperable and machine-understandable format (e.g., RDF, OWL). The contextual metadata produced by the mobile device can be exploited for deriving high level context information. For instance, annotation systems can transform the GPS coordinates into a more suitable spatial representation (e.g., a city name). These system can combine location, orientation, and spatial reasoning in order to infer, for each photo, cardinal relationships (e.g., north of Paris) and 3D spatial relations (e.g., in front of the Eiffel Tower) between the photo shot location and a city tourist sight. The annotation system can also adopt an approach similar to [10] in order to enrich the context description of their photos using external context resources (e.g., weather conditions). In addition, annotation systems can use social network descriptions associated with contextual metadata for inferring the *social context* of a photo (e.g., the nearby user's friends) [9]. Furthermore, high level contexts can be used together with algorithms of image process for increasing the accuracy of automatic content annotation. For example, both the location and camera settings information can be exploited in order to infer whether a photo is an outdoor or an indoor image [3]. All this enriched context and content annotations can be used for developing better personal image organization tools.

3 ContextPhoto Ontology

The W3C Multimedia Annotation group has pointed out the major issues in the development of image annotation tools. The lack of syntactic/semantic interoperability, and the automatic generation of effective content annotation are some of these challenges. Semantic annotation using ontologies can be the first step to solve those hard problems since the use of ontologies for annotation is more suitable for making the content machine-understandable. Ontologies provide a common vocabulary to be shared between people and heterogeneous, distributed application systems. In addition, ontologies reduce the semantic gap between what image search engines are able to find out and what the people expect to be retrieved when they make a search. Moreover, annotations described in a formal and explicit way allow reasoning methods to be applied in order to infer about both the content and the context of a photo. In the field of multimedia annotation, numerous ontologies for image metadata representation have been proposed [2][14][5]. These ontologies are well-suited for extensive content-oriented annotation. In our approach, however, we are more interesting in representing what is the context of the user when she takes her photos. Fig. 2 shows graphical representation of our ontology, called ContextPhoto.

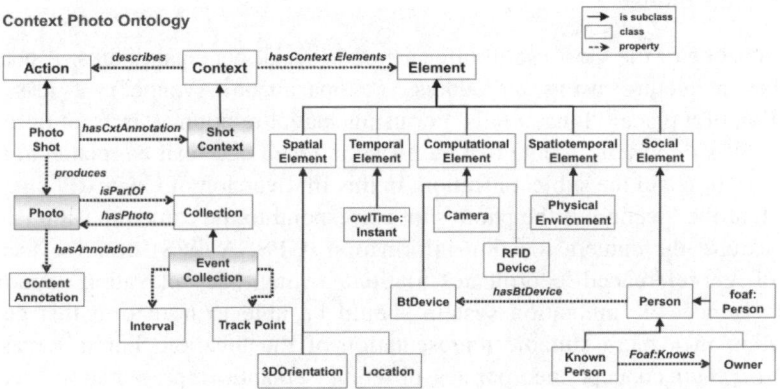

Fig. 2. ContextPhoto Ontology

ContextPhoto is an OWL-DL ontology for annotating photos and photo collections as well with contextual metadata and content descriptions. The ContextPhoto main idea is to associate a spatial itinerary and a temporal interval with a photo collection. The authors in [10] and [7] have pointed out that grouping photos in events (e.g., a vacation, a tourist visit) is one of the most common way people use to recall and to organize their photos. The concept *EventCollection* of ContextPhoto represents the idea of a photo collection associated with a **spatiotemporal event**. This concept is a subclass of the *Collection* concept and it contains a time interval and an ordered list of track points (i.e., timestamp and geographic coordinates) (see Fig. 2). The *Photo* concept contains the basic image properties (e.g., name, width and height) described with the Dublin Core vocabulary. Besides those properties, each photo has two annotation types: content annotation (*Content Annotation*) and contextual annotations (*Shot Context*). An annotation system can use the *Content Annotation* in order to describe the image content using manual textual annotation (e.g., keyword tags) or integrating ContextPhoto with other image annotation ontologies such as the Visual Descriptor Ontology (VDO) [2]. Context is generally defined in order to describe the user situation when she interacts with a system [4][6][12]. The challenges for the context-aware research community are design and develop systems that can self-adapt their behavior and content according to the current situation of the user and her preferences. For example, in a previous work [6] we proposed a context definition and an object-based context representation in order to support adaptation in context-aware collaborative web systems. However, in multimedia annotation systems, ones are more interesting in describing in which circumstances a media was produced. In order to support a context description relevant for annotating photos, we have defined five context dimensions: spatial (*Spatial Element* subclasses), temporal (*Temporal Element* subclasses), social (*Social Element* subclasses), computational (*Computational Element* subclasses), and spatiotemporal (*Spatiotemporal Element* subclasses). These concepts correspond to the major elements for describing a photo (i.e., where, when, who, with what) [10][7][15].

3.1   Spatial Context

Location is one of the most useful information to recall personal photos. A person can remember a picture using an address ("Copacabana Avenue"), a less precise description of a place ("Iguaçu falls") or using spatial relations ("in front of the Eiffel Tower"). It is important to note that the location recall cues can be related to both the camera location and the subject location. In this first version of ContextPhoto, we will assume that the location of the photo shot correspond to the camera position. Systems which acquire the camera location information (GPS, A-GPS) describe location in terms of georeferenced coordinates (latitude, longitude, elevation, coordination system). Hence, an annotation system should be able to transform this geometric information in a more suitable representation of the location. For this reason, the *Spatial Element* concept incorporates different semantic representation levels of a location description. Fig. 3 shows the spatial element *Location,* its subclasses, and its relationships. For instance, in order to describe location places in a geometric way (polygons, lines, points), ContextPhoto imports the NeoGeo[5] ontology. NeoGeo is an OWL-DL representation of the core concepts of GML (Geographic Markup Language), an open interchange format defined by the Open Geospatial Consortium (OGC) to express geographical features. Using the NeoGeo concepts, ContextPhoto define subclasses of *Location* in order to represent geometric regions, pointed locations, and GPS coordinates. The other levels of location representation are

---

[5] http://mapbureau.com/neogeo/neogeo.owl

defined associating the *Location* element with a textual description. This description represents a complete addresses or a hierarchical description for imprecise locations. For instance, America→ Brazil→ Iguaçu falls).

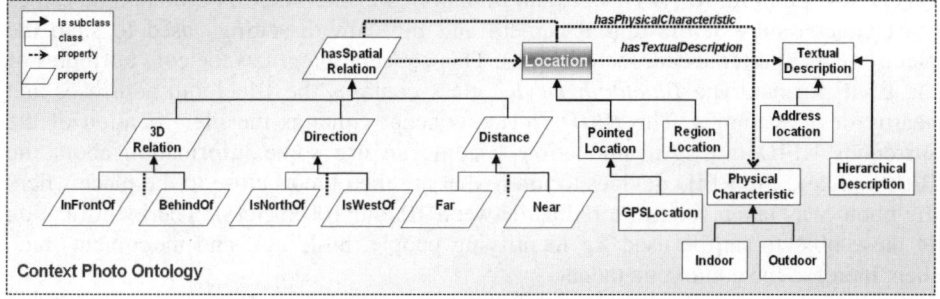

Fig. 3. Spatial elements

We use the spatial ontology proposed in [13] in order to represent the spatial relations between two Location entities. An annotation system can use the GPS coordinates, digital compass data, and spatial databases for inferring spatial relations among the photo location and some point of interests (e.g., a tourist sight). For example, an annotation system can calculate if a photo was taken in the north of Paris or in front of the Eiffel tower. Web image search engines can exploit this information for developing location-based image queries.

3.2   Temporal and Spatiotemporal Contexts

Time is another important aspect to be considered in the organization and the retrieval of personal photos. However, a precise date is not the most temporal attribute used when a person tries to find a photo in a chronological way [10]. When a photo has not been taken at a date that can be easily remembered (e.g., birthday, Valentine's Day), people rather use the month, the day of week, the time of day, and/or the year information when formulating their query. Thus, the *Temporal Element* concept allows the association of an instant (date and time) with a photo, and also of the different time interpretations and attributes listed above (e.g., night, Monday, July). In order to represent time intervals and time instants in ContextPhoto and to allow time reasoning with the photo annotations, we use the concepts designed in the OWL-Time ontology[6]. The *Spatiotemporal Element* concept describes the context elements that depend on time and location data to be calculated. In the ContextPhoto ontology, we have defined only one spatiotemporal class: the physical environment. This concept has properties that describe the season, the temperature, the weather conditions, and the day light status (e.g., day, night, after sunset) when the photo was taken. Usability tests show that these spatiotemporal attributes are useful photo clues that can be both used for search purposes and to increase the described information of a photo [10] [7].

---

[6] http://www.w3.org/TR/owl-time/

3.3 Computational and Social Contexts

Computational Element describes all the digital devices present at the moment of the photo shot (i.e., the camera, and the surrounding Bluetooth devices). We define three subclasses: *Camera*, *RFID Device*, and *Bluetooth Device*. The *Camera* class describes the characteristics of the digital camera and the camera settings used to snap the picture (e.g., shutter speed, focal length). This concept integrates the core attributes of the EXIF format. The *Bluetooth Device* class contains the Bluetooth addresses and nearby devices names. The *RFID Device* concept contains the identification of the surrounds RFID tags. An annotation system can use some information about the Bluetooth and the RFID devices to infer what are the objects close to the place where the photo was taken (i.e., in a radius between 10 and 100 meters). The identification of these objects can be used for identifying people, buildings, and monuments and, then, increase the context metadata.

One of the innovating features of the ContextPhoto ontology is the ability to describe the social context of a photo. The main idea is to represent who were the people present nearby the location where the photo has been taken. The description of social context is based on the proposal of [9] and FoafMobile[7], and uses the Bluetooth address of personal devices in order to detect a person's presence. The main idea is to associate Bluetooth addresses with Friend-Of-A-Friend[8] (FOAF) profiles. FOAF is a RDF ontology that allows the description of a person (name, personal image) and of her social networks (the person's acquaintances). Thus, an annotation system can use the nearby Bluetooth addresses in order to calculate if these addresses identify a user's friend, and afterwards, to annotate the photo with the nearby acquaintances' names. In order to represent the acquaintances and the user, we define in ContextPhoto three classes that can be elements of the photo *social context*: the *Person*, *Owner* and the *Known Person* classes. The *Person* is defined as a subclass of *foaf:Person* that contains a Bluetooth device. The *Owner* describes the owner of the photo collection annotated using ContextPhoto (e.g., the photographer or the annotation system user). The *Known Person* concepts represent people among the *Person* instances who know the *Owner* (i.e., the *foaf:Knows* relation). In order to infer the photo social context, we propose the combination of FOAF profiles and SWRL rules. We suppose that a Web image management system (e.g., Flickr) can describe its users and social networks using FOAF profiles. These Web tools can provide interfaces which allow users to link their FOAF profile with other profiles in the Web. The system can allow a user to associate Bluetooth addresses for identifying her FOAF profile. This association can also be set up automatically. A mobile application can acquire the Bluetooth address of the mobile user device and send this information to the image management system. With the association of Bluetooth devices to people, an annotation system can execute an inference process in order to derivate the social context of a photo determining who was present at the moment of a photo shot. This process reads the photo annotation and gets the *Owner* identification. The system can, then, use the FOAF profile of the photo owner as a start point of a search. The tool navigate through the properties "*foaf:Knows*" and "*rdf:seeAlso*" to get the FOAF profiles of people that the owner knows. All the found profiles are used to instantiate

---

[7] http://jibbering.com/discussion/Bluetooth-presence.1
[8] http:// www.foaf-project.org/

Known Person individuals. After the instantiation of the *Known Person* and *Owner*, the system can use a rule-based engine in order to infer which acquaintances were present when the photo was shot. Formula 1 shows the SWRL rule used to infer the presence of an owner's friend. After the end of the inference process, individuals representing the nearby acquaintances are associated with the photo *social context*.

$$
\begin{aligned}
&\text{Owner}(?\text{owner}) \wedge \text{KnownPerson}(?\text{person}) \wedge \text{SocialContext}(?\text{scctxt}) \\
&\text{ComputationalContext}(?\text{compctxt}) \wedge \text{BTDevice}(?\text{btDv}) \\
&\text{foaf:knows}(?\text{person}, ?\text{owner}) \wedge \text{hasBTDevice}(?\text{person}, ?\text{btDv}) \\
&\text{hasContextElement}(?\text{scctxt}, ?\text{owner}) \wedge \\
&\text{hasContextElement}(?\text{compctxt}, ?\text{btDv}) \\
&\quad \rightarrow \text{hasContextElement}(?\text{scctxt}, ?\text{person})
\end{aligned}
\tag{1}
$$

4 PhotoMap

PhotoMap is a mobile and Web location-based system for photo annotation. We use the image annotation process proposed in the Section 2 (see Fig. 1) for guiding the PhotoMap design. The PhotoMap three main goals are: (1) to offer a mobile application enabling users to take pictures and to group theirs photos in spatiotemporal collections; (2) to propose a Web system that organizes the user's photos using the acquired spatial and temporal data; and (3) to improve the users recall of their photos showing inferred spatial, temporal, and social information. PhotoMap is structured according to a client-server model.

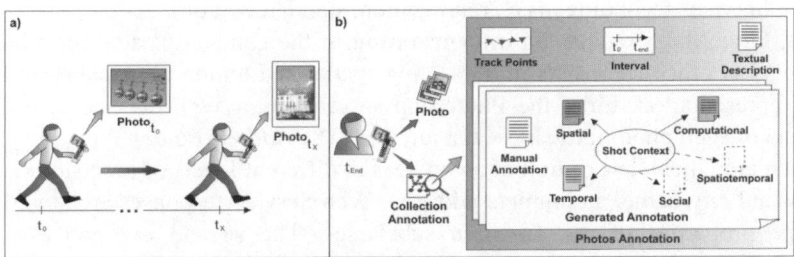

Fig. 4. Photomap use and the generated annotation

The client application runs on J2ME-enabled devices. It allows the users to create photo collections representing events (e.g., tourist visits). The user can give a name and a textual description when she starts a collection. The client application runs a background process that monitors the current physical location of the mobile device (see Fig. 4.). It accesses the device sensors (e.g., built-in GPS, Bluetooth-enabled GPS receiver) via the Location API (i.e., JSR 179) or via the Bluetooth API (i.e., JSR 82).

The gathered coordinates (latitude, longitude, and elevation) are stored in order to build a list of track points. This list represents the itinerary followed by the user to take the pictures. The acquired track list is associated with the metadata of the current photo collection. Moreover, the PhotoMap client application captures the photo shot context when a user takes a picture with her camera phone. The mobile client gets the geographic position of the device, the date and time information, the bluetooth addresses of the nearby devices, and the configuration properties of the digital camera. All these metadata are stored for each photo of the collection. After a photo shot, the user can also add manual annotations. The textual description of the collection, its start and end instants, and the track points list are added to the annotation. For each photo, the gathered context and the possible manual annotation are stored in the ontology instantiation. The taken photos and the generated annotation (i.e., ContextPhoto instances) are stocked in the device file system. Afterwards, the user uploads her photos and the annotation metadata of the collection to the server application. The user can execute this upload process from her mobile phone directly (e.g., via HTTP) or via a USB connection.

4.1   Interpretation, Inference and Indexation Processes

The PhotoMap server is a J2EE Web-based application. Besides acting as an upload gateway, the server application is in charge of the photo indexation, inference, and interpretation processes. After the transmission of a collection and its metadata, the PhotoMap server reads the annotations associated with each photo and then executes some internal processes which enrich the contextual annotation. When a user sends a photo collection to PhotoMap, the annotation contains only information about the computational, spatial and temporal contexts of a photo. Thus, PhotoMap accesses off-the-shelf Web Services in order to augment the context annotation. This approach, proposed by [10], reduces the development cost and benefits from the advantages of the Web Services technology (i.e., reutilization, standardization).

    First, PhotoMap executes an **interpretation** of the gathered spatial metadata. The interpretation process consists in translating spatial and temporal metadata in a more useful representation. First, the PhotoMap server uses a Web Service to transform GPS data of each photo into physical addresses. The *AddressFinder* Web Service [19] offers a hierarchical description of an address at different levels of precision (i.e. only country and city name, a complete address). Web Service responses are stored in the Datatype properties of the *Location* subclasses. The second interpretation phase performs the separation of temporal attributes of the date/time property. The PhotoMap server calculates day of week, month, time of day, and year properties using the instant value. Later, the PhotoMap server gets information about the physical environment at the photo shot time. PhotoMap derives temperature, season, light status, and weather conditions using the GPS data and the date/time annotated by the PhotoMap client. We use the *Weather Underground* [19] Web Service to get weather conditions and temperature information. In addition, PhotoMap uses the *Sunrise and Sunset Times* [21] Web Service to get the light status. The season property is calculated using the date and GPS data. After the interpretation process, the server application executes the **inference** process in order to derivate the photo social context (see section 3.3). At the end of the inference process, individuals representing the present acquaintances are associated with the *Shot Context* of the current photo. Hence, the final OWL annotation of a photo contains spatial, temporal, spatial-temporal, computational and social information. Then, the PhotoMap server executes an **indexation** process in order to optimize browsing and interaction methods. The number of photo collection annotations should increase in our system quickly. To avoid problems of performance linked to sequential searches in these OWL annotations, spatial and temporal indexes are generated for each collection using the PostgreSQL database extended with the PostGIS module. Using the spatial information of the collection track points list, PhotoMap calculates a minimum bounding box that includes the entire itinerary. The generated polygon is stored as a GiST index and it is associated with a string indicating the file path of the OWL annotation. The collection start and end times are used to create temporal indexes which are also associated with the collection file path.

4.2 Browsing and Querying Photos

PhotoMap offers graphical interfaces for navigation and query over the users captured photo collections. We have designed the PhotoMap portal using map-based interfaces for browsing photos purposes. Usability studies [16] show that map interfaces present more interactivity advantages than browsing photos only with location hierarchical links (e.g.; Europe → France → Grenoble). Furthermore, with a map-based interface, we can easily represent the itineraries followed by users when taking their photos. Besides the map-based visualization, users can look into the inferred and manual annotations of their photos and exploit them for retrieval purposes.

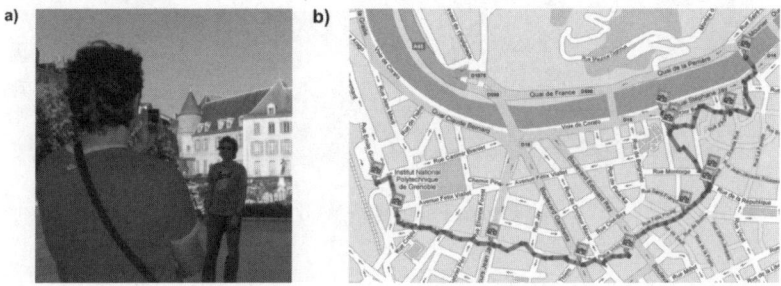

Fig. 5. PhotoMap in a real situation

Fig. 5 illustrates an example of a photo collection taken during a tourist tour in Grenoble (109 GPS data and 13 photos). We have used a Sony Ericsson K750i phone to snap the photos in the city center. The mobile phone had our annotation application installed and was connected to a Bluetooth-enabled GPS. At the beginning of the tour, the user has activated the client application for taking his photos and for tracking his followed path. Fig. 5a shows the user taking a photo with the K750i phone. In the Fig. 5b, the itinerary followed by the user and the placemarks for each taken photo can be seen on a map view of Grenoble.

Fig. 6 shows a screen shot of the PhotoMap Web site. The rectangular region on the left side of the Web Site is the *menu-search view*. This window shows PhotoMap main functionalities, the social network of the user, and a keyword search engine for her photo collections. On the right side, from top to bottom, are shown the *event-collection view*, the *spatial view*, and the *temporal query window*. When a user enters the Web site, PhotoMap uses the temporal index to extract the ten latest collections. PhotoMap then shows, in the *event-collection view*, thumbnails of the first photos of each collection and the collection names annotated by the user. The latest collection is selected and the itinerary followed by the user is displayed in the *spatial view*. Placemarks are inserted in the map for each photo, and the user can click to view the photo and the generated annotation. Fig. 6 illustrates the photo visualization when the user clicks on a placemark.

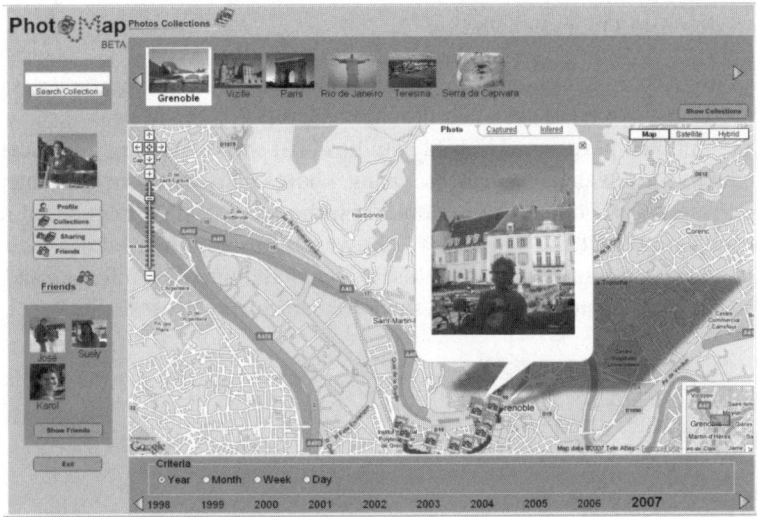

Fig. 6. PhotoMap Web site

Using this web application, the user can read the captured and inferred annotations. Fig. 7a illustrates the visualization of the photo collection over a satellite image of Grenoble. In the "Captured" panel (see Fig. 7b) are described the initial context data annotation: GPS data, date/time, and three physical addressees of the nearby Bluetooth devices. The "Inferred" panel (see Fig. 7c) shows the information derived by the server application: address, temperature, and season. The panel shows the FOAF information about the user's friends present in the photo shot (i.e., name and personal photo). It is important to note that the server application has correctly inferred that only two of the acquired Bluetooth addresses belong to the owner's friends. The other device seems to be a Bluetooth headphone of someone passing by.

Fig. 7. An illustation of PhotoMap annotation panels

The *spatial view* displays maps using the Google Maps API. Navigation on the map, in a spatial query mode, changes the displayed map and also the visible collections in the *event-collection view*. In order to achieve this operation, PhotoMap queries the spatial database that is constrained to return only the collections intersecting a defined view box. PhotoMap uses the bounding box coordinates of the *spatial view* and the current zoom value in order to calculate the view box coordinates. A virtual representation of the view box and the cached collections are presented at the Fig. 8.

Fig. 8. An illustation of PhotoMap spatial cache

5   Conclusion and Future Work

Context metadata associated with Semantic Web technologies can be useful to increase knowledge about photo content. As a matter of fact, this knowledge is essential for developing better image management tools and Web image search engines. We have described, in this paper, a new mobile and Web location-based system that captures and infers context metadata of photos. We have proposed an OWL-DL ontology to annotate spatiotemporal photo collections which can be employed for temporal, spatial, and social reasoning. Finally, PhotoMap Web site offers interfaces for spatial and temporal queries over the user photo collections. In the future, we will investigate ways to calculate spatial relations among the photos and tourist sights of a city. We intend to create a geographic knowledge-base of the tourist attractions of a city by using the ONTOAST tool [8], compass data and location information in order to determine the spatial relations.

References

1. Ames, M., Naaman, M.: Why We Tag: Motivations for Annotation in Mobile and Online Media. In: CHI 2007. Proc. of Conference on Human Factors in computing systems (2007)
2. Athanasiadis, Th., Tzouvaras, V., et al.: Using a Multimedia Ontology Infrastructure for Semantic Annotation of Multimedia Content. In: Proc. of 5th International Workshop on Knowledge Markup and Semantic Annotation, Galway, Ireland (November 2005)
3. Boutella, M., Luob, J.: Beyond pixels: Exploiting camera metadata for photo classification Pattern Recognition 38(6), 935–946 (June 2005)
4. Dey, A.K., Abowd, G.D.: Towards a Better Understanding of Context and Context-Awareness. In: CHI 2000 Workshop of Context-Awareness (2000)
5. Hollink, L., Nguyen, G., Schreiber, G., Wielemaker, J., Wielinga, B., Worring, M.: Adding Spatial Semantics to Image Annotations. In: Proc. of 4th International Workshop on Knowledge Markup and Semantic Annotation (2004)
6. Kirsch-Pinheiro, M., Villanova-Oliver, M., Gensel, J., Martin, H.: Context-aware filtering for collaborative web systems: adapting the awareness information to the user's context. In: Proc. of the ACM Symposium on Applied Computing, Mexico (2005)
7. Matellanes, A., Evans, A., Erdal, B.: Creating an application for automatic annotations of images and video. In: SWAMM 2006. Proc. of 1st International Workshop on Semantic Web Annotations for Multimedia, Edinburgh, Scotland (2006)
8. Miron, A.D., Villanova-Oliver, M., Gensel, J., Martin, H.: Towards the Geo-spatial querying of the Semantic Web with ONTOAST. Submit for the W2GIS, UK (2007)
9. Monaghan, F., O'Sullivan, D.: Automating Photo Annotation using Services and Ontologies. In: Mdm 2006. Proc. of 7th International Conference on Mobile Data Management, pp. 79–82. IEEE Computer Society, Washington, DC (2006)
10. Naaman, M., Harada, S., Wang, Q., Garcia-Molina, H., Paepcke, A.: Context data in geo-referenced digital photo collections. In: MULTIMEDIA 2004. Proc. of 12th ACM international Conference on Multimedia, pp. 196–203. ACM, New York (2004)
11. Pigeau, A., Gelgon, M.: Organizing a personal image collection with statistical model-based ICL clustering on spatio-temporal camera phone meta-data. Journal of Visual Communication and Image Representation 15(3), 425–445 (2004)
12. Petit, M., Ray, C., Claramunt, C.: A Contextual Approach for the Development of GIS: Application to Maritime Navigation. In: W2GIS 2006. Proc. of 6th International Symposium on Web and Wireless Geographical Information Systems Conference, Hong Kong (2006)
13. Reitsma, F., Hiramatsu, K.: Exploring GeoMarkup on the Semantic Web. In: 9th AGILE International Conference on Geographic Information Science: shaping the future of Geographic Information Science in Europe, Visegrád, Hungary, pp. 20–22 (April 20-22, 2006)
14. Schreiber, A.Th., Dubbeldam, B., Wielemaker, J., Wielinga, B.J.: Ontology-based photo annotation. IEEE Intelligent Systems (2001)

15. Sarvas, R., Herrarte, E., Wilhelm, A., Davis, M.: Metadata creation system for mobile images. In: MobiSys 2004. Proc. of 2th International Conference on Mobile Systems, Applications, and Services, pp. 36–48. ACM, Boston, MA (2004)
16. Toyama, K., Logan, R., Roseway, A.: Geographic location tags on digital images. In: MULTIMEDIA 2003. Proc. of 11th ACM international Conference on Multimedia, pp. 156–166. ACM Press, Berkeley, CA (November 2003)
17. Yamaba, T., Takagi, A., Nakajima, T.: Citron: A context information acquisition framework for personal devices. In: Proc. of 11th International Conference on Embedded and real-Time Computing Systems and Applications (2005)
18. Wang, L., Khan, L.: Automatic image annotation and retrieval using weighted feature selection. Journal of Multimedia Tools and Applications, 55–71 (2006)
19. http://ashburnarcweb.esri.com/
20. http://www.weatherunderground.com/
21. http://www.earthtools.org/

Towards the Next Generation of Location-Based Services

Elias Frentzos, Kostas Gratsias, and Yannis Theodoridis[*]

Department of Informatics, University of Piraeus,
80 Karaoli-Dimitriou St, GR-18534 Piraeus, Greece,
Tel.: +30-2104142449; Fax: +30-2104142264
{efrentzo,gratsias,ytheod}@unipi.gr
and
Research Academic Computer Technology Institute,
Patras University Campus, GR-26500 Rio, Greece

Abstract. Location-based services (LBS) constitute an emerging application domain rapidly introduced in modern life habits. However, given that LBS already count a few years of commercial life, the services provided are rather naïve, not exploiting the current software capabilities and the recent research advances in the fields of spatial and spatio-temporal data management. The goal of this paper is to fill this gap by first presenting the next generation of location-based services and then, demonstrating their implementation which takes advantage of both modern commercial software and state-of-the-art spatial network and spatio-temporal databases techniques. Novel techniques are also proposed in fields non-thoroughly addressed by the research community, such as the prediction of future position of objects moving on a road network.

Keywords: Location Based Services, Spatial Network Databases, Spatio-temporal databases.

1 Introduction

The rapid introduction of location-aware mobile devices in modern life, such as GPS-enabled mobile phones and Personal Digital Assistants (PDAs), has triggered the development of an emerging class of e-services, the so-called Location-Based Services (LBS), which provide information relevant to the location of a receiver [15]. LBS are rapidly introduced in modern life habits, influencing the way that people organize their activities, promising great business opportunities for telecommunications, advertising, tourism, etc. [11], also setting the research agenda in several technological fields including spatial and spatio-temporal data management.

From the database management perspective, efficient LBS support request the integration of several research advances in indexing [13], [17] and query processing techniques [5], [12], [14], [16]. The development of such services has also indicated several open research issues such as the processing of forecasting queries (i.e., determining future positions of moving points [4]), the support of continuous location change in query processing techniques [3], knowledge discovery from data collected via LBS [10], etc.

---

[*] Corresponding author.

J.M. Ware and G.E. Taylor (Eds.): W2GIS 2007, LNCS 4857, pp. 202–215, 2007.

On the other hand, although LBS already count some years of commercial life (i-area by NTT DoCoMo was launched in 2001), the services currently provided are rather naïve, not exploiting the current software capabilities and the recent advances in the research fields of spatial and spatio-temporal databases. The goal of this paper is to present the next generation of LBS and, then, demonstrate their implementation taking advantage of modern GIS and DBMS software as well as recent advances in spatial network and spatio-temporal databases.

Investing on the taxonomy of LBS proposed in [1], we present a set of advanced LBS and then sketch up the respective algorithms along with a description of their implementation details. The taxonomy classes provided in [1] are based on the discrimination of mobile vs. stationary reference (query) object, on the one hand, and data objects (e.g., landmarks), on the other hand, resulting in four classes of services (i.e., both being static, *S-S*, one being static and one being mobile, *S-M* and *M-S*, both being mobile, *M-M*), summarized in Table 1.

Table 1. LBS Classification and example services, according to [1]

Reference (Query) Object / Data Objects	Stationary (S)	Mobile (M)
Stationary (S)	*What-is-around* *Routing* *Find-the-nearest*	*Guide-me*
Mobile (M)	*Find-me*	*Get-together*

The driving force behind the LBS classification presented in [1] is about query processing issues: the query processing techniques behind a service involving mobile objects, clearly fall into the domain of Moving Object Databases (MOD), while processing of services concerning stationary objects is likely to be an issue related to the area of Spatial (or Spatial Network) Databases (SDB or SNDB). The services presented in this paper cover the *S-S*, *M-S* and *M-M* classes introduced in [1], employing respective query processing techniques, while, their majority, to the best of our knowledge, is not currently supported by commercial LBS providers.

Outlining the rest of the paper, Section 2 presents the framework on top of which LBS are developed as well as a set of fundamental LBS of the *S-S* LBS class. Section 3 is the core of the paper where a set of novel services is presented, constituting our proposal regarding the next generation of LBS, as well as the algorithmic and implementation issues raised during their development. Section 4 provides technical details regarding the development platforms. Finally, Section 5 summarizes our work providing the conclusions and some interesting directions for future work.

2 Background

In this section, we first introduce the framework used in the rest of the paper, and then describe a set of services already provided by current LBS solutions. We include this set in our discussion since they are fundamental and, thus, used as a basis for the (more advanced) novel LBS set that will follow in Section 3.

Table 2. Table of notations

Symbol	Description
$V = \{V_i\}$	the set of vertices corresponding to road network junctions on a road network
$E = \{E_{i,j}\}$	the set of edges connecting vertices V_i and V_j, corresponding to road segments on a road network
$G(V, E)$	the directed graph that represents the underlying road network on which objects are moving
$L = \{L_i\}$	the set of points of interest (POIs) or Landmarks
T_i, $T_{i,j}$, $T_{i,j}^\dagger$	the trajectory of a mobile user, its actual (sampled) and its predicted spatio-temporal position (i.e., time-stamped spatial point) at timestamp t_j
$T_{i,j}^x, T_{i,j}^y, T_{i,j}^t$	the x-, y-, t- components of the spatio-temporal position $T_{i,j}$. Also note that $T_{i,j}^t \equiv t_j$
$\overline{V_{i,j}}, \overline{V_{i,j}^x}, \overline{V_{i,j}^y}$	the (estimated) velocity vector of object T_i at timestamp t_j and its projection along the x and y axes
$T = \{T_i\}$	the set of trajectories of mobile users
$D_{Eucl}(P, Q)$	the Euclidean distance between the two dimensional points P and Q
$D_{Net}(P, Q)$	the network distance on the graph G between the two dimensional points P and Q
$Buffer(X, D)$	A method that builds a buffer of width D around a path X
$Route(P, Q)$	A method that retrieves a set of bi-connected line segments $\{E_i\}$ of the network graph forming a single path between points P and Q; usually, the result of a routing operation

2.1 The Framework

Before proceeding into describing the LBS, we set the framework on which the services will be based. Specifically, this framework contains:

- A directed graph $G(V, E)$ that represents the underlying road network on which objects are moving. The set of vertices (nodes) $V = \{V_i\}$ of G correspond to road junctions while its edges $E=\{E_{i,j}\}$ (connecting nodes V_i and V_j} represent road segments. Each edge $E_{i,j}$ is associated with two weights (distance metrics): the length of the corresponding road segment and the average time required to travel through that segment, respectively. Movement constraints (one-ways etc.) can be also applied in the graph $G(V, E)$ by appropriately setting the weights of each edge.
- The set $L = \{L_i\}$ containing the points of interest (POIs) or landmarks. POIs can be also categorized into classes (e.g., restaurants, gas stations, ATMs etc); in fact, in our implementation POIs are divided in such categories. We choose however, to restrict our discussion in the case of one class for sake of clarity in presentation.
- The set $T = \{T_i\}$ of trajectories of mobile users. Each trajectory is represented as a set of time-stamped sampled locations <x, y, t>, applying linear interpolation in-between them.

The above definitions along with the notation used in the rest of the paper are summarized in Table 2.

The services presented in this paper are part of an LBS platform following the system architecture illustrated in Fig.1. Among the components contained in the system architecture, we focus on the LBS suite, which encloses the logic and a number of advanced algorithms behind the developed services. Regarding the other components, the architecture includes an extensible Database Management System (*DBMS*) with a *Spatial Extension* communicating with the LBS suite via OLE DB protocol; the *Routing Server* maintains the network graph *G* and provides routing between two points on the network, while its communication with the LBS suite is achieved via the Routing Client middleware. Exploiting the simple functionality provided by these components, we expand it towards many directions. Among others, the developed software supports nearest neighbor search using network (rather than Euclidean) distance, in-route nearest neighbour search, and predictive location queries which are supported by novel techniques.

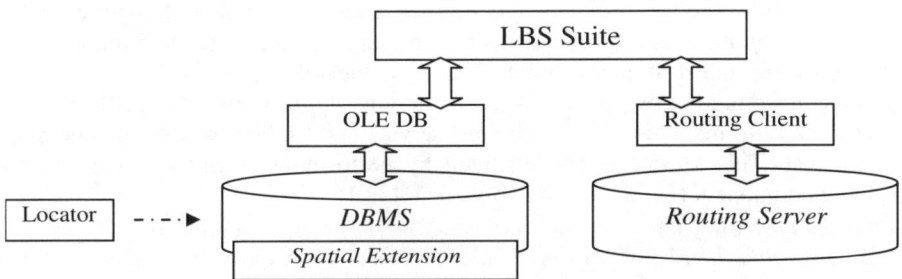

Fig. 1. LBS Suite Architecture

The database maintained in the DBMS of Fig.1 contains, among others, a set of tables fundamental for the advanced services that will be proposed later. Specifically, *Landmarks(object_id, object_name, object_geometry)*, contains the set *L* of POIs while *Trajectories(User_id, Trajectory_id, sampled_position_geometry, timestamp)* contains the set *T* of trajectories of the tracked LBS users. For performance issues, we build a view *Current_Positions(User_id, last_position_geometry)* on top of *Trajectories*, which contains the current positions of all tracked LBS users. The table *Trajectories* (and its view *Current_Positions*) are assumed to be periodically updated from an external Locator component. We also point out that *geometry* type columns in tables assume the Spatial Extension component in the involved DBMS illustrated in Fig.1.

2.2 A Set of Fundamental LBS

The basic set of LBS usually operated by current LBS solutions consists of three fundamental services, namely *What-is-around*, *Routing*, and *Find-the-Nearest*, which evidently fall into the *S-S* class of the taxonomy presented in [1]. All three services (and the respective algorithms) assume the presence of a graph *G* and/or a set of POIs

L. Moreover, all services involving graph operations (e.g., routing between two points), can be evaluated with any of the two optimization criteria presented, either traveled distance or traveling time, choosing between the two weights set for each edge of the graph (length and time, respectively); in the following sections, for sake of simplicity, we restrict our discussion to the distance (rather than time) optimization. The following paragraphs describe the functionality of each of the above fundamental LBS:

- *What-is-around:* The simplest service is the one that retrieves and displays the location of every POI being located inside a rectangular area (Q, d), where Q is the location of the user (or simply a user-defined point) and d is a selected distance (i.e., the half-side of the query rectangle). The input of the corresponding algorithm consists of the point Q and the distance d, while it returns the set $L' \subseteq L$ containing all POIs inside the rectangular area (Q, d). This LBS is fundamental for the user in order to know where he/she is located and what he/she can find nearby, while in terms of spatial database operations it involves a simple *spatial range query*.
- *Routing:* This service provides the optimal route between a departure and a destination point, P and Q, respectively. The input of the respective algorithm is the departure and destination points P and Q; the service returns a route on the graph connecting the two points. Apart from other applications, this LBS gives the user the tool to make use of the previous service and find the way to the landmark of interest. This service is implemented by performing a simple request to the routing component (i.e. the *Routing Server* in Fig.1).
- *Find-the-Nearest:* This service retrieves the k nearest landmarks (POIs). For example, *"find the two restaurants that are closest to my current location"* or *"find the nearest café to the railway station"*. The underlying algorithm takes as input the query point Q (for example, calling user's current location), and returns the set of points $L' \subseteq L$, which are the k nearest to Q members of L.

Regarding the third service, it is important to note that the conventional nearest neighbor (NN) search supported by current LBS solutions retrieves the k-NN objects based on the Euclidean distance between the reference (query) and the data objects stored in the database. However, the proper functionality of this service requires finding the nearest neighbor based on the *network distance* between the two points (i.e., the distance traveled by an object constrained to move on the network edges). On the other hand, a recent solution in the field of SNDB includes the *"Euclidean Restriction"* algorithm described in [12]. The algorithm is based on the observation that $D_{Eucl}(Q, P) \leq D_{Net}(Q, P)$ holds for each pair of two-dimensional points Q and P; as such, every object P' with Euclidean distance from Q greater than the respective network distance of another object P can be safely pruned without further considering its network distance, which by definition is greater than the respective Euclidean distance.

Based on the above three fundamental LBS, in Section 3 we propose a set of advanced LBS, covering also the *M-S* and *M-M* classes of services, according to the taxonomy introduced in [1].

3 Next Generation LBS

In this section we describe a set of novel services constituting our proposal regarding the next generation of LBS, which can be also considered as extensions of the three fundamental services discussed in Section 2.2. In particular, we focus on the following three services, named *Guide-me* (or, *Dynamic Routing*), *In-Route-Find-the-Nearest*, and *Get-together*. Once again, all services (and the respective algorithms) assume the presence of a graph G and/or a set of POIs L:

- *Guide-me (Dynamic Routing):* A first extension of the (static) *Routing* described in Section 2.2 is the so-called *Guide-me* service, illustrated in Fig. 2(a). Likewise, the system determines the best route between the calling user's current location (point P) and a destination point Q, and, then, keeps track of the user's movement (by simply updating its position) towards the destination point, allowing him/her to deviate from the 'optimal' route, as long as his/her location does not fall out of a predefined safe area (buffer) built around this route. The user is notified of his/her deviation every time he/she crosses out of the buffer's border and he/she is given the option of re-routing from that current location (point R). The input of *Guide-me* algorithm contains the *id* of the calling user, the destination point Q, and the distance D, which defines the buffer width.

- *In-Route-Find-the-Nearest:* It is a combination of *Routing* and *Find-the-Nearest* services which, given a departure and a destination point, P and Q, respectively, finds the best route between them, constrained also to pass through one among the specified set of candidate points (e.g., one of the points contained in Landmarks). For example, a request for this service could be, *"provide me the best route from my current location to city A constrained to pass from a gas station"*. Once again, the input of the respective algorithm contains the departure and destination points, P and Q respectively. This problem can also be considered as a special case of the so-called *Trip Planning Query* (*TPQ*) [5], with the number of different classes requested set to one.

- *Get-together:* With this service a moving user T_k 'attracts' a set of other users $S_k \subseteq T$ (let us assume, members of a community), also moving in the same area, to converge at a meeting point not known in advance. This point is periodically (every Δt seconds) calculated by the system based on the future projection of the calling user's trajectory. As an example, consider Fig.2(b), where the calling user T_k is at (the spatiotemporal) location $(T_{k,j0})$ and moves towards the direction shown. The 'attracted' users T_{i1} and T_{i2} are located at $(T_{i1,j0})$ and $(T_{i2,j0})$ respectively. The system predicts that by the time $t_j + \Delta t$ user T_k will most probably be located at location $T_{k,j+\Delta t}^\dagger$ (point P in Fig.2(b)) and, therefore, routes T_{i1} and T_{i2} towards this point. Eventually, at $t_j + \Delta t$ the (recorded) location of T_k is $T_{k,j0+\Delta t}$. Likewise, the system projects the trajectory of T_k to the future time $t_{j0} + 2\Delta t$, predicting that at that time T_k will be at $T_{k,j+2\Delta t}^\dagger$ (point P' in Fig.2(b)), so it routes the rest of the users accordingly. The service terminates when at least one or all the members of S_k reach T_k. The input of the respective algorithm is the id of the 'master' user T_k, the set of the ids of the 'attracted' users S_k, and a time delay Δt. The algorithm returns a set of (dynamically updated) paths between each $T_i \in S_k$ and the periodically predicted location of T_k.

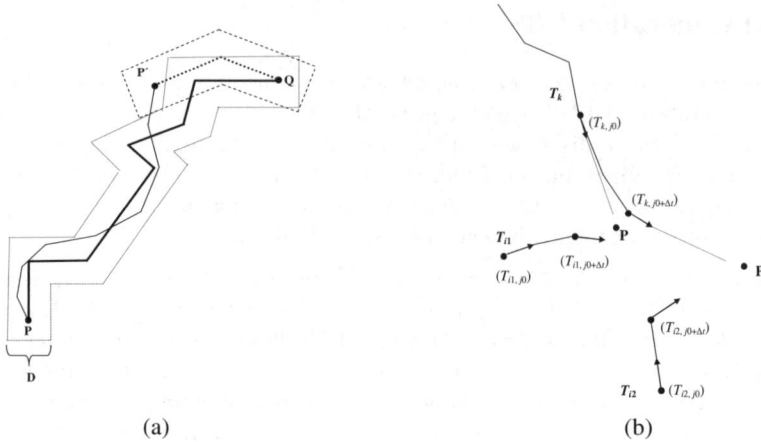

(a)   (b)

Fig. 2. (a) Guide-me and (b) Get-together examples

Among these services, the first evidently falls into the *M-S* class since it involves a moving user and several static data objects, while, the second is classified as *S-S*. Finally, the third service belongs to *M-M* class since both the reference (query) and the data objects are moving. The algorithms supporting the above three services are presented in the sections that follow.

3.1 Guide-Me (Dynamic Routing)

As already discussed (and illustrated in Fig. 2(a)), this service requires the DBMS to keep track of the user's current position. As such, it exploits the previously introduced view *Current_Positions* over the table *Trajectories* (containing the current position of each user T_i). The algorithm developed to support the *Dynamic Routing* service is illustrated in the pseudo-code of Fig.3.

Apparently the first step of the `Dynamic_Routing` algorithm, involving the retrieval of the current position of object T_i, is a simple selection on *Current_Positions*. Regarding the second step, `Route` is performed by a simple request to the routing component (i.e., the Routing Server of Fig.1). Finally, in step 9, it is requested to check whether the object's current location $T_{i,j}$ lies on a buffer of the route R with distance D; this operation is performed by compiling the `Buffer` and `Contains` spatial methods provided by the Spatial DBMS extensions:

```
SELECT * FROM CurentPositions
WHERE User_id=UId AND
Contains(Buffer(R,D),last_position_geometry)
```

The `Contains(A,B)` function returns `true` when the spatial object A contains the spatial object B, while the `Buffer(A,D)` function constructs a spatial object representing the buffer of the A with distance D.

```
Algorithm Dynamic_Routing(User Id •ᵢ, destination point Q,
distance D, time period •t)
1    •ᵢ,ⱼ = Retrieve current position of •ᵢ
2    R = Route(•ᵢ,ⱼ,Q)
3    DO WHILE •ᵢ,ⱼ has not reached Q
4       Wait •t; j = j+•t
5       •ᵢ,ⱼ = Retrieve current position of •ᵢ
6       IF NOT •ᵢ,ⱼ lies in the buffer Buffer(R,D) THEN
7          R = Route(•ᵢ,ⱼ,Q)
8       ENDIF
9    LOOP
```

Fig. 3. Algorithm `Dynamic_Routing`

3.2 In-Route-Find-the-Nearest

This service retrieves the best route one has to follow in order to travel from a departure to a destination point, P and Q, respectively, also constrained to pass via a landmark among the ones contained in the Landmarks table. The developed algorithm, based on the TPQ solutions provided in [5], is illustrated in the pseudo-code of Fig.4.

```
Algorithm In_Route_Find_the_Nearest(departure point P,
destination point Q)
1    Nearest.Dist=∞
2    Retrieve route R=Route(P,Q)
3    Find the Euclidean nearest object N to the route object R
4    Calculate D_Net(P,N) and D_Net(N,Q)
5    Retrieve all POIs Nᵢ having
                D_Eucl(P,Nᵢ)+D_Eucl(Nᵢ,Q)< D_Net(P,N)+ D_Net(N,Q) and sort
         them incrementally according to D_Eucl(P,Nᵢ)+ D_Eucl(Nᵢ,Q)
6    DO WHILE D_Eucl(P,Nᵢ)+ D_Eucl(Nᵢ,Q)≤ Nearest.Dist
7       IF Nearest.Dist > D_Net(P,Nᵢ)+ D_Net(Nᵢ,Q) THEN
8          Nearest.Point = Pᵢ
9          Nearest.Dist = D_Net(P,Nᵢ)+ D_Net(Nᵢ,Q)
10      ENDIF
11   NEXT Nᵢ
12   Return Nearest
```

Fig. 4. Algorithm `In_Route_Find_the_Nearest`

The `In_Route_Find_the_Nearest` algorithm is based on the same principle with the Euclidean restriction algorithm [12], that is, the Euclidean distance between two points serves as a lower bound for their network distance. As such, the algorithm initially produces the optimal route R between P and Q by performing a request to the Routing Server, while subsequently, uses R as a query object in order to retrieve the Euclidean NN among the records contained in the Landmarks table (Lines 2-3 in pseudo-code, also illustrated in Fig.5(a)). In this step, we exploit the R-tree-based nearest neighbor operator provided from the DBMS Spatial Extension, which retrieves the nearest to the query object R among those that are contained in the Landmarks table.

Then, the algorithm performs requests to the Routing Server in order to retrieve the best route between P, N and Q calculating the network distance between them (Line 4 in pseudo-code, also illustrated in Fig.5(b)), while afterwards uses their sum in order to retrieve candidate objects with a total distance from both P and Q upper bounded by it (Line 5 in pseudo-code, also illustrated in Fig.5(c)). It is also important to note that these objects are contained inside an elliptical region with P and Q as foci.

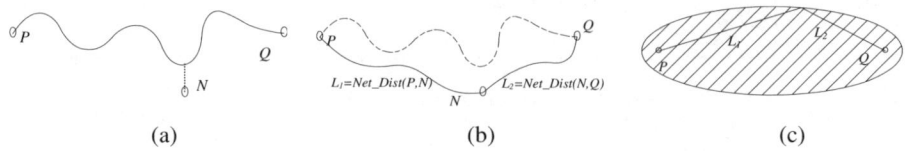

(a)   (b)   (c)

Fig. 5. An illustration of the In-Route-Find-the-Net-Nearest functionality

In its final step, the algorithm calculates the network distances between P, Q and the candidate points N_i, until the sum of the Euclidean distances of N_i from P and Q is greater than the respective network distance of the candidate nearest (lines 6-11). Finally, the algorithm reports the candidate nearest as the answer to the query.

3.3 Get-Together

This service requires the DBMS to keep track of a user's current position (we name him/her 'master'), along with the positions of the users in the set S_k 'attracted' by user T_k; recall that all current positions are maintained in *Current_Positions* view. Moreover, in order to 'project' the master user's trajectory in a future position, it also employs the *Trajectories* table containing the history of each tracked object. The algorithm developed to support the *Get-together* service is illustrated in the pseudo-code of Fig.6.

The `Get_together` algorithm iterates until the position of all users contained in set S_k have converged to the master user's current position $T_{k,j}$ at timestamp t_j. Inside each iteration, the algorithm projects the trajectory of object T_k into the future timestamp $t_j + \Delta t$ calculating the respective position $T_{k,j}^\dagger$ on which T_k is estimated to be found at this future timestamp (Line 2). Subsequently, the algorithm computes and reports the routes needed for all objects of set S_k, in order to reach position $T_{k,j}^\dagger$ (Line 3), and finally, updates the actual positions of all objects involved in the algorithm at timestamp $t_j + \Delta t$ (Lines 5-6) - these positions are used in order to determine whether the algorithm may or not continue with the next iteration (Line 1).

Evidently, the most interesting operation involved in the algorithm is the `Project_trajectory` function. This routine estimates the future location of a moving object by projecting its trajectory to a future timestamp $t_j + \Delta t$. Such a computation requires some extra knowledge about the objects movement, such as the velocity and the direction of each object. Although several methods have been proposed for processing and indexing the current and future location of objects moving without network constraints, including [13] and [16], the problem of

Algorithm Get_together(User Id $\bullet_k$, destination point Q, distance D, time period $\bullet t$)

1	DO WHILE $T_{i,j}$ has not reached $\bullet_{k,j}$ ($\forall\ T_i \in S_k$)
2	$T_{k,j}^\dagger$ =Project_trajectory($\bullet_k$, $\bullet t$)
3	FOR EACH $T_i \in S_k$ Report Route($\bullet_{i,j}$, $T_{k,j}^\dagger$)
4	Wait $\bullet t$; $j = j + \bullet t$
5	FOR EACH $T_i \in S_k$ Update current position $\bullet_{i,j}$
6	Update current position $\bullet_{k,j}$ of $\bullet_k$
7	LOOP

Fig. 6. Algorithm Get_together

predicting the position of an object moving in a network, is very challenging by its nature, and therefore, still remains open. Specifically, the majority of the proposed strategies consider only the object's current position [4], while an efficient Project_trajectory function should make use of a trace (of dynamic length) of the trajectory for the estimation of the object's future location. Moreover, since our need is to estimate a single point on which the members of S_k will be routed, the Project_trajectory function should produce a 'winner' (i.e., single) projected point. As such, existing approaches which produce matrices with probabilities for each candidate network edge [4] cannot be used; instead we must estimate a single destination point for all 'attracted' objects.

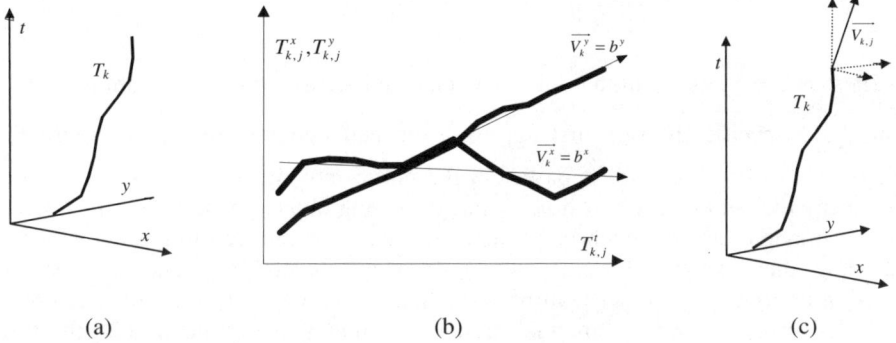

(a)   (b)   (c)

Fig. 7. The Project_trajectory function: (a) a trajectory in the 3-dimensional space, (b) the x and y trajectory coordinates expressed as functions of time, and, (c) the estimated vector $\overrightarrow{V_{k,j}}$

We address this problem based on a simple regression model, which calculates the projection of the velocity vector along the x- and y- axes. Specifically, assuming that the object follows a 'general' direction, our goal is to determine this direction; we therefore employ the *simple linear regression* along each axis, assuming that the values of $T_{k,j}^x$ and $T_{k,j}^y$ at each timestamp $t_j = T_{k,j}^t$, linearly depend on time, i.e.,

$$T_{k,j}^x = a^x + b^x \cdot T_{k,j}^t, \text{ and } T_{k,j}^y = a^y + b^y \cdot T_{k,j}^t,\qquad(1)$$

and, the velocity vector of this 'general' direction at timestamp t_j will be determined as the vector $\overrightarrow{V_{k,j}} = \overrightarrow{V_{k,j}^x} + \overrightarrow{V_{k,j}^y}$. The magnitude for both $\overrightarrow{V_k^x} = b^x$ and $\overrightarrow{V_k^y} = b^y$ can be calculated using *linear regression analysis techniques* given the values of time-stamped sampled positions of T_k. Consider, for example, Fig.7(a) illustrating a trajectory T_k in the 3-dimensional space; then, Fig.7(b) demonstrates the corresponding x and y trajectory coordinates as functions of time, along with the values of $\overrightarrow{V_k^x} = b^x$ and $\overrightarrow{V_k^y} = b^y$ calculated by the 'mean' trajectory direction along each axis. Then, their combination in Fig.7(c) produces the vector $\overrightarrow{V_{k,j}}$ standing for the general trajectory direction.

Finally, a first approximation of the projected location $T_{k,j+\Delta t}^\dagger$ can be calculated by the simple formula $T_{k,j+\Delta t}^\dagger = T_{k,j} + \overrightarrow{V_{k,j}} \cdot \Delta t$, which leads to

$$T_{k,j+\Delta t}^{\dagger x} = T_{k,j}^x + \left| V_{k,j}^x \right| \cdot \Delta t \quad \text{and} \quad T_{k,j+\Delta t}^{\dagger y} = T_{k,j}^y + \left| V_{k,j}^y \right| \cdot \Delta t \tag{2}$$

It is also worth to note that the velocity vectors produced by linear regression already include the mean object speed along the two axes, which, consequently, is incorporated in the above formulas. Anyway, its value $\overline{V_{k,j}}$ can be easily calculated as the magnitude of vector $\overline{V_{k,j}}$, thus:

$$\overline{V_{k,j}} = \left| V_{k,j} \right| = \sqrt{\left| V_{k,j}^x \right|^2 + \left| V_{k,j}^y \right|^2} \tag{3}$$

There is one more issue that has to be clarified regarding the calculation of $\left| V_{k,j}^x \right|$ and $\left| V_{k,j}^y \right|$. Specifically, one first approach on their computation via the regression analysis, is to proceed with it based on the entire moving object trajectory history (i.e., using the complete set of time-stamped moving object positions $T_{i,j}$ for $j = 0, ..,$ *now*). Although, this approach may sound reasonable, it can lead to false conclusions regarding their estimated values since, moving objects may change travelling direction arbitrarily, perform U-turns, vary their speed etc. Consider, for example, a truck delivering several commercial products around a metropolitan area; the truck starts its trip from a departure point (e.g., the company warehouse), delivers the products to several locations around the city (stopping at each one of them), and returns back to its original location (i.e., the company warehouse). Thus, the values of $\left| V_{k,j}^x \right|$ and $\left| V_{k,j}^y \right|$ calculated from the regression model will include all this arbitrarily directed trajectory history.

On the other hand, the direction criterion applied in *local circumstances* can provide more accurate results; for example, when this delivery truck moves between two delivery points, it actually follows a general direction. As such, we propose that *the calculation of* $\left| V_{k,j}^x \right|$ *and* $\left| V_{k,j}^y \right|$ *could be safely performed based on the Δt last timestamps*. Formally, in our implementation, $\left| V_{k,j}^x \right|$ and $\left| V_{k,j}^y \right|$ are determined by the following formulas (which are directly derived from the linear regression analysis of the last Δt time-stamped trajectory sampled positions):

$$\left|V_{k,j}^{x}\right| = \frac{\sum_{l=j-\Delta t}^{j} \left(T_{k,l}^{x} - \overline{T_{k,j}^{x}}\right) \cdot \left(T_{k,l}^{t} - \overline{T_{k,j}^{t}}\right)}{\sum_{l=j-\Delta t}^{j} \left(T_{k,l}^{x} - \overline{T_{k,j}^{x}}\right)^{2}}, \text{ and } \left|V_{k,j}^{y}\right| = \frac{\sum_{l=j-\Delta t}^{j} \left(T_{k,l}^{y} - \overline{T_{k,j}^{y}}\right) \cdot \left(T_{k,l}^{t} - \overline{T_{k,j}^{t}}\right)}{\sum_{l=j-\Delta t}^{j} \left(T_{k,l}^{y} - \overline{T_{k,j}^{y}}\right)^{2}} \quad (4)$$

where $\overline{T_{k,j}^{x}} = \frac{1}{\Delta t + 1} \cdot \sum_{l=j-\Delta t}^{j} T_{k,l}^{x}$, $\overline{T_{k,j}^{y}} = \frac{1}{\Delta t + 1} \cdot \sum_{l=j-\Delta t}^{j} T_{k,l}^{y}$ and $\overline{T_{k,j}^{t}} = \frac{1}{\Delta t + 1} \cdot \sum_{l=j-\Delta t}^{j} T_{k,l}^{t}$.

Subsequently, in order to support more realistic, network-constrained projected location, in the proposed algorithm we search for the nearest network link based on $T_{k,j+\Delta t}^{\dagger}$, and we route object T_k to it, thus producing route R. Finally, we determine the point on this route, on which the object will be found after the Δt time period, i.e., after traversing a distance of $\overline{V_k} \cdot \Delta t$ (which is performed by simply traversing it and counting the distance so far). The developed algorithm is illustrated in the pseudo-code of Fig.8.

Algorithm Project_Trajectory(User Id •ₖ, time period •t)

1	Calculate $\left|V_k^x\right|$ and $\left|V_k^y\right|$ based on Eq.(4)
2	$T_{k,j+\Delta t}^{t} = T_{k,j} + \overline{V_k} \times \Delta t$
3	R = Route($T_{k,j}$, $T_{k,j+\Delta t}^{t}$)
4	Determine point P on R with $D_{Net}(T_{k,j}, P) = \left|V_k\right| \cdot \Delta t$
5	Return P

Fig. 8. Algorithm In_Route_Find_the_Nearest

A final point to be discussed is that the calculation of $\left|V_{k,j}^{x}\right|$ and $\left|V_{k,j}^{y}\right|$ through Eq.(4) (i.e., the first algorithm step) requires only to query the *Trajectories* table based on the master user's id and the last Δt timestamps; the results of this query will retrieve the set of all time-stamped trajectory positions involved in Eq.(4) which will be subsequently used to produce $\left|V_{k,j}^{x}\right|$ and $\left|V_{k,j}^{y}\right|$.

4 Implementation Details

All the services presented in Sections 2 and 3 have been implemented on top of three basic components. The first one is the *Microsoft SQL Server* [9], which is a relational database management system (RDBMS) that does not natively support spatial objects such as points, lines etc; consequently, the employment of an extension component, which enables SQL Server to support spatial data is an obligatory action, in order for the LBS suite to be properly developed.

The Spatial Extension component is the *MapInfo SpatialWare* [6], which enables the DBMS to store, manage, and manipulate location-based data. It allows therefore spatial data to be stored in the same place as traditional data, ensuring data accessibility, integrity, reliability and security through the mechanisms of the SQL

Server. SpatialWare includes a variety of non-traditional data types, such as points, lines, polyline, regions (polygons), supports numerous spatial functions, and it is Open GIS compliant [11]. However, the most important SpatialWare feature is the support it provides for R-tree indexing [2], making it able to support huge volumes of spatial data; R-tree indexing allows pruning the search space when a spatial query is executed. Otherwise (i.e., in the case where no spatial index is present), the execution of each spatial query would lead to linear scans over the entire dataset, which is a very expensive operation.

The *Routing Server* component used in our implementation is the *MapInfo Routing J Server (RJS)* [7], which is a street network analysis tool for finding a route between two points, the optimal or ordered path between many points, the creation of drive time matrices, and the creation of drive time polygons. RJS calculates either the shortest distance or quickest timed route between any two points, returning text-based driving directions and spatial points to the parent application. This functionality is achieved by *XML requests* over a continuously running server: the client queries the RJS with an XML file containing information such as, the departure and the destination point, and after processing the request, RJS returns another XML file containing the optimal route in terms of its lines segments (i.e., edges of the respective directed graph). Mapinfo provides also the *RJS .NET client* middleware, which undertakes the tasks of composing the XML file used for making the request, and subsequently, interpreting the server's answer to a set of comprehensive objects implemented in the form of .NET objects.

Finally, the LBS suite, requesting data from all the above components, is implemented with Microsoft Visual Studio .NET [8], while the connection to the DBMS is realized by an *OLE DB connection* [9].

5   Summary

In this paper we presented typical examples of the next generation of location based services, sketched up the respective algorithms and provided some interesting details on their implementation. The majority of the presented services are not currently supported by commercial LBS providers, or are available in an inefficient and inaccurate manner. Among others, the developed services involve nearest neighbor queries using network (rather than Euclidean) distance, optimal route finding between a set of user-defined landmarks, and in-route nearest neighbor queries. The developed algorithms and their implementation are supported by recent advances in the field of Spatial Network Databases [5],[12],[14], and novel techniques originally proposed in this paper.

The solutions provided are not only focused on LBS; actually, many of them are directly applicable in the context of route planning, which is a task usually performed via web services. Therefore, our plan is to employ our LBS suite in the framework of a web-based application providing users with advanced functionality regarding their travelling needs.

Acknowledgments

Research partially supported by FP6/IST Programme of the European Union under GeoPKDD project (2005-08) and EPAN Programme of the General Secretariat for

Research and Technology of Greece under NGLBS project (2004-06). The services proposed in this paper are part of a commercial LBS platform; they designed and developed in cooperation with Telenavis S.A. in the context of NGLBS project. Telenavis S.A. is a commercial LBS provider providing both GSM-based and web-based private and corporate solutions (see for example http://www.navigation.gr).

References

1. Gratsias, K., Frentzos, E., Delis, V., Theodoridis, Y.: Towards a Taxonomy of Location-Based Services. In: Li, K.-J., Vangenot, C. (eds.) W2GIS 2005. LNCS, vol. 3833, Springer, Heidelberg (2005)
2. Guttman, A.: R-Trees: a dynamic index structure for spatial searching. In: Proceedings of ACM SIGMOD Conference (1984)
3. Jensen, C.S., Christensen, A., Pedersen, T., Pfoser, D., Saltenis, S., Tryfona, N.: Location-Based Services: A Database Perspective. In: Proceedings of Scandinavian GIS (2001)
4. Karimi, H., Liu, X.: A Predictive Location Model for Location-Based Services. In: Proceedings of ACM-GIS (2003)
5. Li, F., Cheng, D., Hadjieleftheriou, M., Kollios, G., Teng, S.-H.: On Trip Planning Queries in Spatial Databases. In: Bauzer Medeiros, C., Egenhofer, M.J., Bertino, E. (eds.) SSTD 2005. LNCS, vol. 3633, Springer, Heidelberg (2005)
6. MapInfo Corporation, MapInfo SpatialWare, (accessed May 18, 2007) (2007a), Available at http://extranet.mapinfo.com/products/ Overview.cfm?productid=1141
7. MapInfo Corporation, MapInfo Routing J Server (accessed May 18, 2007) (2007b), Available at http://extranet.mapinfo.com/ products/Overview.cfm?productid=1144
8. Microsoft Corporation: Microsoft Visual Studio.NET (accessed May 18, 2007) (2007a), Available at http://msdn.microsoft.com/ vstudio/
9. Microsoft Corporation, Microsoft SQL Server (accessed May 18, 2007) (2007b), Available at http://www.microsoft.com/sql/
10. Nanni, M., Pedreschi, D.: Time-focused clustering of trajectories of moving objects. J. Intell. Inf. Syst. 27(3), 267–289 (2006)
11. Open GIS Consortium, OpenGIS® Location Services (OpenLS): Core Services (accessed May 18, 2007) (2007), Available at http://www.opengis.org
12. Papadias, D., Zhang, J., Mamoulis, N., Tao, Y.: Query Processing in Spatial Network Databases. In: Aberer, K., Koubarakis, M., Kalogeraki, V. (eds.) DBISP2P 2003. LNCS, vol. 2944, Springer, Heidelberg (2004)
13. Saltenis, S., Jensen, C.S., Leutenegger, S., Lopez, M.: Indexing the Positions of Continuously Moving Objects. In: Proceedings of ACM SIGMOD Conference (2000)
14. Sankaranarayanan, J., Alborzi, H., Samet, H.: Efficient Query Processing on Spatial Networks. In: Proceedings of ACM-GIS (2005)
15. Theodoridis, Y.: Ten Benchmark Queries for Location-based Services. The Computer Journal 46(6), 713–725 (2003)
16. Tao, Y., Papadias, D., Shen, Q.: Continuous Nearest Neighbor Search. In: Bressan, S., Chaudhri, A.B., Lee, M.L., Yu, J.X., Lacroix, Z. (eds.) CAiSE 2002 and VLDB 2002. LNCS, vol. 2590, Springer, Heidelberg (2003)
17. Tao, Y., Papadias, D., Sun, J.: The TPR*-Tree: An Optimized Spatio-Temporal Access Method for Predictive Queries. In: Bressan, S., Chaudhri, A.B., Lee, M.L., Yu, J.X., Lacroix, Z. (eds.) CAiSE 2002 and VLDB 2002. LNCS, vol. 2590, Springer, Heidelberg (2003)

Automated Schematization for Web Service Applications

Jerry Swan[1], Suchith Anand[1], Mark Ware[2], and Mike Jackson[1]

[1] Center for Geospatial Science, University of Nottingham, UK
{Jerry.Swan, Suchith.Anand, Mike.Jackson}@nottingham.ac.uk
[2] Faculty of Advanced Technology, University of Glamorgan, UK
jmware@glam.ac.uk

Abstract. For the purposes of this paper, a schematic map is a diagrammatic representation based on linear abstractions of networks. With the advent of technologies for web-based delivery of geospatial services it is essential to develop map generalization applications tailored for the same. This paper is concerned with the problem of producing automated schematic maps for web map applications. The paper looks at how previous solutions to the spatial conflict reduction can be adapted and applied to production of automated schematic maps for web services.

Keywords: Schematic maps, Web services, Map generalization, Simulated annealing.

1 Introduction

Maps can be treated as graphical representations of spatial structure (i.e. they show the locations of objects in a geographic space and the attributes to which they are linked). Mapping can be viewed as an abstraction process by which real world objects are measured and then subjected to both simplification and reduction, and then stored on a medium. This leads to the important concepts of scale, classification, symbolization and generalization in cartography (Robinson et al., 1995).

The process of simplifying the form or shape of map features, usually carried out when the map is changed from a large scale to a small scale, is referred to as generalisation. Map generalisation is a process of extracting the important and relevant spatial information from reality. It is essential when the content of the map to be displayed exceeds the capability of graphic representation. Monmonier (1996) makes the point that reality is three-dimensional, is detail rich and is far too factual to allow a complete yet uncluttered two-dimensional graphic scale model. Figure 1 shows how map generalisation might be applied when there is a change from a larger to a smaller scale. The process involves the use of generalisation operations such as simplification, selection, displacement and amalgamation of features (Ware and Jones, 1998).

Generalization is a complex process in cartography involving considerable decision making. As a result there has not been much success in attempts to fully automate the process, though partly automated systems have been developed. In the foreseeable future the process will be a semi-automated collaboration between cartographer and machine (GIS Files, OS 2007). Map schematization is one subset in the domain of map generalization.

J.M. Ware and G.E. Taylor (Eds.): W2GIS 2007, LNCS 4857, pp. 216–226, 2007.
© Springer-Verlag Berlin Heidelberg 2007

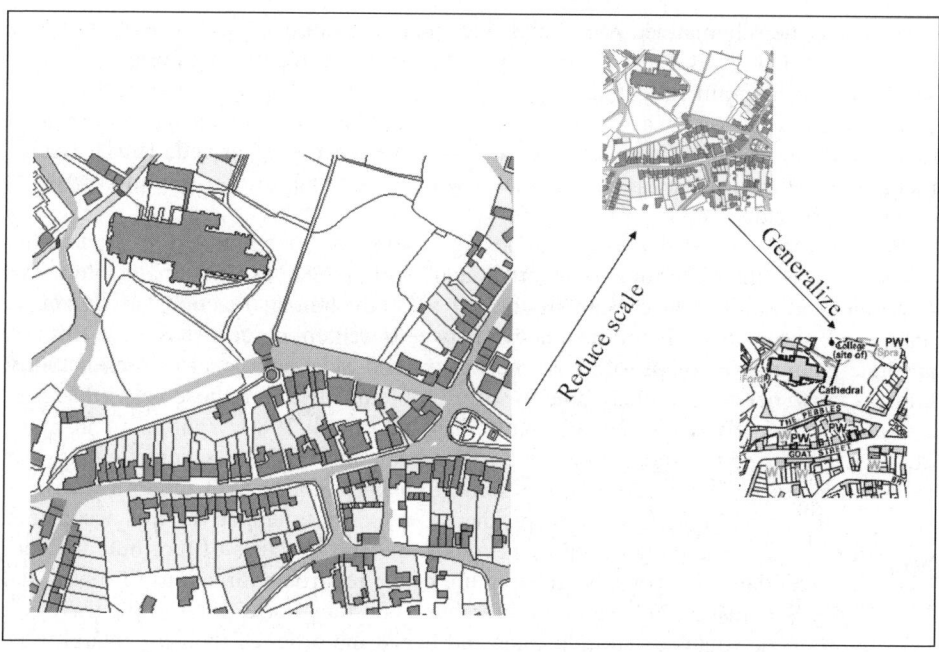

Fig. 1. Example of generalisation brought about by a change in scale. Data sets shown are OS Mastermap (large scale) and OS 1:50,000 raster for the St. David's area, West Wales.

This paper is concerned with the problem of producing automated schematic maps suitable for web based applications. Prototype software has been developed to derive schematic maps on the fly with reduced linear information from the detailed network dataset using web services. The results are promising. Future work will concentrate on refining the technique through the use of additional constraints to enhance visualization and usability for different applications.

2 Background of the Study

A schematic map is a diagrammatic representation based on linear abstractions of networks. Typically transportation networks are the key candidates for applying schematization to help ease the interpretation of information by the process of cartographic abstraction. Schematic maps are built-up from sketches that usually have close resemblance to verbal descriptions about spatial features (Avelar, 2002).

The best example of a modern day schematic map is the London Tube map designed by Harry C Beck in 1931. An electrical engineer, he based his design on a circuit diagram and used a schematic layout. The map locally distorted the scale and shape of the tube route but preserved the overall topology of the tube network. It took two years of persistent efforts by Beck before the Underground Publicity Department of London Tube network accepted the first publication. Their fear was that the total abandonment of geographical accuracy of the map would render it incomprehensible to the majority of underground travelers. But the public appreciated the helpful character of the schematized routes and since then its usability grew. The only surface feature included in Beck's Tube map is the stylized representation of the river Thames. The diagram reflected this in this unemphathic display of the central area, where no single feature was dominant. Also the central area was enlarged in relation to the outlying regions for purposeful and skilful distortion (Garland, 1994). Beck's method has since been adapted and used in different forms in transportation systems around the world.

Perhaps the best justification for choosing to adopt a schematic map in a particular application is their current popularity and widespread use in many domains. Psychological studies have been performed to study human reactions to schematic maps and show that given the choice between written descriptions of networks, accurate planimetric maps of those networks, and schematic map representations, humans grasp networks and are able to solve problems involving those networks faster and more accurately using the schematic maps (Bartram, 1980; Agarwala and Stole, 2001). Considering transport modes, it has also been observed that schematic maps can be more appropriate for representing some transport modes than others (Morrison, 1996). In the case of underground maps, there is a special tolerance for, and acceptance of, geographical inaccuracy to represent routes, but the structure of the route network has to be essentially the same as the structure of the real route directions (Tversky and Lee, 1999). Schematic maps can be preferable when they correspond to the traveler's perception of the routes as straight lines and hence the order of their acceptability is: underground railways; surface railways; trams and light rails; and, buses.

3   Basic Process in Schematization

Not only is it easy to lie with maps, it's essential. To portray meaningful relationships for a complex, three-dimensional world on a flat sheet of paper or a video screen, a map must distort reality (Monmonier, 1996). Schematic maps are good examples of this and are an ideal means for representing specific information about a physical environment. They play a helpful role in spatial problem solving tasks such as way finding. This section introduces the concepts of schematic maps, its relevance in transportation maps, the key process in generating schematic maps and previous work carried out in the area of automated schematic mapping techniques.

Avelar (2002) observes and lists the following characteristics common to most schematic maps:

- Easy to follow diagrammatic representation;
- Highly generalized lines;
- Geometrically inaccurate (e.g. node positions not necessarily correct, line lengths do not correspond to actual distances);
- Routes usually drawn as straight lines, with lines varying only via fixed, stylized angles (commonly 45 and 90 degrees);
- Do not have a constant scale factor, with the scale being relatively large in congested or important regions.

For one-off, relatively static maps (such as city metro maps) manual and computer assisted generation is acceptable. However, for dynamic applications, where layout is likely to change more frequently (e.g. bus routes or pedestrian routes), these techniques are not so useful, and there exists a definite need for automated solutions. In the automatic production of schematic maps vector based source datasets are simply input to suitable algorithm/software application, together within any control parameters that might be required.

The basic steps for generating schematic maps are to eliminate all features that are not functionally relevant and to eliminate any networks (or portions of networks) not functionally relevant to the single system chosen for mapping. All geometric invariants of the network's structure are relaxed except topological accuracy. Routes and junctions are symbolized abstractly (Waldorf, 1979).

Elroi (1988) refined the process by adding three graphic manipulations, although implementation details and results were not given. First, lines are simplified to their most elementary shapes. Next, lines are re-oriented to conform to a regular grid, such that they all run horizontally, vertically or at a forty-five degree diagonal. Third, congested areas are increased in scale at the expense of scale in areas of lesser node density. These three steps are illustrated in Figure 2. Though Elroi in this paper has listed the theory on schematization, the actual real-life implementation was not given.

The first step in the process is line simplification, which can be achieved using an algorithm such as that of Douglas and Peucker (1973). Care must be taken when performing this step to avoid the introduction of topological errors; this can be achieved most easily by making use of topology preserving variants of the Douglas-Peucker algorithm, such as that presented by Saalfeld (1999). Steps two and three are the key components of the process, and their automation has been the focus of previous work by several researchers.

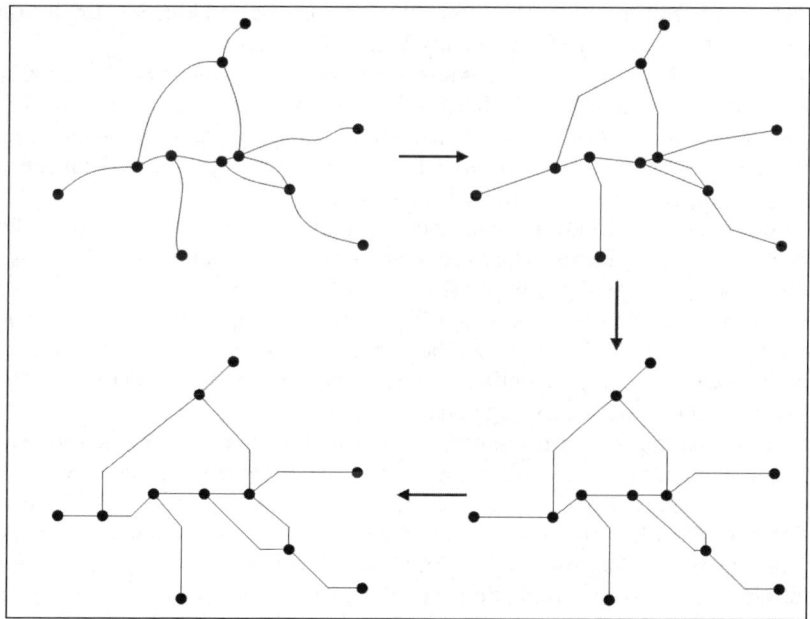

Fig. 2. Key processes for producing a schematic map

Avelar (2002) presents an algorithm for the automatic generation of schematic maps from traditional vector-based route networks, which are pre-generalized using the Douglas-Peucker algorithm. The algorithm makes use of gradient-descent based optimization in an attempt to force the network to conform to orientation and minimum separating distance constraint. By using an optimization technique, the lines of the original route network are modified to meet geometric and aesthetic constraints in the resulting schematic map. Map modifications are achieved by the iterative displacement of map vertices. At each iteration, the best single vertex displacement for each vertex is calculated and applied. The displaced location of a particular vertex v is obtained by initially calculating the arithmetic mean of the displacements required in order to correctly orient each of the edges to which v belongs. The displacement is then adjusted in order that minimum separating distances are maintained between features. It could be the case that none of the calculated vertex displacements offer any improvement to the current map. A problem with this is that, since only map improving displacements are accepted, there is the risk of getting caught in locally optimal solutions.

As part of the algorithm, special emphasis is placed on preserving topological line structure of the line network during the transformation process. This is achieved using simple geometric operations and tests. The author also analyses the convergence of the generalization approach and suggests suitable stopping criteria. The work was applied to transport network data for the city of Zurich. The results appear very promising but are likely to be non-optimal due to the use of gradient descent.

Agrawala and Stolte (2001) also make use of iterative improvement optimization, in their case simulated annealing, to produce what they refer to as sketch route maps for vehicle navigation. These sketch maps are simplified versions of road networks with the particular characteristic of having a relatively high level of detail near the route start and end points. The algorithm works by making a single random displacement of a randomly chosen vertex at each iteration.

Cabello et al (2001) present a combinatorial algorithm for the schematization of road networks. The algorithm produces maps in which every path has two or three links and edge orientation is restricted (horizontal, vertical or diagonal). It is guaranteed to find a correct solution if one exists, but has the disadvantage of not providing an output if such a solution does not exist.

Barkowsky et al. (2000) present their discrete curve evolution algorithm for simplifying geometric shapes. They use a stepwise elimination of kinks that are least relevant to the shape of the polygonal curve in the simplification. An algorithm is presented, which on the one hand simplifies spatial data up to a degree of abstraction intended by the user and which on the other hand does not violate local spatial ordering between cartographic entities, since local arrangement of entities is assumed to be an important spatial knowledge characteristic.

The metro map algorithm presented by Stott and Rodgers (2005) is a graph based method for generating schematic maps from a set of nodes and connecting edges. Nodes represent stations and these are connected by one or more edges. Starting with an initial map the algorithm makes use of node displacement and gradient descent based optimization to improve the overall layout with respect to a set of schematic map defining criteria (i.e. orientation, angular resolution, line straightness and edge length). The approach is interesting in that the graph is embedded on an integer square grid in such a way that nodes (i.e. stations) can only ever be centered on grid intersections. This has the effect of greatly reducing the total number of possible alternative map solutions when compared to methods that make use of a continuous search space, such as that presented by Avelar (2002) and the algorithm developed for this project. The work presented by Stott and Rodgers is still at an early stage and the limited results available are difficult to assess. A potential advantage of their discrete approach is reduced processing times (when compared to continuous approached). However it also thought likely that the quality of result will suffer - depending on the resolution of the integer square grid.

Iterative improvement algorithms are used widely for solving large optimization problems. A well known cartographic example of an optimization problem is the point-label placement problem. The most common solution to this problem is to make use of trial positions, in which each point-label is assigned a fixed number of alternative positions in-and-around its point. If there are 'n' point features and each is assigned 'k' label trail positions, then there are a total of k^n alternative map labellings available; the assumption is that some of the labellings provide better solutions with less graphic conflict than others. Generating and evaluating all realizations is not practical even for relatively small values of n and k. Christensen et al (1995) suggest the use of iterative improvement algorithms, including discrete gradient descent and simulated annealing, as a means of limiting the number of realizations examined.

Ware and Jones (1998) adapt this trial position technique for automated building object displacement for the purpose of graphic conflict resolution between the building objects and road features (the conflict arising due to road symbolization that is often required subsequent to scale reduction). In their work, each of n building objects is assigned k trial positions, giving rise to k^n alternative maps; some of these maps will contain less graphic conflict than others.

4   Web Services in Schematic Map Production

Neun and Burghardt (2005) work highlights the capabilities of web services in web based generalization. Foester and Stoter (2006) have implemented a specification for web processing service for generalization as a proof of concept , to establish the WPS as a research platform for generalization.

A *service* can be considered as a functional "black box", together with a "service contract" – essentially a description of the pre- and post-conditions applicable to that function. Services are the atomic elements of the Service-Oriented Architecture (SOA) paradigm.

Web services are the emerging *de jure* mechanism for achieving interoperability between remotely-hosted data and applications. At the most fundamental level, W3C standardization mandates XML as the data format and HTTP as the transfer protocol.

SOA may be considered to take the longstanding software-engineering practices of loose-coupling and component re-use to their logical extreme. The tightest coupling between components is their service contracts, which are agnostic as regards physical location and execution environment (Windows\Linux\C#\Java\C++ etc). Re-use is facilitated by service composition mechanisms (such as BPEL), in which applications are constructed (possibly dynamically) by orchestrating sequences of service invocations.

Layered on top of W3C standardization, the Open Geospatial Consortium (OGC) specifies a range of services that facilitate the exchange and processing of geospatial data. The OCC Web Map Service (WMS) is predominantly concerned with raster maps, whereas the Web Feature Service (WFS) is concerned with collections of geometric features. The Web Processing service (WPS) provides for generic transformations, without prior restriction on the data formats it can process.

A schematizing web-service enjoys the advantages of web services in general: it can be made available for discovery (e.g. via a standard OGC cataloging mechanism) and incorporated directly into a workflow, whether as an intermediate step or to visualized as a map layer by a WFS client.

In order to be of general applicability, it is desirable that a geospatial web service has a relatively small number of parameters and produces intuitive output for a range of different ground-scales.

In order to meet these criteria, the operation of the schematizing web service therefore differs from the work of (Anand, 2006) in a number of respects:

Grid alignment is performed with respect to a coordinate system derived from the geometry envelope, as opposed to aligning to the input. SRS. The number of grid points is an optional service parameter. It was empirically determined that between 20 and 40 grid points works well for a variety of feature collections and ground scales, and hence the default parameter value is 30 grid points.

Users of the schematization service can elect to trade processing time against output quality via a "schematization quality" parameter having a small number of discrete values. Specific decision points in the optimization algorithm then make use of the quality value, most notably the choice of cooling schedule and its associated parameters.

5   The Algorithm and Software Implementation

The simulated annealing schematic map algorithm is presented in Figure 3. The approach used here is similar to that used by Agrawala and Stolte (2001) to produce sketch route maps. At the start of the optimisation process the algorithm is presented with an initial approximate solution (or state). In the case of the schematic map problem, this is the initial network (line features made up of edges, which in turn are made up of vertices). The initial state is then evaluated using a cost function C; this function assigns to the input state a score that reflects how well it measures up against a set of given constraints. If the initial cost is greater than some user defined threshold (i.e. the constraints are not met adequately) then the algorithm steps into its optimisation phase. This part of the process is iterative. At each iteration the current state (i.e. the current network) is modified to make a new, alternative approximate solution. The current and new states are said to be neighbours. In simulated annealing algorithms the neighbours of any given state are generated usually in an application-specific way. In the algorithm presented here, a new state is generated by the function RandomSuccessor, which works by selecting a vertex at random in the current state and subjecting it to a small random displacement, subject to some maximum displacement distance. This compares to the random displacement methods favoured by Agrawala and Stolte (2001) and Thomas et al (2003) and is in keeping with the random approach inherent to simulated annealing approaches. An alternative method would be to calculate displacements in a way similar to Avelar (2002).

The new state is also evaluated using C. A decision is then taken as to whether to switch to the new state or to stick with the current. Essentially, an improved new state is always chosen, whereas a poorer new state is rejected with some probability p, with p increasing over time. The iterative process continues until stopping criteria are met (i.e. a suitably good solution is found or a certain amount of time has passed or a certain number of iterations have taken place without improvement).

```
    Algorithm SchematicMap
input: Initial, Annealing_Schedule, Stop_Conditions
    D_Current← D_Initial
t←GetInitialTemperature(Annealing_Schedule)
Cost_current=C(Current)
while NotMet(StopConditions)
        New←RandomSuccessor(D_Current)
        Cost_new=C(New)
        ΔE←Cost_current-Cost_new
        if ΔE >0 then
            Current←New
            Cost_current=Cost_new
        else
                    -ΔE/t
            p = e
            r = Random(0,1)
            if (r<p) then
                    Current←New
                    Cost_current=Cost_new
            end
        end
        t←UpdateTemperature(t, Annealing_Schedule)
end
Return(D_current)
```

Fig. 3. Simulated Annealing Algorithm

As with other simulated annealing solutions, at each iteration the probability p is dependant on two variables: ΔE (the difference in cost between the current and new states); and t (the current temperature). p is defined as:

$$p = e^{-\Delta E / t}$$

The variable t is assigned a relatively high initial value; its value is decreased in stages throughout the running of the algorithm. At high values of t higher cost new states (large negative ΔE) will have a relatively high chance of being retained, whereas at low values of t higher cost new states will tend to be rejected. The acceptance of some higher cost new states is permitted so as to allow escape from locally optimal solutions. In practice, the probability p is tested against a random number r ($0 \leq r \leq 1$). If r < p then the new state is accepted. For example, if $p = 1/3$, then it would be expected that, on average, every third higher cost new state is accepted. The initial value of t and the rate by which it decreases is governed by what is called the annealing schedule. Generally, the higher the initial value of t and the slower the rate of change, the better the result (in cost reduction terms); however, the processing overheads associated with the algorithm will increase as the rate of change in t becomes more gradual (Ware et al, 2006).

Based on the characteristics of a schematic map identified, the schematic map algorithm presented here considers seven constraints. Additional constraints can be added if required.

- Topological: The original network and derived schematic map should be topologically consistent;
- Orientation: If possible, network edges should lie in a horizontal, vertical or diagonal direction;
- Length: If possible, all network edges should have length greater than or equal to some minimum length (to ensure clarity);
- Clearance: If possible, the distance between disjoint features should be greater than or equal to some minimum distance (to ensure clarity);
- Angle: If possible, the angle between a pair of connected edges should be greater than or equal to some minimum angle (to ensure clarity).
- Rotation: An edge's orientation should remain as close to its starting orientation as possible
- Displacement: Vertices should remain as close to their starting positions as possible (Anand, 2006)

A software prototype has been implemented based on the simulated annealing algorithm described in the previous section. The algorithm is implemented in Java as a Web Feature Service. A Java-client application (Figures 4 and 5) has also been created to act as a workbench for demonstration and research purposes.

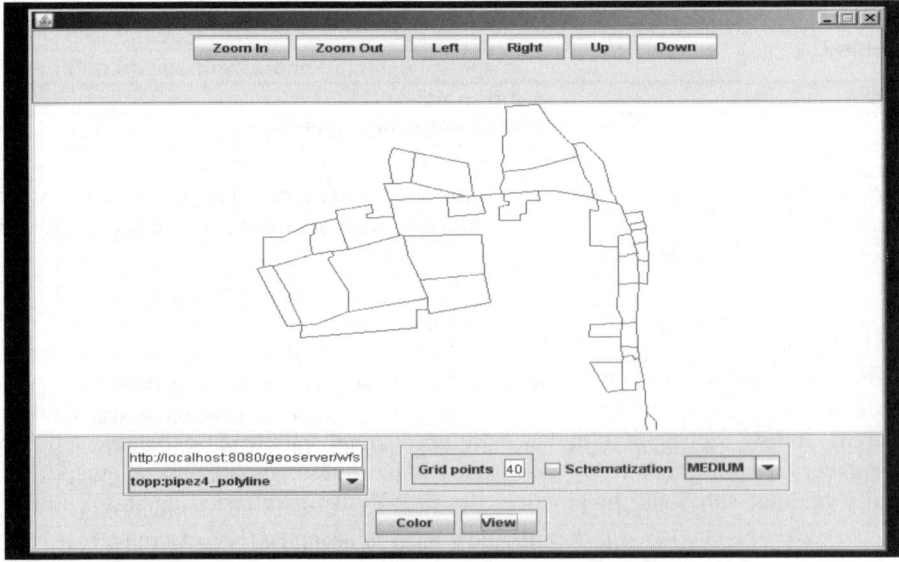

Fig. 4. Client application showing original featureset

The client application allows a user to specify the URL of remotely-hosted data from an external WFS source. In the present implementation, input feature geometry is required to be of polyline format – an error message is displayed if the input data does not conform. As discussed above, the user can additionally select the number of grid points and the schematization quality. These parameters are then passed to the schematizing feature service, which obtains the feature collection from the remote datasource and transforms them using the above algorithm. When optimization is complete, the schematized features are transferred to the client for visualization.

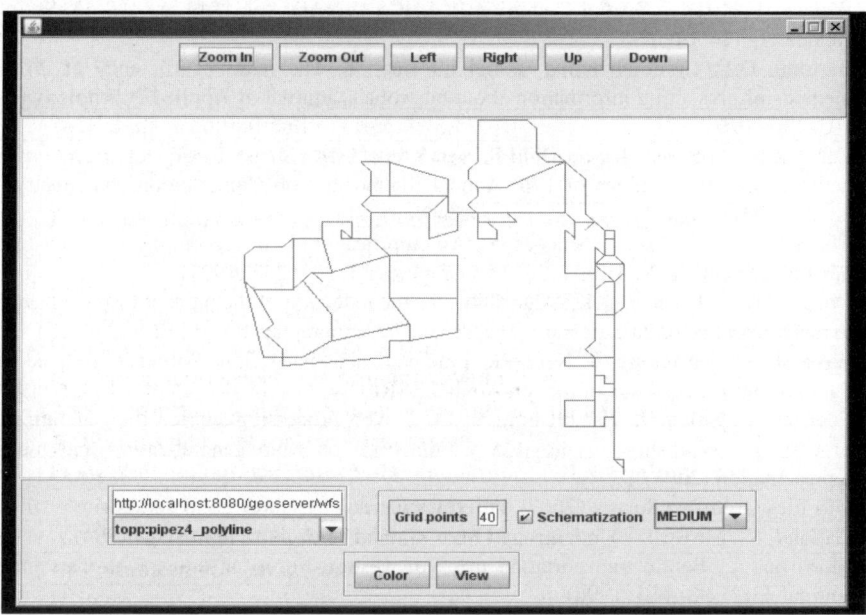

Fig. 5. Client application showing the results of schematization

6 Conclusions and Future Work

This paper looks into the application of web services in generating schematic maps. Despite the immense potential of a schematizing web-service, there are very few examples of actual implementations and this study has been successful in developing tools for the same. A web based schematization tool will be of practical use to many application domains like transportation, water network modelers, utility engineers etc. A prototype Simulated Annealing technique has been developed to derive schematic map from various polyline datasets for web service based applications. The results are promising but more work need to be done in refining the process. Future work will concentrate on refining the technique through the use of additional constraints to enhance visualization and usability.

References

1. Avelar, S.: Schematic Maps On Demand: Design, Modeling and Visualization, Zurich. Swiss Federal Institute of Technology (2002)
2. Anand, S.: Automatic derivation of schematic maps from large scale digital geographic datasets for MobileGIS, PhD Thesis. University of Glamorgan (2006)
3. Agrawala, M., Stolte, C.: Rendering effective route maps: improving usability through generalization. In: Proceedings of SIGGRAPH 2001, Los Angeles, pp. 241–250 (2001)
4. Barkowsky, T., Latecki, L.J., Richter, K.-F.: Schematizing maps: Simplification of geographic shape by discrete curve evolution. In: Habel, C., Brauer, W., Freksa, C., Wender, K.F. (eds.) Spatial Cognition II. LNCS (LNAI), vol. 1849, pp. 41–53. Springer, Heidelberg (2000)
5. Bartram, D.J.: Comprehending spatial information: The relative efficiency of different methods of presenting information about bus routes. Journal of Applied Psychology 65(1), 103–110 (1980)
6. Cabello, S., de Berg, M., van Dijk, S., van Kreveld, M., Strijk, T.: Schematization of road networks. In: Proceedings of 17th Annual Symposium on Computational Geometry, pp. 33–39 (2001)
7. Christensen, J., Marks, J., Shieber, S.: An empirical study of algorithms for point-feature label placement. ACM Transactions on Graphics 14, 203–232 (1995)
8. Douglas, D.H., Peucker, T.K.: Algorithms for the reduction of the number of points required to represent a line or its caricature. The Canadian Cartographer 10(2), 112–122 (1973)
9. Elroi, D.S.: Designing a Network Linemap Schematization Software Enhancement Package. In: Proceedings of the 8th Annual ESRI User Conference, Redlands, CA (1988)
10. Foerster, T., Stoter, J.: Establishing an OGC Web processing service for generatlization process. In: Workshop of the ICA Commission on Map generalization and multiple representation (2006)
11. GIS files Ordnance Survey (2007), http://www.ordnancesurvey.co.uk/oswebsite/gisfiles/
12. Garland, K.: Mr Beck's Underground Map. Capital Transport Publishing (1994)
13. Morrison, A.: Public transportation maps in western European cities. The Cartographic Journal 33(2), 93–110 (1996)
14. Monmonier, M.: How to lie with maps. The University of Chicago Press (1996)
15. Neun, M., Burghardt, D.: Web services for an open generalization research platform Workshop of the ICA Commission on Map generalization and multiple representation (2005)
16. Saalfeld, A.: Topologically consistent line simplification with the Douglas-Peucker algorithm. Cartography and GIS 26(1), 7–18 (1999)
17. Stott, J.M., Rodger, R.: Automated Metro Map Design Techniques. In: Proceedings of the 22nd International Cartographic Conference, International Cartographic Association (2005)
18. Thomas, N., Ware, J.M., Jones, C.B.: Continuous search space and building grouping: improvements to the simulated technique. In: 5th ICA Workshop on Progress in Automated Map Generalisation, Paris (2003)
19. Tversky, B., Lee, P U.: Pictorial and verbal tools for conveying routes. In: Freksa, C., Mark, D.M. (eds.) COSIT 1999. LNCS, vol. 1661, pp. 51–64. Springer, Heidelberg (1999)
20. Ware, J.M., Jones, C.B.: Conflict Reduction in Map Generalization Using Iterative Improvement. Geoinformatica 2(2), 383–407 (1998)
21. Ware, J.M, Anand, S., Taylor, G.E, Thomas, N.: Automatic Generation of Schematic Maps for Mobile GIS Applications. Transactions in GIS 10(1), 25–42 (2006)

A Continuous Spatial Query Processing for Push-Based Geographic Web Search Systems

Ryong Lee[1] and Yong-Jin Kwon[2]

[1] Samsung Advanced Institute of Technology,
P.O. BOX. 111, Suwon, Kyunggi, Korea
ryong.lee@samsung.com
[2] Department of Telecommunication and Information Engineering,
Korea Aerospace University, Koyang, Kyunggi, Korea
yjkwon@tikwon.hangkong.ac.kr

Abstract. With the explosive growth of the web, people often need to monitor fresh information about their areas of interest by browsing the same sites repeatedly. Especially, even for a local area, so many pages are created every day that might be a greatly helpful knowledge in daily decision supports. In order to reduce such monitoring work and not to miss chances to meet critical information for a local area, we are developing a **Continuous Geographic Web Search System** with *Push-based* web monitoring services. This paper will describe the system architecture to deal with multiple user queries to represent users' geographic attention over multiple data pages incoming from geographic web crawlers. If a newly incoming data page is relevant to a pre-registered query, the user who previously registered the query will be informed spontaneously about the new information by our notification service. This paper especially focuses on the problem of how to match multiple data pages and multiple queries as quickly as possible. For the purpose, we will propose a fast matching algorithm based on a spatial join and show a primitive experimental result with synthesis data.

Keywords: Continuous Query Processing, Geographic Web Search Systems.

1 Introduction

The integration of Web Information Systems (WIS) and Geographic Information Systems (GIS) has been very attractive because of their interesting applications. Especially, for the purpose of geographic web search and mining, we have studied critical issues of the integration of WIS and GIS [11, 10, 9] and developed an integrated search system that can explore local information of interest on the web. Recently, people have been faced with the new problem of the explosively ever-increasing size of the web, even for a local area. It is very hard for personal users to monitor and analyze for their geographic interests because the information exploration method of current web search engines is mainly based on the *Pull-based* approach. That is, users need to represent their query explicitly with keywords and

J.M. Ware and G.E. Taylor (Eds.): W2GIS 2007, LNCS 4857, pp. 227–238, 2007.

submit them to search engines to get results every time when they want to find information. In order to reduce this boring work to monitor newly updated contents on web pages, the *Push-based* approach has been widely used. In contrast, with the *Pull-based* approach, a server needs to perform an active and iterative search task instead of users following a pre-registered query, which can define general search terms, trigger conditions and notification methods [19, 21]. While a *Pull-based* query can ask for past and current data from search engines, a *Push-based* query can ask for future data that will appear on the web. This will greatly reduce users' efforts to devote their attention to monitoring updated information repeatedly, since users will be informed about updated information on the web adaptively when the information appears and is related to the users' specified interests. However, major efforts of the *Push-based* web search systems [14, 15, 12, 6] have focused on extending traditional Relational DBMS to deal with a continuous query for monitoring, not for ceaselessly updating web resources and notifying users about updates when they are available and satisfying user specifications.

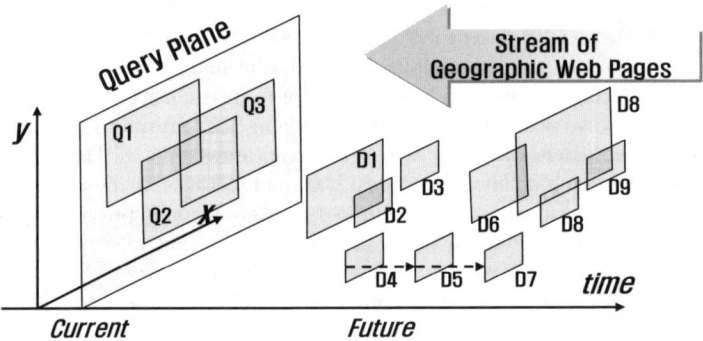

Fig. 1. Continuous Query Processing over Stream of Geographic Web Pages

In this paper, we propose a **Continuous Spatial Query Processing** for Geographic Web Search Systems, where users can register continuous queries for newly updated web information about a geographic area of interest. In order for this extension in geographic web search systems, we also introduce our model to deal with continuous spatial queries for *Push-based* services. To monitor newly created web content about a geographic area, the system should be able to crawl web content as new as possible and needs to monitor whether the incoming stream of crawled content satisfies the conditions of pre-registered queries. For these goals, we applied a continuous query processing approach popularly adopted for stream databases [3, 8, 20]. Their query types, however, focused mainly on search terms and temporal dimensions. That is, in their proposals, users can register a query such as '*notify me if there are new pages including the keyword <Seoul> for the next week.*' The geographic area of interest still has to be given in keyword terms in ambiguous forms. However, a geographic scope represented by keyword terms is hard to specify its exact location or area and difficult to use geographic relationships such distance or other topological relations such as *near* or *inclusion*. This incurs poor quality search results since they could not use *map semantics* such as a spatial relationship between page contents and their real relevant geographical scope.

To improve this problem in geographic web searches, we proposed a method to compute the geographic scope of a page in a geometric representation on a map. For that, geographic keywords such as building names (geowords) that appear in web content are transformed into locations or areas with the help of map semantics such as geowords-to-area correspondence on the map. This approach could greatly reduce false negatives or false positives, since it uses map semantics to replace the keyword matching problems by relatively logical and reasonable geometric relationships. Thus, in our web stream monitoring, we use a geographic scope as a user query for Push-based Geographic Web Searches. However, the simple geographic scope comparison between users' queries and incoming web pages will have another problem if the system has to deal with a lot of queries concurrently. Generally, tens of thousands or millions of queries are pre-registered by a large number of users. For new updated pages crawled by the system, the page should be compared with every pre-registered query in their geographic scopes. But the number of scope comparisons for every incoming new page is still the same with the number of the queries. If the number of queries is large, the response time to meet the user requests would be degraded. In order to solve the problem, stream database systems usually adopt a query indexing method to reduce the time to filter out unrelated queries. In our system, we will also index continuous spatial queries registered by users in a two-dimensional index. In our research, incoming pages are also indexed using the two-dimensional index. Instead of finding each page's corresponding queries using the query index, a instant data index made by a group of pages in a short time will be used to take advantage of reduction in comparing the scopes in an efficient manner.

The remainder of this paper is organized as follows. Section 2 describes the system architecture of *Push-based* Geographic Web Search Systems and reviews our previous work to compute the geographic scope of web pages using web content analysis. Section 3 explains our strategy to match incoming data streams of geographic web pages with continuous spatial queries pre-registered by users. Section 4 presents the continuous spatial query processing algorithm with an improved spatial join method. Section 5 shows an experimental result to evaluate a proposed algorithm using synthesis data. Section 6 concludes the paper and discusses some future work.

2 Push-Based Geographic Web Search System

Before the detailed discussion later, we introduce the system architecture for continuous geographic web searches and review our research as to how a web page including geowords is projected on a map.

2.1 System Architecture

The proposed system we have been developing has five major parts shown in Fig. 2 in terms of their functionalities. At first, in the part of *User Clients*, users can specify their geographic area of interest, instead of geowords, on a graphic user interface having a map viewer as shown in Fig.3. On the interface, an interested region can be specified by drawing a rectangular area. With push-based query registration, the client will also support Pull-based queries to search information over previously stored databases. Next, in the ***Registry Part***, queries made by user clients are sent to this part and managed in a user/query database. In contrast to clients, this part can exist as a separated server. This part manages user accounts and extant queries that are out-dated according to the period of user specification. In the ***Monitoring Part***, the computation of matching incoming pages and pre-registered queries is resolved in this part. Actually, there are two kinds of index: the first, the 'query index,' has query elements from the Registry part. The second is called 'data index,' and it simply manages the geographic scopes of incoming pages. The data index is also composed of the 'instant data index' and the 'regular data index.' While the instant index is for streamed data that have recently been imported in the system, the regular index manages the geographic scopes all the data accumulated in local web databases. In order to answer *Push-based* queries, we use the instant data index and query index to find matches. But, for *Pull-based* queries, the regular index is usually used to search for past data in the databases. In the ***Crawling Part***, for the purpose of geographic web searches, the system has a multi-threaded crawling engine that can visit many web sites concurrently. Especially, in order to find geographic web pages that are relevant to our target geographic area, we use some heuristic methods to determine the geographic scope of each site, as is discussed in the next subsection. If a page is relevant to a target area, the page will be revisited at the scheduled time. For the scheduling of revisits, we will ignore the details because that is not the main focus of this paper. However, managing web resources as quickly as possible is also an important issue in continuous geographic web search systems.

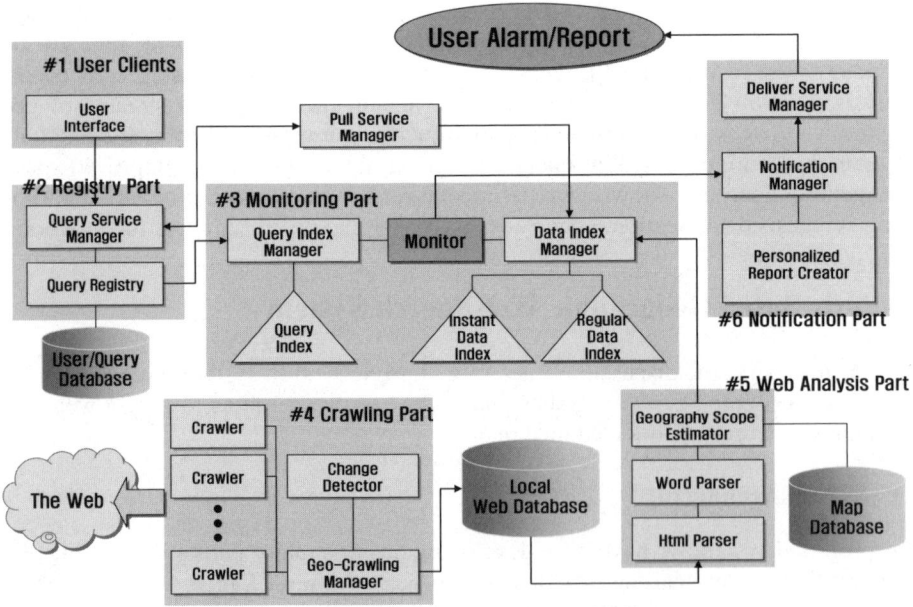

Fig. 2. System Architecture

In the **Web Analysis Part**, incoming web pages from above the crawling part must be analyzed to know the new links and geowords in the contents. The links are extracted to extend the list of target sites and given to the crawling part for further web navigation. The geowords are used to compute a representative geographic scope of a page. In order to find geowords in a page, we need a list of geowords for a target geographic area and each geographic scope as a geographic location or a range on the map. Actually, we made a list using a map made by the 'National Geographic Information Institute' of Korea. For a complex type of geographic area such as 'town' or 'ward,' we simply represent its area as an MBR covering the relevant polygons on the map. The resulting list is on the map database as a table that has fields for 'geoword' and 'scope of coverage.' Finally, in the **Notification Part,** if the monitoring part finds matching cases over the two indexes, it sends an alarm to this part with the pairs of (relevant page IDs, relevant query IDs). This means that found pages should be sent to a user who has registered the matching query. For the users' convenience, we create a report to summarize the newly appearing pages in a document. In the notification, actually, reports must be made to the same number of users. However, if two or more users have the same set of queries, the system can reduce an effort to make the same reports.

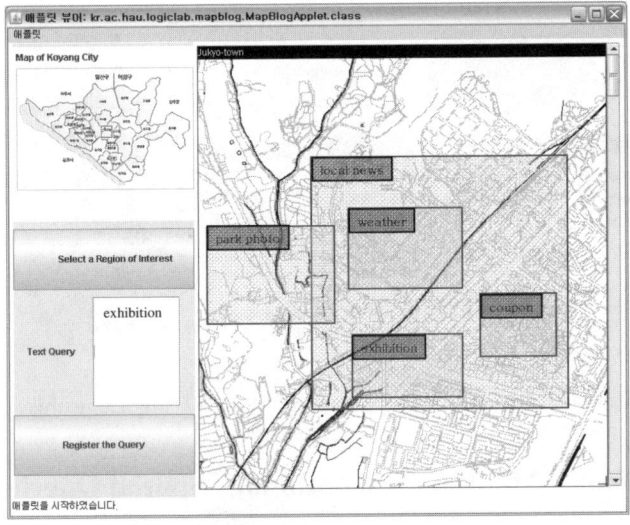

Fig. 3. A User Interface for Map-based Query Registration

2.2 Computation of Geographic Scope of a Web Page

In order to compute the geographic area of a web page, some interesting approaches have used. A common problem is the deficiency of map semantics in web contents. In order to estimate geographic scopes, there were two kinds of approaches as follows:

1) Page Contents Analysis [2, 18, 17]: A web page has its own geographic area information in its contents such as addresses, phone number, etc. While this internal

information is very informative to determine what locations are related to, only a few pages include such information.

2) Linkage Analysis [4, 7]: Internal information can be inaccurate, for example [17]: the NY Times site has global coverage, while the name represents only a relatively small local area. The geographic scope of the 'NY Times' would be larger than just the New York area, if we consider a geographic distribution of the users who access the site.

In our previous work[10], we took the first approach of using geowords in web page contents. In order to project a web page on a map, we could set a representative area from geowords in pages, as shown in Fig. 4. In this figure, the geowords on the left page are projected to simple MBRs instead of complex polygons for geographic objects. Then, these projected elementary MBRs are summarized by a larger covering MBR. Actually, there is a problem in the determination of corresponding area for a geoword; if a geoword is mapped to two or more areas (that is, homonymic names such as towns having same names), we need to choose one of them. In our heuristic method, we select the one that has a minimum distance with other geowords appearing in the same page.

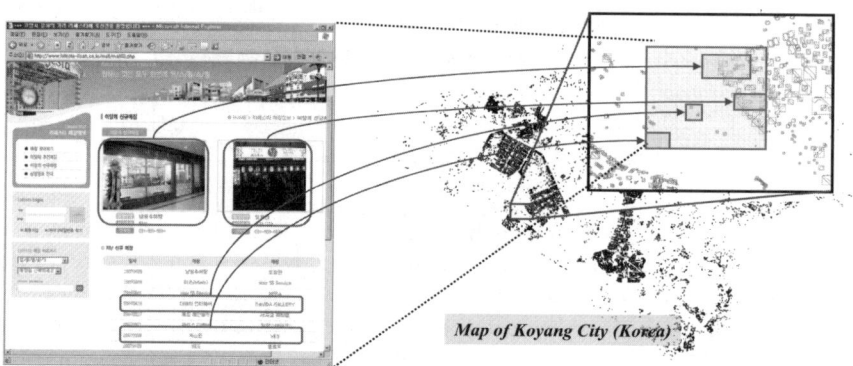

Fig. 4. Computation of Geographic Scope Using Geowords

3 A Fast Search Method for Continuous Spatial Queries

As described in previous sections, our system is targeted to many users on the web. Thus, we must prepare a QoS problem, especially for delays occurring throughout all the processing stages from query registration to final notification. First, each user can make multiple queries on the system. Then, the queries are generally processed one by one when every new page comes into the system. If the total number of queries that should be managed by the system is more than tens of thousands or millions, there would be a linearly increasing computation burden. In order to solve such problems, grouping the registered queries has been proposed[15]. However, the proposed methods [14, 15, 12, 6] focus on non-spatial dimensions about keywords or time domain. In our research, we extend such queries to spatial domains for geographic web searches. Thus, we should consider a method to search for registered queries on the streamed geographic web contents we modeled in this paper.

3.1 Query Processing Strategy

To search for registered queries on incoming pages, we will assume that a registered query is relevant to an incoming page if there is a geographic intersection (overlap) over the geographic scopes. Thus, we need to find the intersecting queries as soon as possible. In our system, we apply a two-dimensional R-Tree[1, 16] to index the registered queries based on spatial relationships. In the index, the search key is a geographic area of an incoming page, and the target values are a set of registered query identifiers that overlap with a given search area. The index is a dynamic structure that can be inserted or deleted according to the lifetimes of the queries. That is, we should manage the index autonomously by deleting due-dated queries. For the period of query lifetime, we represent it as a time window of the start time and the end time. Actually, this is easily adoptable by extending only one dimension of time from 2d spatial to 2d+1d spatio-temporal index. However, the temporal dimension used for the notification period will not be discussed here to focus on the spatial dimension. Thus, we consider only cases of permanent spatial queries on the web.

For query indexing, we could find the user propensity for geographic searches over the map interface; first, the geographic scopes drawn by many users have focused on common areas or buildings that are usually called "Landmarks." Second, the aggregated areas are often different in all their coordinates, since drawing the area of interest on the map by using a rectangle often has a broader range rather than keywords-based query specification. That is, the keyword specification, the data are distinct and discrete. But coordinate specifications are somewhat imprecise and relatively continuous data. For these kinds of properties on the query set, we need to develop a practical method to find their common range of interest as quickly as possible.

Likewise, a geographic scope registered by a user is inserted into this tree as an element. Basically, we assume that all scopes have different coordinates, but are highly overlapped with another. The query index manager in Section 2 has this index and performs a search traversal over the tree. In the other side, we also represent a geographic scope of a page simply as an MBR; the query index manager needs to perform a range search that travels multiple paths from the root to intersecting elements. However, for every incoming page or a group of pages, finding overlapped queries still has high costs. In order to reduce the computation of overlapping comparisons between many geographic scopes of registered queries and geographic scopes of concurrently incoming pages (by multi-threaded crawlers), we need a much faster comparison method. This problem can be converted into the Spatial Join problem in database systems. In order to reduce the complexity of join computation, we take two approaches as follows:

1) Group search over a query index

If we want to monitor incoming pages concurrently, searching at the level of the bulk of the pages can have advantages of reducing the cost to do common search tasks. For example, if we group many pages into one MBR as their overall geographic scope, this united area can restrict the search area in the query index, often into a smaller intersecting search area. For this approach, we introduce a data indexing method for geographic scopes of incoming pages, already named the 'instant data index' in Section 2.

2)   Using the overlapping condition to reduce a traversal in the R-Tree
As we described above, we found that user queries often have a commonality with each other. If we deal with such information over the R-tree and use them over the tree traversal, this can greatly reduce the traversal efforts and time. This means that the common area should be indexed differently.

3.2   Indexing for Multiple Spatial Data

Concurrent queries that should be answered as possible as soon are processed in a grouped form. We first make an index of multiple data pages. Each page is represented as a geographic scope by an MBR. These multiple pages are also grouped with an R-tree as a data index. Contrary to the above query index, this index is relatively smaller in its size, since we index some pages for a short time. This data index is compared with the query index to find corresponding queries. The grouped index has an abstract form rather than its elements. The root node of the data index represents the whole geographic range of such incoming pages in a short time. Instead of a distinctive comparison, this abstract coverage can reveal relevant queries in a short time. If there is an overlap between the abstract coverage and the query index, the child elements of the query index are retrieved. Next, the retrieved queries are also used to select which data contribute to the answers. Finally, pairs of incoming data and their relevant query set are made as a result. This data index is then trashed in the memory to load the next incoming data.

3.3   Indexing for Multiple Overlapped Spatial Queries

In this subsection, we propose an index structure to improve a spatial search for multiple data streams over multiple queries. In order to index multiple queries, we adopted an R-Tree index, which is often used in spatial databases. In the index, registered queries by users become elements in the lowest level. However, different from the conventional R-Tree, we should consider the characteristics of user queries, which are highly overlapped. To make the index faster, we can use such a pattern of overlapping. If a non-leaf node has elements that have a commonly intersected part as in Fig. 5, the intersected area could be a key to determine whether the child elements in a node are all relevant to compare a data area.

If a data area intersects (or covers) this area, it is certain that all the child nodes are relevant without computing the intersection. This can reduce the comparing time between the data index and the highly-overlapped query index. To embed the intersect information in the R-Tree, we extend nodes of the tree to have an intersected MBR in the header of every node. For the common intersection, we define an Common MBR as $Common\_MBR = Q_1 \cap Q_2 \cap Q_3, \cdots, Q_k$ (k: the number of child elements in a node). However, the higher the number of elements, the lower the possibility of overlaps. In contrast, the fewer the number of elements, the less the effect of computation reduction. In our assumption that the registered user queries are highly overlapped, we could take advantage of this new criterion instead of the original R-Tree's Union representation.

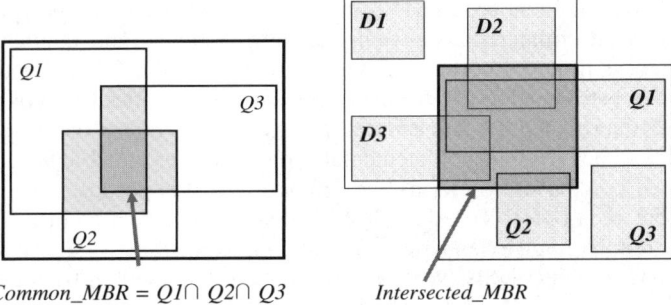

$Common\_MBR = Q1 \cap Q2 \cap Q3$ $Intersected\_MBR$

Fig. 5. Common_MBR   **Fig. 6.** Intersected_MBR

Algorithm RJ_Recursive(TypeNode *RN1*, TypeNode *RN2*)

01. **if** *RN1* is leaf node **and** *RN2* is leaf node
02.   **if** *RN1.mbr* ∩ *RN2.mbr* ≠ φ
03.   **output**(e_1, e_2)
04. **else if** *RN1* is leaf node **and** *RN2* is non-leaf node
05.   **foreach** child_node∈RN2
06.   Call **RJ_Recursive**(*RN1*, *child_node*)
07. **else if** *RN1* is non-leaf node **and** *RN2* is leaf node
08.   **foreach** child_node∈*RN1*
09.   Call **RJ_Recursive**(*child_node*, *R2*)
10. **else if** *RN1* is non-leaf node **and** *RN2* is non-leaf node
11.   *intersected_mbr = RN1.mbr* ∩ *RN2.mbr*
12.   **if** *intersected_mbr = φ*
13.   return
14.   **foreach** *r2_child_node∈RN2*
15.   **if** *intersected_mbr* ∩ *RN2.common_mbr* ≠ φ
16.   **if** *intersected_mbr* ∩ *r2_child_node.mbr =φ*
17.   **continue**
18.   **endif**
19.   **endif**
20.   **foreach** *r1_child_node∈RN₁*
21.   **if** *intersected_mbr* ∩ *RN1.common_mbr* ≠ φ
22.   **if** *intersected_mbr* ∩ *r1_child_node.mbr =φ*
23.   **continue**
24.   **endif**
25.   **endif**
26.   Call **RJ_Recursive**(*r1_child_node, r2_child_node*)
27.   **endfor**
28.   **endfor**

Fig. 7. Overlap-based Spatial Join Algorithm

4   A Fast Algorithm for Overlap-Based Spatial Join

As described above, we need to find matched pairs of the geographic scopes of the data and the queries over *Push-based* stream systems together. If we do not use the data index and the query index, each data element should be traversed over the index. It will take a linear time over the number of incoming data and the number of registered queries. If there are m pages incoming and there are n queries not indexed, we need an m* n comparison between the data elements and queries. In order to reduce the access time to queries, indexing queries with the R-Tree would be a good solution. But we still need to ask for the query index in m times for m pages. Thus, we also grouped data using the same index to ask for the query index in rather large representative ones. Instead of elementary data, the grouped data can share the information of comparing MBRs in two indexes. For this purpose, we first applied a spatial join between two R-Trees[5, 13] to deal with our matching problem. However, we need to modify and extend the original join algorithm for the case when two indexes have different heights. We also use the heuristic method by Intersected MBR shown in Fig. 6, where the intersected area can determine the next comparison of intersected elements in each MBR as D2/D3 and Q1/Q3. In this figure, D1 and Q3 are excluded because there is no overlap with the intersected MBR. Furthermore, in our extension, nodes in the R-Tree have a common MBR, as described in Section 3.3. This common area in a node can reduce the comparisons over all the child elements Q1~Q3, if a search range intersects this area. Finally, we describe the overlap-based spatial join algorithm in Fig. 7.

5 Experiments and Evaluation

In order to evaluate our overlap-based spatial join algorithm in various distributions of data and queries, we worked on a synthesis data set. We compared three kinds of searches; 1) *Linear*: we don't use either index and only compare above m * n pairs. 2) *One-by-One*: we index only the queries and ask *m* times for *m* data pages. 3) *Common_MBR*: this is the result of the proposed join algorithm. In our experimental setting, the data or queries exist in a two-dimensional area of U={(x,y):0≤x≤500, 0≤y≤500}. Each datum or query is given as an MBR in a range from 100 to 200 together, since we want to look at the overlap effect. In order to test an objective data set, we generated two random sets for data and queries. We control only the number of elements in each index and the number of child items in the R-Tree.

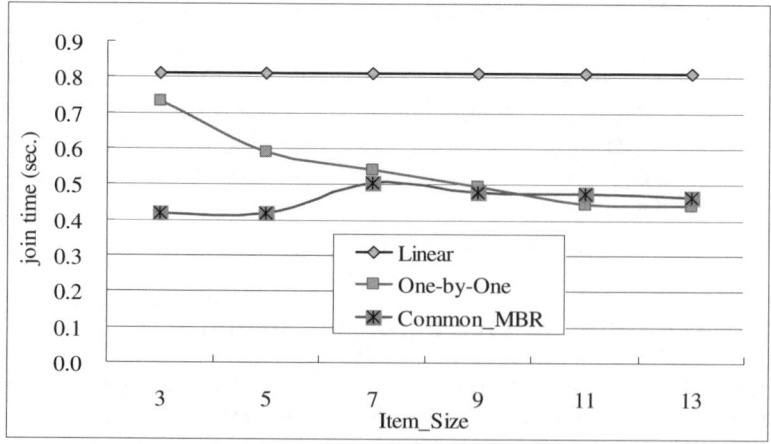

Fig. 8. Experimental Results

From iterative tests, we had some interesting results to increase the effect of spatial join; first, in our algorithm, the possibility of which Common_MBR would occur in elements in a node of the R-Tree will decrease if the number of nodes increases. However, the reduction effect to survey other nodes on behalf of the Common_MBR will increase if the number of nodes increases. These two conditions should meet in a trade-off point to maximize the join reduction. Furthermore, the ratio of join reduction is proportional to a similarity of heights of the two indexes. That is, in each index, the reduction has an impact on the other index. We could have an experimental result shown, as in Fig. 8; with increasing item_size, the join time of one-by-one cases improves, not under the Common_MBR. As we said, the large item_size can't have the benefit of Common MBR. But in small item_sizes or highly overlapped queries, join time would be greatly reduced to more than half the time of linear comparison.

6 Conclusion and Future Work

In this paper, we propose a new kind of geographic web search system to deal with *Push-based* monitoring. Especially, as the most important function in such systems, we introduce a comparison problem over multiple queries and multiple data pages. We also present a spatial join algorithm to reduce the time to find matched pairs and show its effects in an experiment. As future work, we will study a method to reduce spatial joins for highly overlapped data and to maintain a local web database as fresh possibilities by a well-scheduled revisiting strategy for the web.

Acknowledgments. This research was partly supported by the Internet information Retrieval Research Center(IRC) in Korea Aerospace University, and was partly supported by 2005 Korea Aerospace University faculty research grant.

References

1. Antonin, G.: R-TREE: A Dynamic Index Structure for Spatial Searching. In: Proceeding of ACM SIGMOD, pp. 47–57 (1984)
2. Arikawa, M., Okamura, K.: Spatial Media Fusion Project. In: Proceeding of Kyoto International Conference on Digital Libraries: Research and Practice, pp. 75–82 (November 2000)
3. Babcock, B., et al.: Models and Issues in Data Stream Systems. In: Babcock, B., et al. (eds.) Proc. The 21st ACM SIGACT-SIGMOD SIGART Symp. on Principles of Database Systems (PODS), pp. 1–16. Madison, Wisconsin (June 2002)
4. Buyukkokten, O., Cho, J., Garcia-Molina, H., Gravano, L., Shivakumar, N.: Exploiting geographical location information of web pages. In: Proceeding of the ACM SIGMOD Workshop on the Web and Databases, WebDB (1999)
5. Brinkhoff, T., Kriegel, H.-P., Seeger, B.: Efficient Processing of Spatial Join Using R-trees. In: Proc. Int'l Conf. on Management of Data, ACM SIGMOD, Washington, DC, pp. 237–246 (May 1993)
6. Chen, J., et al.: NiagaraCQ: A Scalable Continuous Query System for Internet Databases. In: Proc. Int'l Conf. on Management of Data, ACM SIGMOD, Dallas, Texas, pp. 379–390 (June 2000)

7. Ding, J., Gravano, L., Shivakumar, N.: Computing Geographical Scopes of Web Resources. In: VLDB 2000, pp. 545–556 (2000)
8. Golab, L., Ozsu, M.T.: Issues in Data Stream Management. ACM SIGMOD Record 32(2), 5–14 (2003)
9. Kwon, Y.J., Kim, D.H.: Mobile SeoulSearch: A Web-based Mobile Regional Information Retrieval System Utilizing Location Information. In: Kwon, Y.-J., Bouju, A., Claramunt, C. (eds.) W2GIS 2004. LNCS, vol. 3428, pp. 206–220. Springer, Heidelberg (2005)
10. Lee, R., Shiina, H., Tezuka, T., Yokota, Y., Takakura, H., Kwon, Y.J., Kambayashi, Y.: Map-Based Range Query Processing for Geographic Web Search Systems. Digital Cities, 274–283 (2003)
11. Lee, R., Takakura, H., Kambayashi, Y.: Visual Query Processing for GIS with Web Contents. In: The 6th IF Working Conference on Visual Database Systems, pp. 171–185 (May 29-31, 2002)
12. Leopold, J.L., Heimovics, M., Palmer, T.: Webformulate: a web-based visual continual query system. In: WWW 2002, pp. 221–231 (2002)
13. Lim, H., Lee, J., Lee, M., Whang, K., Song, I.: Continuous query processing in data streams using duality of data and queries. In: Proceedings of the 2006 ACM SIGMOD international Conference on Management of Data, June 27 - 29, Chicago, IL, USA (2006)
14. Liu, L., Pu, C., Tang, W.: WebCQ: Detecting and Delivering Information Changes on the Web. In: CIKM 2000. The Proceedings of International Conference on Information and Knowledge Management, ACM Press, Washington, D.C. (November 7-10, 2000)
15. Liu, L., Pu, C., Tang, W., Han, W.: CONQUER: A continual query system for update monitoring in the WWW. International Journal of Computer Systems, Science and Engineering (1999)
16. Manolopoulos, Y., Nanopoulos, A., Papadopoulos, A.N., Theodoridis, Y.: Rtrees: Theory and Applications. Series in Advanced Information and Knowledge Processing. Springer, Heidelberg (2005)
17. Matsumoto, C., Ma, Q., Tanaka, K.: Web Information Retrieval Based on the Localness Degree. In: Hameurlain, A., Cicchetti, R., Traunmüller, R. (eds.) DEXA 2002. LNCS, vol. 2453, pp. 172–181. Springer, Heidelberg (2002)
18. McCurley, K.S.: Geospatial Mapping and Navigation of the Web. WWW10 (2000)
19. Paton, N.W., Díaz, O.: Active database systems. ACM Comput. Surv. 31 1, 63–103 (1999)
20. Tao, Y., Yiu, M.L., Papadias, D., Hadjieleftheriou, M., Mamoulis, N.: RPJ: Producing Fast Join Results on Streams through Rate-based Optimization. In: SIGMOD Conference 2005, pp. 371–382 (2005)
21. Widom, J., Ceri, S.: Active Database Systems. Morgan Kaufman, Seattle (1996)

OGC Web Processing Service Interface for Web Service Orchestration

Aggregating Geo-processing Services in a Bomb Threat Scenario

Beate Stollberg and Alexander Zipf

i3mainz - Institute for Spatial Information and Surveying Technology,
University of Applied Sciences FH Mainz,
Holzstraße 36, Mainz, Germany
{stollberg, zipf}@geoinform.fh-mainz.de

Abstract. The aggregation of web services in order to achieve a common goal is a basic concept in *Service Oriented Architectures* (SOA). This paper demonstrates *OGC Web Services* (OWS) aggregation based on a simulated bomb threat scenario which is a possible disaster management example. Web services are aggregated based on the OGC *Web Processing Service* (WPS) interface. Here the concept of a "Composite-WPS" is introduced and used to invoke all other services involved. The presented approach is compared to the use of BPEL which is an orchestration language with some technical pitfalls when applied to the orchestration of OWS.

Keywords: geo-processing, geographical web services, orchestration, Web Processing Service (WPS), service chain, composition, SDI, disaster management.

1 Introduction

Spatial Data Infrastructures (SDIs) support discovery, access, and use of geographic information in the decision-making process [1]. Traditionally, *Geographic Information Systems* (GIS) have been used for the extraction of information from spatial data in order to answer spatial questions. Today, the transition from the traditional monolithic GIS to an interoperable *Service Oriented Architecture* (SOA) is taking place. Increasingly research on the consequences of this paradigm shift is carried out, e.g. in resolving some of the key interoperability problems among GI services [2] or the chaining of these services [3,4]. SDIs are based on the assumption that users are usually not interested in data but in a piece of information that can be generated by Geographic Information (GI) services which are main components of SDIs [5]. Therefore, Spatial Data Infrastructures provide an important basis in the field of disaster management where geographic information must be discovered, processed and visualized quickly to provide critical assistance in emergencies and in support decision makers and rescue workers [6].

This is also one of the main goals within our project "OK-GIS". OK-GIS is the acronym for "Open Disaster Management using free GIS" (http://www.ok-gis.de).

J.M. Ware and G.E. Taylor (Eds.): W2GIS 2007, LNCS 4857, pp. 239–251, 2007.
© Springer-Verlag Berlin Heidelberg 2007

Within OK-GIS we aim at the development of a flexible toolbox for the administration, application, visualization and mobile creation of geodata using standard based services in a disaster management environment. Basic principles are derived from the *Open Geospatial Consortium* (OGC) standardized *OGC Web Services* (OWS) such as *Web Map Service* (WMS), *Web Feature Service* (WFS), *Web Coverage Service* (WCS), and *Catalog Service for the Web* (CSW). Furthermore, OK-GIS incorporates services according to the OpenLS OGC initiative, including the *Route Service* (RS) [7], which was extended into a 3D Emergency Route Service (3D-ERS) [8]. The *OpenLS Presentation Service* and the OpenLS *Utility Service* (Geocoder / Reverse Geocoder) were also implemented and the *OpenLS Directory Service* is currently under development.

Visualization, vector and raster data access, along with the ability to search for spatial data is mostly covered by the standards mentioned above. However, sophisticated and recognized standards for distributed spatial data processing, leading to user specific information preparation, are still missing [9] and were taken into consideration in an OGC discussion paper that recommended a *Web Processing Service* (WPS) specification via a standardized interface.

The described WPS interface specification is also used within the OK-GIS project for realizing example scenarios in the field of disaster management. This will be demonstrated in this paper on a specific example - finding a bomb. A first version of this bomb scenario which was based on [10] has already been discussed by [11,12] concerning the orchestration of existing OWS using the *Business Process Execution Language* (BPEL). In this article the scenario was elaborated and focuses on the geo-processing aspect using the OGC WPS interface. While [11,12] only presented this aspect in form of a "black box" through calls of generic WPS processes that were specified in detail, we will discuss the interaction of several processes and present the implementation of the scenario using the existing WPS framework [13] of the *deegree* project (http://www.deegree.org). Since [12] uncovered some technical problems when chaining OCG Web Services with BPEL we will introduce and discuss a new approach for composing OWS without an orchestration language but rather by using the WPS interface itself.

2   OGC Web Processing Service

The idea behind a WPS is to offer any kind of GIS processing functionality. It may provide simple calculations (e.g. the calculation of a buffer) as well as complex computations (e.g. the generation of a climate model). Thus, in principle there are no restrictions on what can be implemented based on the WPS interface.

A WPS is able to handle more than a single process and there are three mandatory operations performed by a WPS, namely *GetCapabilities*, *DescribeProcess* and *Execute*. The response to a *GetCapabilities* request is an XML-document containing metadata of the WPS and all available processes. A detailed process description as well as input and output parameters are provided for every process as response to a *DescribeProcess* request, also in form of a XML-document. The final process execution is carried out when an *Execute* request is send to the WPS.

3 Example: Search for Adequate Evacuation Shelters in a Bomb Threat Scenario

In our scenario a bomb was found. This can be either an "old" bomb from World War II or a "ticking" bomb from a current attack. All persons within a specified radius around the bomb, namely within the "danger zone", must be evacuated. A second, larger buffer must also be defined around the bomb, namely the "security zone": Within this security zone only persons inside buildings of special interest (such as kindergartens, schools, etc.) will be evacuated. The scenario is illustrated in figure 1.

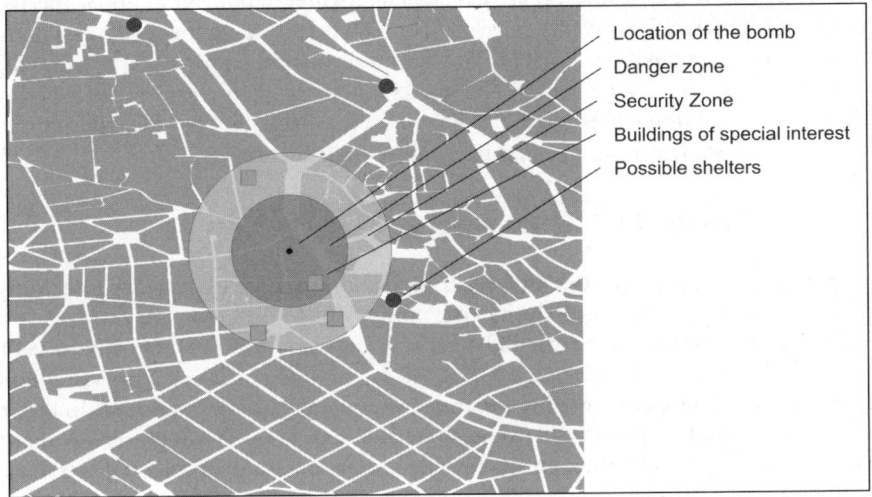

Fig. 1. Controlled Areas around a discovered Bomb

First the danger and the security zones must be defined. This is done by using the well-known GIS base function "Buffer", which is available as a WPS process. In order to calculate the number of persons to be evacuated, population numbers are available per building block. The building block information has to be clipped with the danger zone and then the population information has to be summed up. This is achieved by using the WPS process *PolygonIntersectsPolygonJoinAggregation*. Additionally the number of persons within the security zone who are inside buildings of special interest must be calculated. These building and respectively the number of people inside are provided by a WFS. Using the WPS process *PointInPolygonJoin* all the buildings of special interest inside the security buffer zone are determined. Then, by using the WPS process *Aggregation*, the number of persons inside those buildings is calculated [14] (we assume here that the needed data is available or can be found through a search in the CSW and retrieved from a WFS). On top of this, a WFS provides information on possible shelters, where the evacuees can be safely assembled together, with all relevant attributes for these shelters such as address and building capacity in order to find the best suitable shelter. A buffer around the bomb area is calculated and then a decision can be made where to send the affected people by determining which shelters are closest for the calculated number persons seeking refuge. If no shelter is found, the bounding box is automatically extended until at least one appropriate shelter is found. In case there is more than one shelter within the bounding box, a route will be calculated between the location of the bomb and all possible shelters by an OpenLS RS. The resulting shortest route is then suggested to the decision makers. Finally all routes between the buildings of special interest and the allocated shelter are calculated by an *Emergency Route Service* (ERS) [15]. The ERS is a RS that conforms to the OpenLS specification but automatically by-passes restricted zones or streets. These restrictions need to be made available within the disaster management SDI, e.g. by a WFS, and automatically considered by the ERS. Here the danger zone is defined as an area to be avoided and is by-passed in the route calculation.

For representing the whole bomb scenario as a single application all of the listed services (WPS, RS, ERS, WFS) must be combined to one aggregated service, also referred to as a composition. For the description of business processes in general the orchestration model can be used. The background of *Web Service Orchestration* (WSO) and the option of using the WPS interface itself as possible alternative of using the BPEL standard is presented in the following sections.

4   Orchestration of OGC Web Services

The aggregation of Web Services is the composition of a set of services to achieve a common goal [15]. The possibility of compositing different services is often perceived as the greatest value of the web service paradigm [16] and it is a central concept in the framework of *Service Oriented Architectures* (SOA). Various complex tasks can be solved by compositing and reusing several "simple" services. In order to represent the complete process logic orchestration can be used which means that the sequence and implementation terms for activating external or internal services is presented through the eyes of a single participant [17].

4.1   Using BPEL for the Orchestration of OGC Web Services

An established standard concerning WSO is the use of BPEL which uses the *Web Service Description Language* (WSDL). WSDL is a standard for the description of a web service interface and acts as the link between the *Orchestration Engine* (OE) and the services involved [12]. The WSDL interface offers the possibility of providing a defined functionality for a client where the underlying implementation is not transparent. This leads to one of the problems related to the use of BPEL for the orchestration of OGC Web Services, namely the lack of WSDL documents describing the OWS. These documents have to be created manually for every service used in order to orchestrate them using BPEL. Furthermore the *Simple Object Access Protocol* (SOAP) support is missing in OWS. At the moment OGC Web Services are invoked via HTTP POST and/or GET which is not supported by all Orchestration Engines tested or at least not directly supported. The OGC is aware of these problems and there are ongoing discussions which will end in concrete solutions in the near future.

On the other hand there is also a technical problem with the OE and the transfer of raw binary data. This kind of data is served in response to a WMS *GetMap* or WCS *GetCoverage* request which cannot for this reason be orchestrated using the BPEL approach. All these difficulties are described in detail in [12]. The mentioned pitfalls forced us to look for alternatives and lead us to the use of the WPS interface itself which is an alternative to the *World Wide Web Consortium* (W3C) compliant WSDL interface in OWS architectures.

4.2 Using the WPS Interface Itself – Three different Approaches

The OGC WPS interface provides the technical possibilities for a orchestration of web services, although this may be in contrast to the original WPS idea. Therefore our focus is on the implementation and aggregation of WPS processes. At the same time, the possibility of using an WPS interface to aggregate complex geo-processing services is investigated. In principle there are no restrictions on what can be implemented using the WPS interface. The specification provides plenty of room for interpretation because it does not describe the problem clearly. The specification primarily describes the concrete implementation of geo-processing methods and does not entirely exclude its use as an orchestration service. Therefore it is possible to use a WPS for aggregating participating OWS. The interface is specified in such a way that any kind of processing can be hidden, if desired.

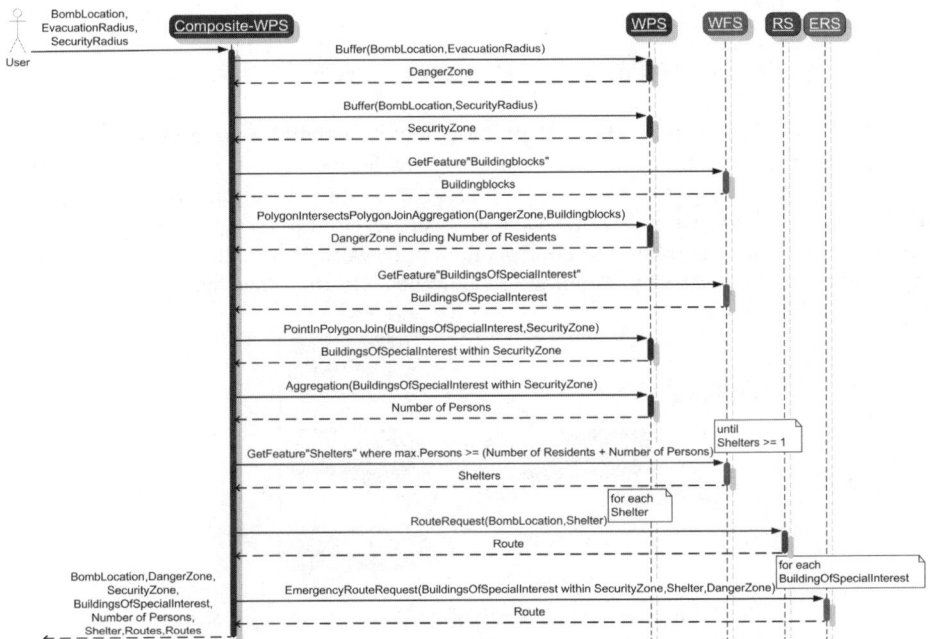

Fig. 2. Composite-WPS calling all other Services in the Bomb Threat Scenario

In the first approach of the implementation of our scenario we define a "Composite-WPS" as a WPS that can serve as an orchestration service for activating other services. The Composite-WPS activates all OWS needed for our scenario, including other WPSs as well as RS, ERS (which is an RS with special behavior) and WFS according to the defined application logic. The Composite-WPS can also be addressed as an independent service. The sequence diagram of the bomb threat scenario is presented in figure 2.

Such a solution requires large amounts of data to be passed between the different services. E.g. the security zone is calculated by the WPS process *Buffer*, sent back to the Composite-WPS and then sent again to the process *PointInPolygonJoin*. This can be avoided by calling specific services directly from other services. The *PointInPolygonJoin* process calls the *Buffer* process directly and the *Aggregation* process then calls the *PointInPolygonJoin* process. This possibility is explicitly enabled through the WPS interface by anticipating a *Key-Value-Pair* (KVP) encoded *Execute* request transported by the HTTP GET method. The use of "nested" KVP encoded *Execute* requests is a second option for the implementation of our bomb threat scenario and is shown in figure 3 for a part of the whole scenario.

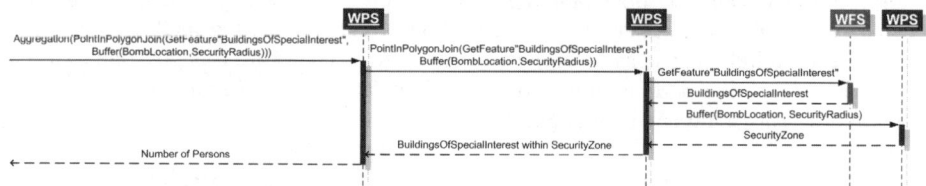

Fig. 3. Example of possible nested Requests in the Bomb Threat Scenario

These two alternatives show that there are in general two different concepts concerning the chaining of services: „*Centralized Service Chaining*" and „*Cascading Chaining*" according to R. Lucchi [personal communication, May 07, 2007]. *Centralized Service Chaining* means that a single central service invokes all other services, one after the other and controls the entire work flow. This approach is state-of-the-art and supported by common technologies. On the other hand, the *Cascading*

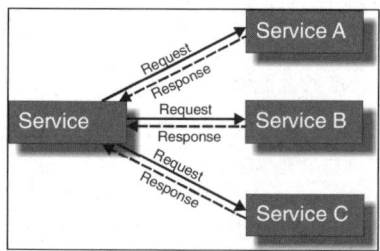

Fig. 4. Illustration of the Centralized Service Chaining Concept

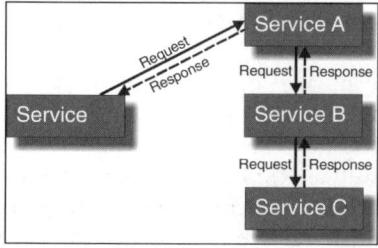

Fig. 5. Illustration of the Cascading Service Chaining Concept

Chaining method exchanges data directly because the individual services communicate with each other. Both concepts are presented in figure 4 and figure 5.

The WPS interface is one of the few specifications which support *Cascading Chaining* and the concept is described by looking closer at a KVP encoded *Execute* request and calling other services out of a called service. We assume for our simplified bomb threat scenario that the location of the bomb is stored in a GML (*Geograpy Markup Language*) encoded file "bomblocation.gml" which is accessible at a server and the danger zone will be calculated for a radius of 1000 m around this bomb location. Here is a possible implementation of the KVP encoded *Execute* request for invoking the WPS process *Buffer*:

```
http://localhost:8080/wps?
request=Execute&
service=WPS&
version=0.4.0&
Identifier=Buffer&
DataInputs=
InputGeometry,http://localhost:8080/bomblocation.gml,
BufferDistance,1000
```

The information about buildings of special interest (e.g. schools, kindergartens etc.) is provided by a WFS and can be requested in the following way:

```
http://localhost:1979/geoserver/wfs?
service=wfs&
version=1.0.0&
request=GetFeature&
typename=buildings_of_special_interest
```

These two KVP encoded requests deliver the data inputs for the WPS process *PointInPolygonJoin*, namely the buildings of special interest (point data) and the danger zone (polygon data). Both can be directly integrated in the request for invoking the WPS process *PointInPolygonJoin* which could then look like that:

```
http://localhost:8080/wps?
request=Execute&
service=WPS&
version=0.4.0&
Identifier=PointInPolygonJoin&
DataInputs=
PointFeatures,
http://localhost:1979/geoserver/wfs?
service=wfs&
version=1.0.0&
request=GetFeature&
typename=os:buildings_of_special_interest,
PolygonFeatures,
http://localhost:8080/wps?
request=Execute&
service=WPS&
version=0.4.0&
Identifier=Buffer&
DataInputs=InputGeometry,http://localhost:8080/bombloca
tion.gml,
BufferDistance,1000
```

This complete request could then be integrated into an WPS *Execute* request to invoke the process *Aggregation*.

Such nested requests become quite complex, however the advantage is obvious: data is requested were it is processed and not passed unnecessarily. When large amounts of GML encoded data are passed around, a performance increase is expected. On the other hand, it is more difficult to trace the outputs of the various services which were called. Therefore, tracking errors is difficult because they are not easily matched to a particular service.

KVP encoded *Execute* requests are optional according to the WPS interface specification and unfortunately, to the best of our knowledge, are not yet implemented within the existing *deegree* WPS framework nor any other framework.

A third approach would be to combine multiple WPS functionalities into a single implementation by calling only classes and the respective methods of the classes in order to run the complete workflow. The whole functionality would be exposed as a single WPS process but at the same time the single building blocks are still available externally, as independent processes. This alternative is purely a question of implementation and shown in figure 6.

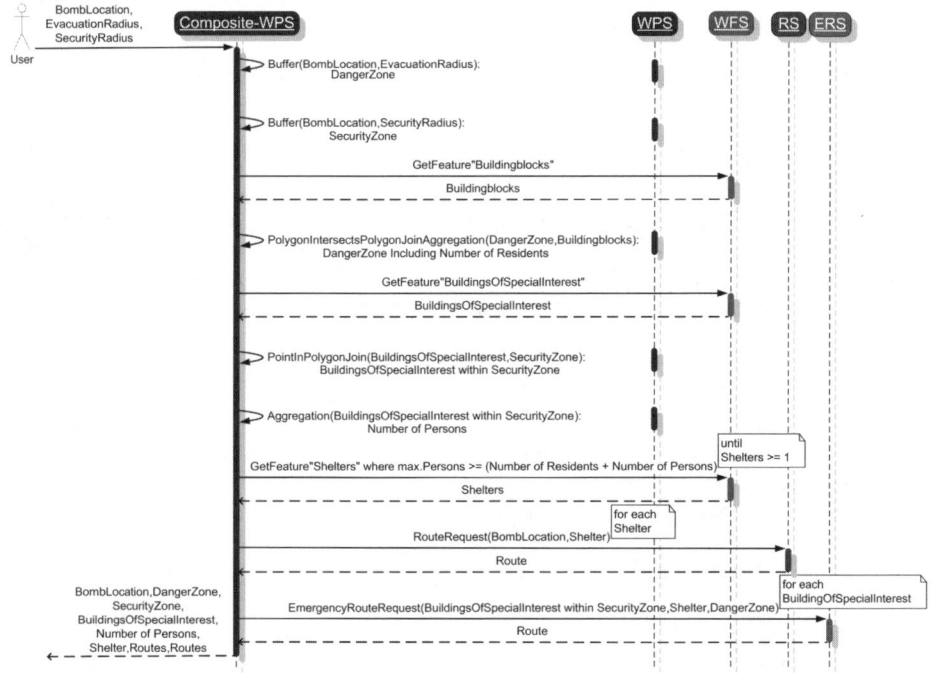

Fig. 6. A third Alternative of Invoking Services in the Bomb Threat Scenario

The advantage of this alternative is the performance boost because no data is exchanged between individual services more than once. The disadvantage is that it is not possible to replace WPS process instances by other instances which offer the same functionality and this does not make sense in terms of a distributed architecture. Thus, it is assumed that only one WPS instance is used to provide the whole scenario functionality as even though another service provider may have developed a different WPS instance with e.g. a process *Buffer* which is faster and achieves a higher accuracy. It would not be possible to address this WPS process within the third presented alternative. In reality however, the described scenario should use distributed WPS processes provided by different organizations. And this would be possible in the both alternatives described above, namely introducing a Composite-WPS or using nested KVP encoded requests. In these two approaches it is quite easy to replace WPS instances and this demonstrates the strength of distributed architectures.

4.3 Towards a Composite-WPS

Currently, we are implementing the scenario by using a Composite-WPS that invokes all other services and respective processes in our scenario in a sequential manner. This is in accordance with the *Centralized Service Chaining* concept. We did not inspect the concept of *Cascading Service Chaining* further because the support of KVP encoded *Execute* requests is at the moment still missing in the existing WPS implementations as mentioned above. We are demonstrating successfully that it is possible to represent complex scenarios completely in accordance with OGC standards, which was our main aim in the first place. The question which of the three approaches described above is preferable for a particular scenario cannot be answered in general. It depends on the number of services involved and especially on the amount of data in every single case.

However, all three presented alternatives offer the possibility to provide the final application as an independent OGC compliant service through the WPS interface. In OK-GIS, these aggregated services are called "virtual services". In comparison to the generic OGC services, the idea within OK-GIS is, that these "virtual services" are scenario or domain specific similar to the described bomb threat scenario. They represent the end-user view who is not interested in technical implementations of OGC service chains, but rather on the task to be performed in order to manage the disaster situation.

Furthermore, it would be possible to create a WSDL document describing this domain specific service in order to achieve W3C compliance. This may be relevant for a developer who is not familiar with OWS and who does not expect an OGC compliant interface but instead, a in the general IT community well established standard like WSDL.

4.4 Using an Accessibility Analysis Service

A part of the described service chain, namely the finding of an optimal evacuation centre could also be realized by using the *Accessibility Analysis Service* (AAS) [18]. This would avoid re-implementing this part of the service chain and we shortly discuss this alternative even though the AAS is not a standardized OWS or even a discussion paper, but its interface is modeled according the OpenLS specification. Therefore it has to be initially hidden behind a WPS facade for being OGC conform.

Based on a street network and a defined time distance, the AAS calculates accessible areas and other information.

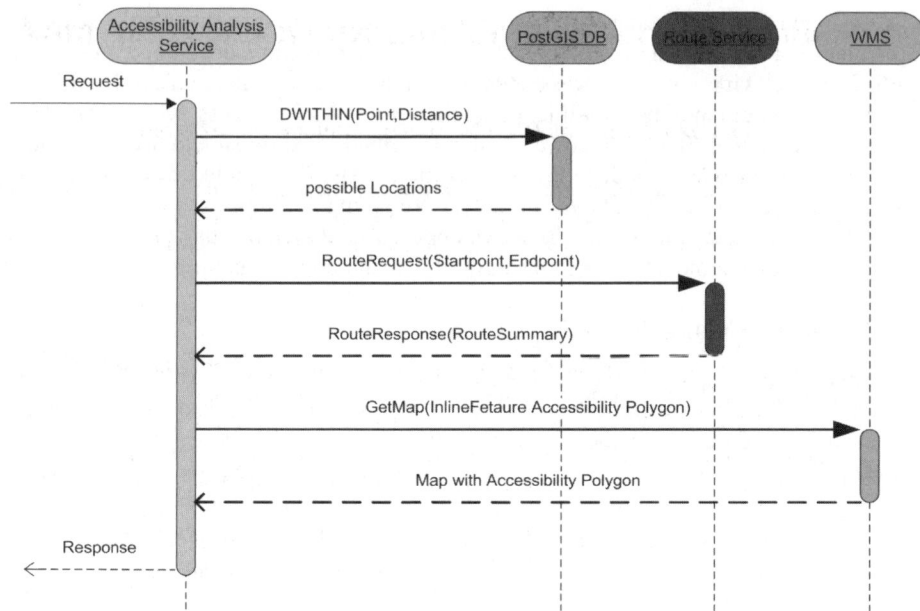

Fig. 7. UML Sequence Diagram of the AAS interacting with other Services

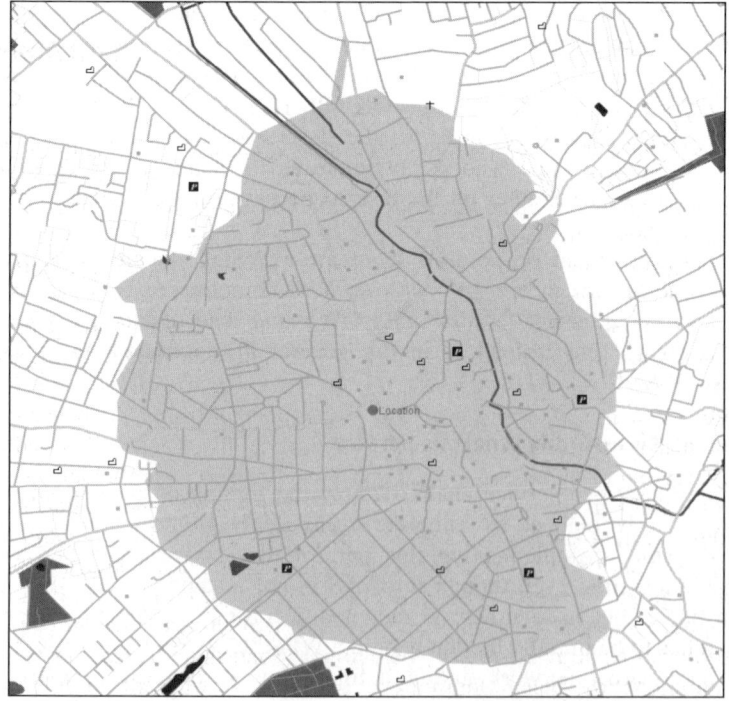

Fig. 8. Example Response Map of the AAS

For using the AAS in the bomb scenario, the evacuation shelters need to be added and respectively the different "types" of places have to be made selectable. This means that the AAS needs slightly to be extended. Additionally, the already existing AAS has several other options, which are not needed in our scenario. However, the use of this already available service would save some implementation work but the advantages and disadvantages of using the AAS need to be assessed. The current AAS implementation is shown in figure 7. Instead of a WFS, currently a PostGIS database is accessed for performance reasons. This database uses the *Simple Feature Specification* for SQL (SFS4SQL), which offers analog functionalities like a WFS and therefore it could be easily replaced by a WFS in principle.

An example of an AAS response is illustrated in figure 8. The convex polygon represents the area which is reachable within three minutes from the marked "Location".

As mentioned above the AAS is not an OGC standard but it would be possible to hide the service behind the WPS interface. This makes clear that the quite open WPS specification offers also the possibility to use any kind of web service "packed" as a Web Processing Service.

5 Implementation

For the implementation of the described bomb threat scenario we use the existing WPS framework of the *deegree2* project (www.deegree.org) [19]. This and other projects such as the *52°North* project (http://52north.org) and the *Agriculture and Agri-Food Canada* project (www.agr.gc.ca) are in accordance with the current version 0.4.0 of the WPS specification from the year 2005 [20].

The *deegree2* framework offers the possibility to implement the application logic of new processes within a separate class. This class is derived from the *org.deegree.ogcwebservices.wps.execute.Process.java* class. In addition an XML configuration document has to be created for the process and integrated into the framework along with the process class. After that the new process is then accessible via the WPS interface.

While the implementation of a WPS from scratch would be time-consuming, simple processes can be rapidly implemented when using frameworks such as the *deegree* WPS. The frameworks relieve programmers of much of the overhead and they can focus on implementing custom processes.

Because the WPS interface is that open, there is no restriction to the process type and process complexity. Therefore, separate processes can also be aggregated into one new WPS process. We have taken advantage of this in our approach by orchestrating all scenario steps via the Composite-WPS but still providing this WPS as an independent WPS process.

The first implementation step based on the Composite-WPS concept is to realize the aggregation of the involved services by direct coding. This will be replaced in a second step by designing a more flexible process class which is fed by configuration files. A future goal is the aggregation of several services within one WPS without touching the implementation itself but only configuring some XML documents.

6 Discussion

Modularization and the reuse of existing modules are the key to Service Oriented Architectures and to object oriented programming in general. Reusing services is supported by applications like BPEL designer which make it possible to orchestrate single services using graphical tools. Unfortunately, difficulties exist with respect to the orchestration of OGC Web Services based on BPEL which compelled us to explore alternatives such as the OGC WPS interface for orchestrating OWS. Consequently, programming languages must be used instead of the simpler graphical configuration tools provided by BPEL designer. The use of BPEL would be more likely in case of existing WSDL documents for all OWS, but with respect to the WPS interface, such a WSDL document is redundant because the response to a *DescribeProcess* requests contains more or less the same information as a WSDL description and would only be necessary for technical reasons, namely the use of BPEL for service orchestration.

The use of the WPS interface itself for the orchestration of OGC Web Services is a "misuse" of the interface and contrary to the original idea. But only the lack of restrictions and the openness of the interface gave us the idea to "abuse it" in order to aggregate several processes to achieve our goal. It is no secret that the current WPS concept is not fully developed and any complex service can be realized, no matter if this service provides geo-processing functionality or not. Alternatively, it would seem to make more sense, if it offered a well known amount of relevant geo-processing procedures which could serve as building blocks for more complex geo-processing functions. This was taken into account for the new 1.0 WPS interface specification which is currently developed within a working group. In this specification, *Profiles* are introduced which should be further developed by user communities to agree on defined WPS processes. Within particular domains, it is imperative that standardized geo-processing functionalities are agreed upon in order to solve interoperability problems. It is important to remember that final aggregated services (like the bomb threat scenario) are usually domain specific and do not really belong to an unspecified domain OGC service. Luckily, other existing OWS are mostly domain-neutral. On the other hand, these processes do represent geo-processing functions and this is what the WPS was actually designed for. That is why the term "complex" needs to be further defined considering aspects such as: what can still be categorized under geo-processing base functions and when do they start becoming domain specific functions which then may not belong to a WPS? However, the WPS interface closes an open gap in the area of OWS and respectively SDIs and this paper demonstrated that complex scenarios can be completely represented in accordance with OGC standards if needed. But neither the use of the WPS interface nor using WSDL for orchestrating OWS give an answer to interoperability problems in SDIs and this challenged has yet to be solved.

References

1. GSDI, Developing Spatial Data Infrastructures: The SDI Cookbook. Version 2.0. Retrieved May 29, 2007 (2004), http://www.gsdi.org/gsdicookbookindex.asp
2. Granell, C., Gould, M., Poveda, J.: Incremental Composition of Geographic Web Services: An Emergency Management Context. In: Proceedings 7th AGILE Conference on Geographic Information Science, Heraklion, Greece (2004)

3. Alameh, N.: Chaining Geographic Information Web Services. IEEE InternetComputing 7(5), 22–29 (2003)
4. Peng, Y., Di, L., Yang, W., Yu, G., Zhao, P.: Path Planning for Chaining Geospatial Web Services. In: Carswell, J.D., Tezuka, T. (eds.) W2GIS 2006. LNCS, vol. 4295, pp. 214–226. Springer, Heidelberg (2006)
5. Bernard, L., Einspanier, U., Haubrock, S., Hübner, S., Kuhn, W., Lessing, R., Lutz, M.: Ontologies for Intelligent Search and Semantic Translation in Spatial Data Infrastructures. In: Photogrammetrie - Fernerkundung - Geoinformation (PFG), pp. 451–462 (2003)
6. Li, J., Zlatanova, S., Fabbri, A. (eds.): Geomatics Solutions for Disaster Management, p. 444. Springer, Heidelberg (2007)
7. Neis, P.: Routenplaner für einen Emergency Route Service auf Basis der OpenLS Spezifikation. Diploma Thesis. University of Applied Sciences FH Mainz (2006)
8. Neis, P., Schilling, A., Zipf, A.: 3D Emergency Route Service (3D-ERS) based on OpenLS Specifications. In: GI4DM 2007. 3rd International Symposium on Geoinformation for Disaster Management, Toronto, Canada (2007)
9. Kiehle, C., Greve, K., Heier, C.: Standardized Geoprocessing - Taking Spatial Data Infrastructures one step further. In: Proceedings of the 9th AGILE International Conference on Geographic Information Science, Visegrád, Hungary (2006)
10. Biermann, J., Gervens, T., Henke, S.: Analyse der Nutzungszenarien und Aktivitäten der Feuerwehr. Internes Projektdokument. Projekt OK-GIS (2005)
11. Weiser, A., Neis, P., Zipf, A.: Orchestrierung von OGC Web Diensten im Katastrophenmanagement - am Beispiel eines Emergency Route Service auf Basis der OpenLS Spezifikation. In: GIS - Zeitschrift für Geoinformatik 09/2006, pp. 35–41 (2006)
12. Weiser, A., Zipf, A.: Web Service Orchestration (WSO) of OGC Web Services (OWS) for Disaster Management. In: Joined CIG/ISPRS Conference on Geomatics for Disaster and Risk Management, Toronto, Kanada (2007)
13. Heier, C., Kiehle, C.: Geodatenverarbeitung im Internet - der OGC Web Processing Service. GIS 2005 6, 39–43 (2005)
14. Stollberg, B.: Geoprocessing in Spatial Data Infrastructures - Design and Implementation of a Service for Aggregating Spatial Data. Diploma Thesis. University of Applied Sciences FH Mainz (2006)
15. Dustdar, S., Schreiner, W.: A Survey on Web Services Composition. International Journal of Web and Grid Services, Inderscience 1(1) (January 2005)
16. Einspanier, U., Lutz, M., Senkler, K., Simonis, I., Sliwinski, A.: Toward a process model for gi service composition. In: GI-Tage (GI Days) 2003, Münster, Germany (2003)
17. Reichert, M., Stoll, D.: Komposition, Choreographie und Orchestrierung von Web Services - Ein Überblick. EMISA Forum, Band 24, Heft 2, 21–32 (2004)
18. Neis, P., Zipf, A.: A Web Accessibility Analysis Service based on the OpenLS Route Service. In: AGILE 2007. International Conference on Geographic Information Science of the Association of Geograpic Information Laboratories for Europe (AGILE), Aalborg, Denmark (2007)
19. Fitzke, J., Greve, K., Müller, M.: Building SDIs with Free Software - the deegree Project. In: Fitzke, J., Greve, K. (eds.) Proceedings of GSDI- 7, Bangalore, India (2004)
20. OGC, Web Processing Service. OGC Discussion Paper, Document Reference Number 05-007r4, Version 0.4.0 (2005)

A Client for Distributed Geo-processing on the Web

Theodor Foerster[1] and Bastian Schäffer[2]

[1] International Institute for Geo-Information Science and Earth Observation
P.O. Box 6, 7500AA Enschede, the Netherlands
foerster@itc.nl
[2] Institute for Geoinformatics
Robert-Koch-Str. 26-28, D-48149 Münster, Germany
schaeffer@uni-muenster.de

Abstract. The OGC Web Processing Service specification provides a means to perform web-based processes on distributed geodata and to produce thereby web-based geoinformation. However a client which integrates different data and geo-processing services is missing but would enable fully web-based information integration. This was the starting point for this study to build a client, which is able to integrate multiple web-based data and processing services. The applicability of the client for web-based information integration will be demonstrated for a real-time risk management scenario. The client is realized on top of uDig and available under Open Source license (GPL) at the 52° North Open Source initiative.

Keywords: Web Processing Service, Web Processing Service Client, web-based processing.

1 Introduction

As computational power and network capabilities mature, processing of distributed data towards information becomes one of the main interests in the IT world. This aspect is addressed by the evolving concept of Web Services and Service-oriented Architecture (SOA, [1]). Distributed processing using Web Service technology is also a main interest in the geoinformation (GI) domain, as most of the GI applications involve large amounts of globally distributed data and as the demand for distributed available geo-information increases. The web-based geodata access was addressed by the concept of Spatial Data Infrastructures (SDI, [2]) and the invention of Web Service and geodata standards as for instance developed by the Open Geospatial Consortium (OGC[1]). Web-based geoinformation has been identified as a key factor for SDIs in the future, therefore sufficient concepts for web-based processing are required. Such processes have to be able to access globally distributed data and to provide the information in line with the already available standards. Regarding such a standardized

---

[1] OGC website: www.opengeospatial.org

J.M. Ware and G.E. Taylor (Eds.): W2GIS 2007, LNCS 4857, pp. 252–263, 2007.

service for web-based processing the OGC proposed the Web Processing Service (WPS) specification as a discussion paper[2] [3].

In the last two years several communities started to implement this specification [4,5] and demonstrated its applicability by several use cases, such as generalization processing [6] or groundwater vulnerability measurement [7]. We developed a processing client based on the *Java Uniform Mapping Platform* (JUMP[3]) [6]. Additionally [8] built a small demo application, but his demo did not provide flexible access to WPS nor integration with other services. So a flexible client application, which utilizes the capabilities to process distributed data services over the web and present the result in a pleasing way to the user has been missing. This was the starting point for this study to investigate the benefits of such a client and to implement a ready-to-use client solution. We decided to implement it as a desktop application as such a breed of application provides more flexibility to the user such as advanced data querying and visualization capabilities. We chose the *User-friendly Desktop Internet GIS* (uDig[4], [9]) like [8] already did for a small demo application on WPS as the appropriate implementation platform, because it already allows connecting to distributed data services such as Web Map Service (WMS) and Web Feature Service (WFS). Thereby it provides many advantages against JUMP, which does not allow to connect to any OGC Web Service and which only provides limited styling capabilities to the user. The client is realized as a plug-in for uDig and is available under Open Source license (GPL) at the 52° North Open Source initiative[5]. It is important to note, that the presented work is unique in the evolving field of web-based geo-processing, as it focuses on a flexible processing client, which allows integrating data and geo-processing services.

In Section 2, the paper describes the technical background of WPS and the idea of processing distributed services as utilized in the proposed client. Then we provide in Section 3 some insights into the client architecture. Section 4 demonstrates a real-time risk management scenario incorporating generalization functionality and buffer analysis. Both processes demonstrate the applicability of the client and the benefit of distributed data processing. Finally we discuss some problems we encountered during this study and give an outlook about possible extensions of the client. We also draw a conclusion about the practical experience with the uDig client in comparison to JUMP. The paper ends with a conclusion.

2   Technical Background

The Web Processing Service specification describes a way how to publish and perform processes on the web. Such a process can range from a simple buffer calculation to a complex process of vector analysis for generalization purposes.

---

[2] The discussion paper is known as WPS version 0.4.0.
[3] JUMP website: www.jump-project.org
[4] uDig website: udig.refractions.net
[5] 52° North website: www.52north.org

The communication is based on three operations (Figure 1), which can be called via HTTP-GET and key-value pair (KVP) encoding or HTTP-POST and a XML-encoding. The *GetCapabilities* operation (step I. in Figure 1) provides service metadata and some brief description of the process. The process metadata including the parameter descriptions is accessible through the *DescribeProcess* operation (step II.). Via the *Execute* operation it is possible to call the desired process (step III.).

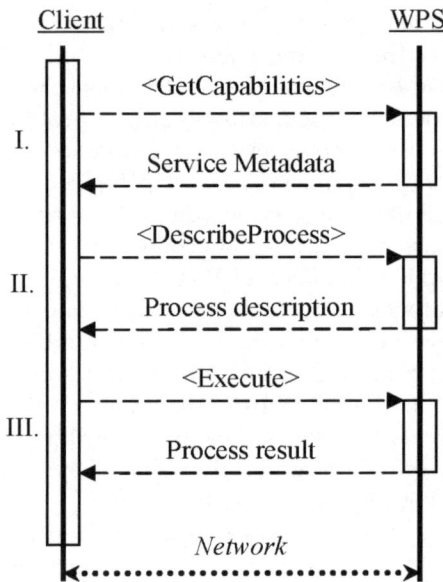

Fig. 1. Basic client-service communication pattern to perform a desired process on WPS

Besides capabilities to perform long-term processes and to store process results on service side, WPS provides the capability to link distributed resources into the process via an Uniform Resource Locator (URL). This URL might identify a location at which a GML file is stored or even be a WFS call for a specific feature type. This reference will be incorporated in the Execute request to the WPS as a *ComplexValueReference*[6] element (Example 1).

Example 1 (Sample ComplexValueReference element linking WFS road data).
<wps:ComplexValueReference ows:reference="**http://geoserver:8080/ geoserver/wfs?SERVICE=WFS&VERSION=1.0.0&REQUEST= GetFeature&typename=roads**" schema="http://schemas.opengis.net/ gml/2.1.2/feature.xsd" />

---

[6] This paper is based on the version 0.4.0 of the WPS specification, so the name of the ComplexValueReference element might differ in the coming versions.

By using this ComplexValueReference element, the client does not have to take care about sending the plain data to the WPS, but only the reference. Additionally such references provide a certain degree of freedom to the WPS implementation about how to retrieve the data or enable even to apply some caching mechanism based on comparing the references over multiple requests. Overall this feature improves the performance of web-based processes on client and service side drastically, as it decreases the workload to prepare the data for sending at the client side and as it enables service-side caching as already mentioned. Additionally looking at the topology of the network, most services are able to retrieve data faster than clients can send them, because services are mostly located in faster network environments.

Besides improved performance linking resources via ComplexValueReference incorporates the notion of distributed data access, as the WPS is able to incorporate multiple data from different locations. This distributed notion guarantees real-time data access. However from the perspective of caching, this might cause some trouble, as some data will be cached but has been updated at the original source in the meanwhile. So a trade-off will always remain between caching (i.e. for preformance purpose) and real-time data processing. Section 4 including Figure 7 demonstrates the application of processing referenced data services in WPS.

3   Client Architecture

The client is realized as a plug-in in uDig, which is organized as a set of plug-ins on top of the eclipse rich client platform (RCP[7]). The eclipse RCP implements the concept of inversion of control and allows thereby only specific parts of the workflow to be customized. The customization within the eclipse RCP is achieved via so called extension points, at which the intended functionality can be inserted into the RCP framework. The concept of extension points is inherited by uDig to insert intended functionality at designated points of the uDig workflow.

So in order to register a new service type, like the WPS, in uDig, the net.-refractions.udig.catalog.ServiceExtension extension point and the net.-refractions.udig.catalog.ui.ConnectionFactory extension point have to be implemented. The ServiceExtension allows the introduction of new service types to uDig. The ConnectionFactoy requires a custom wizard and a class extending net.refractions.udig.catalog.ui.UDIGConnectionFactory as a binding between the user interface and the internal model of uDig. The simplified architecture is presented in Figure 2.

The WPSServiceExtension is a net.refractions.udig.catalog.Service-Extension, which is the major hook to make a WPS available as a new datasource within uDig. The class acts mainly as a factory to create new WPSServiceImpl instances, which inherit from net.refractions.udig.catalog.IService. Such an instance will only be created, if the entry URL of the WPS instance is not yet available in uDig. Meta information about the represented WPS such as the

---

[7] Eclipse RCP website: www.eclipse.org/rcp/

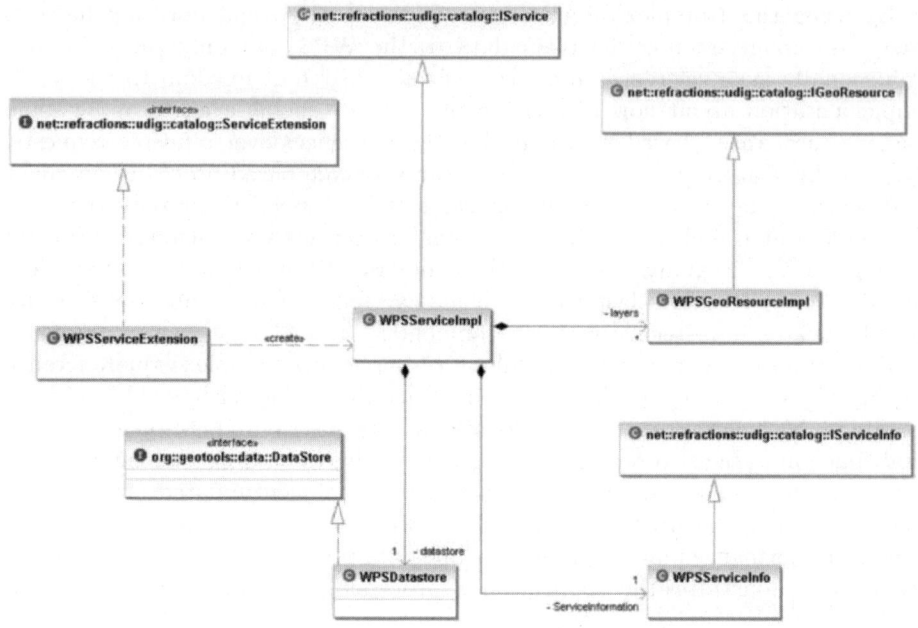

Fig. 2. Simplified class diagram of the developed uDig plug-in

URL, the description and the name are stored in `WPSServiceInfo`, which inherits from `net.refractions.udig.catalog.IServiceInfo`. The WPS does not provide layers like most of the common OGC Web Services. Since uDig applies this common OGC Web Services layer concept, the plug-in has to provide the process results as layers. Therefore, the process results are available to the layer view of uDig through `WPSGeoResourceImpl`.

The integrated execution of a WPS process and the rendering of the process results as layers in uDig involves several steps. This procedure goes beyond the current wizard concept of uDig, because it is mainly designed for a simple datasource import, which is handled by a single wizard page. Therefore the concept of wizards in uDig was extended by enabling the wizard to consist of more than one custom page. This allows separating the complex WPS execution procedure into smaller steps.

Overall, the basic interaction between the user and uDig is realized as follows (Figure 3): First the designated type of service has to be chosen - in this case WPS (step I.). This kicks-off the previously mentioned `WPSConnection-Factory`, since it was registered with the WPS service. The `WPSConnection-Factory.createConnectionpage()` method creates and returns the first custom wizard page to the enclosing wizard (step II). The page requests the user to enter the entry URL to the WPS. After the user has pressed the wizard's next button, the client requests automatically the capabilities for the entered URL and for each of the processes the process description. Thereupon, the second wizard page is displayed and now the user can select the process (step III.), he/she wants to use. The selected process is internally stored and after pressing the next button, the third custom wizard page is shown. This page allows the user to configure the selected process with the necessary parameters (step IV.). The form for the parameters is generated on-the-fly, based on the process description. For every required literal input, the user is presented with a corresponding text box. Additionally, for every complex input, the user can choose from a drop down list containing currently available layers inside of uDig. If the user chooses a WFS layer, the layer can be send by reference according to the general wizard configuration. This allows the WPS to fetch the data and saves up time, since not the whole data has to be prepared on the client-side and transferred to the service (Section 2). In the last step (step V.), the wizard shows a final layer selection wizard page, which allows the user to select the layer(s) he/she wants to add to uDig.

To get more insight into the internal workflow of the client and how the classes cooperate to perform according to the user's action in the wizard, the remainder of this section will summarize the course of action in more detail. So after the user has pressed the next button in step IV., uDig recognizes that there are no more custom wizard pages provided and tries to fetch all layers that can potentially be added. Therefore, a `WPSServiceImpl` instance is created through the `WPSServiceExtension.createService()` method. Figure 4 presents the course of action taken to get all potential layers.

First, the wizard, represented by the `ConnectionState`, calls the `members()` method on the `WPSServiceImpl` instance. This method obtains a `WPSDataStore` instance through the `WPSDataStoreFactory` class. The `WPSDataStore` receives all parameters collected during the previous wizard pages and fetches all process descriptions from the 52° North WPS Client API, represented by the WPSClient class in Figure 4. Next, the `WPSServiceImpl` instance calls the `getTypeNames()` method on the previously created `WPSDataStore` instance. This method analyzes each stored process description and returns the name for each complex output from the requested process. Now, the `WPSServiceImpl` instance is able to create a `WPSGeoResourceImpl` object for each returned process output, which is finally returned back to the calling `ConnectionState` object.

The wizard page in step V. is based on the results returned from the `WPSServiceImpl.members()` method. Finally, the user has to press the finish button to start the actual WPS process execution, which is illustrated in Figure 5. For each selected layer, the corresponding `WPSGeoResourceImpl` object is requested to return the associated features. Therefore, the `resolve()` method is called, which obtains the `WPSDataStore` from the corresponding `WPSServiceImpl` object and fetches the features from the `WPSDataStore` instance by calling the `getFeatureSource()` method. This method constructs the actual WPS execute request based on the stored parameters collected during the wizard page process and sends the request to the `WPSClient`, which finally forwards it to the WPS over the wire.

The process results are cached and passed into a newly created `WPS FeatureReader` object, which parses the data and makes it available. The

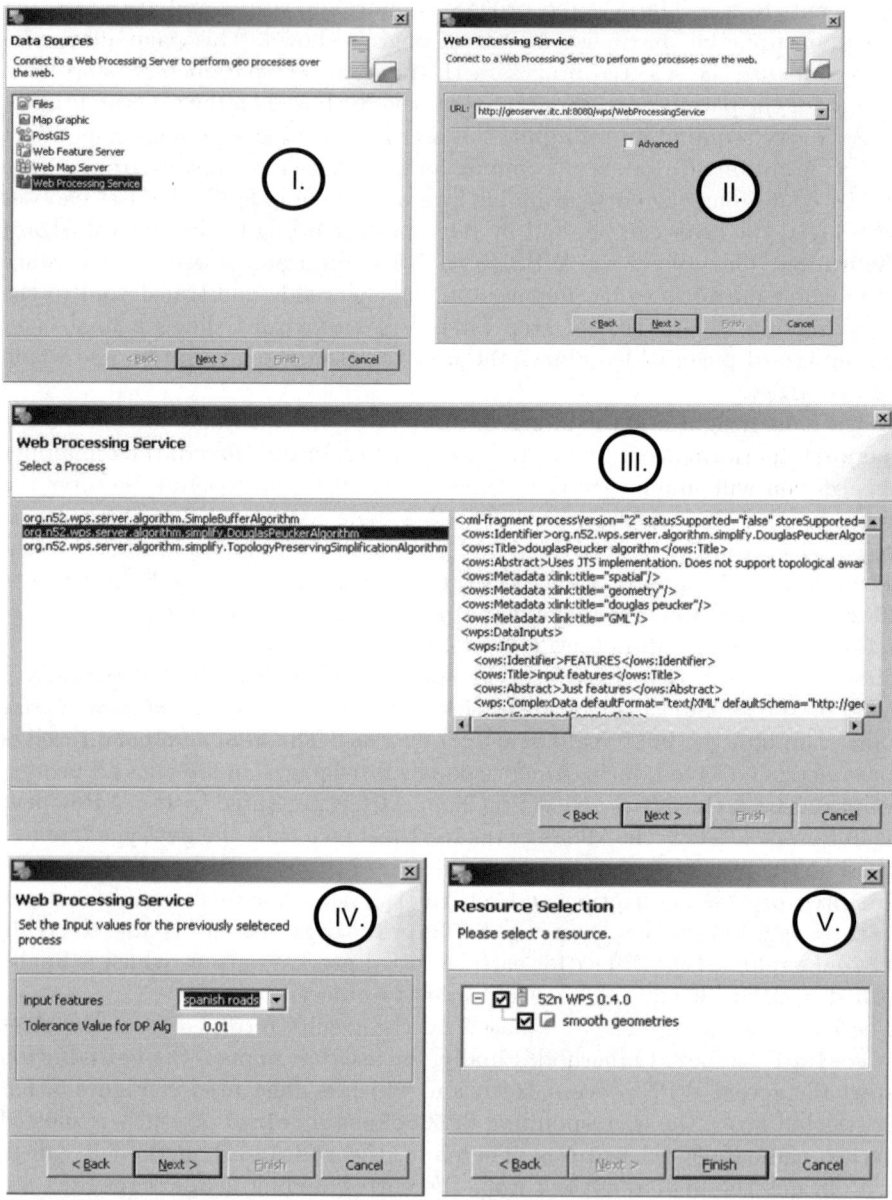

Fig. 3. The sequence of wizard pages to add a WPS process in uDig. I.) Select service type, II.) Enter WPS entry URL, III.) Select process, IV.) Configure process parameters, V.) Select process output.

WPSFeatureReader object is returned back to the WPSGeoResource and finally forwarded to the calling ConnectionState object to be displayed in uDig. As a result, a new layer is added to the layer list and the associated features are

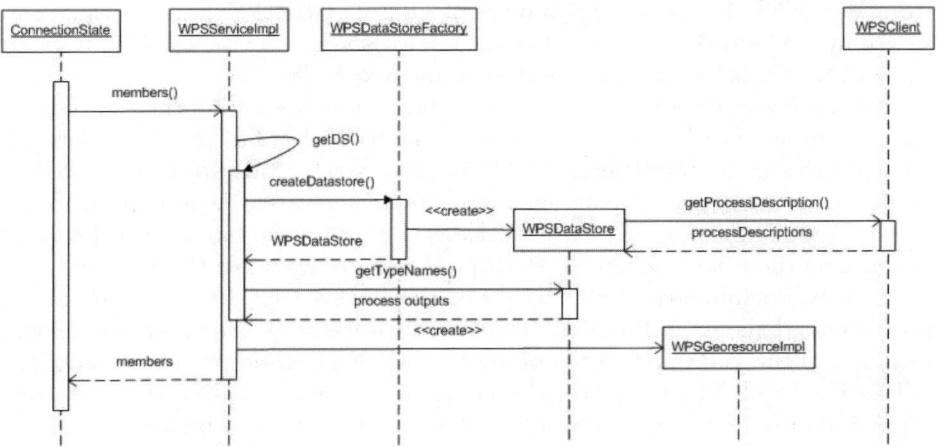

Fig. 4. Sequence diagram describing the creation of uDig layers representing WPS results

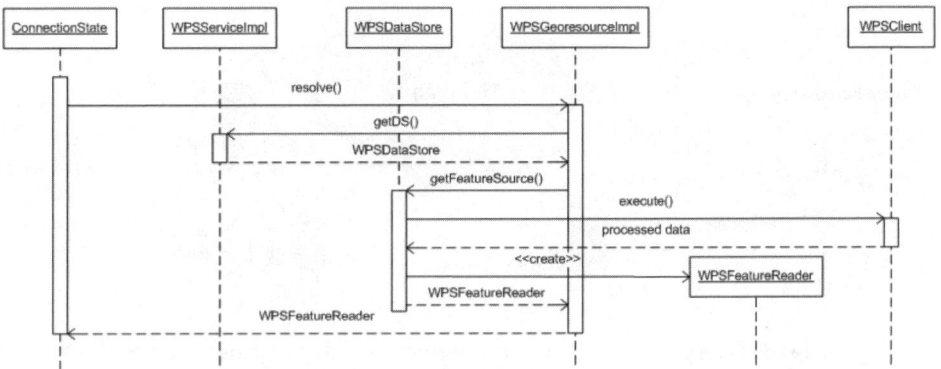

Fig. 5. Sequence diagram describing the execution of a WPS process inside of uDig

displayed. From now on, this layer can be treated as any other uDig resource and especially serve as input for another WPS process.

4   A Distributed Processing Scenario

The scenario aims at producing a readable map to indicate recent fire threats to transport infrastructure. Thus this scenario involves data about fire threat areas (i.e. areas where fire has been reported) and road data. It inherits aspects of a real-time risk-management application, as different data source have to be integrated and processed in order to improve decision making. The chosen location for this scenario is the North-West of Spain. The data for the burnt areas are provided by the courtesy of the Joint Research Center in Italy and served through a WFS. The roads of Spain are taken from CORINE data, provided and served by ITC's WFS. In order to produce a satisfying map we also incorporated some CORINE landcover data, served through a WMS.

However these data sources do not contain the required information per se. In order to analyze, which roads are at stake by such a fire threat, buffers are created around the burnt areas. Additionally, as the roads are too detailed to be displayed on a general overview map, simplification has been applied to the roads. Overall, this involves two processes: Buffering the burnt area data and simplifying the road geometries. Both processes are available through the standardized WPS interface. An overview of the involved services, their data and processes is depicted in Figure 6. The WFS instances are geoserver[8] implementations and the WPS instance is based on the 52° North implementation of the WPS. The result of the described scenario with some CORINE data as background information (served through a WMS) is presented in Figure 7.

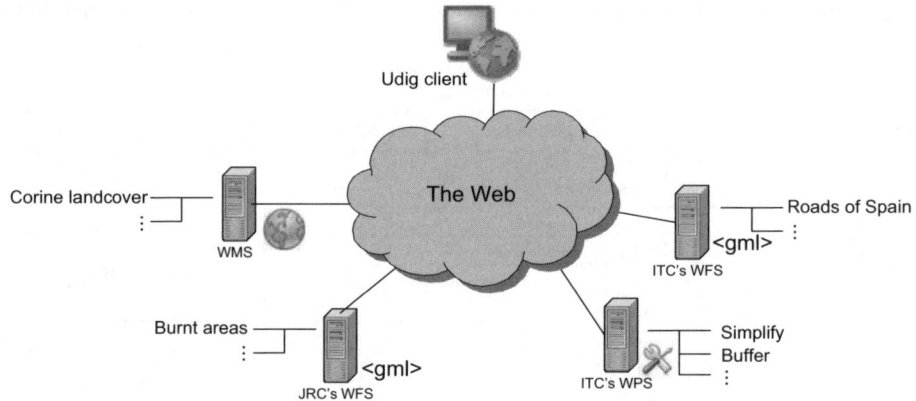

Fig. 6. Overview of the involved services in the described scenario

It is important to note, that this scenario is completely based upon distributed services. Especially incorporating the burnt areas data demonstrate the applicability of this client for real-time data processing and visualization in a risk management scenario. Such burnt areas could be collected from sensors and published on-the-fly to a WFS, as it is described in [10] for a fire alert information system in South Africa. Thus, using ComplexValueReferences allows us to dynamically process real-time data (from WFS) and extract real-time information (by WPS).

The client resolves the references of the data to be processed automatically and incorporates them into the Execute request just as depicted in Example 1 of Section 2 (and Figure 8). Therefore the user has to register the different services into the client by indicating the endpoints of the suggested services. Based on this manual registration the client is able to retrieve the metadata of

---

[8] Geoserver website: www.geoserver.org

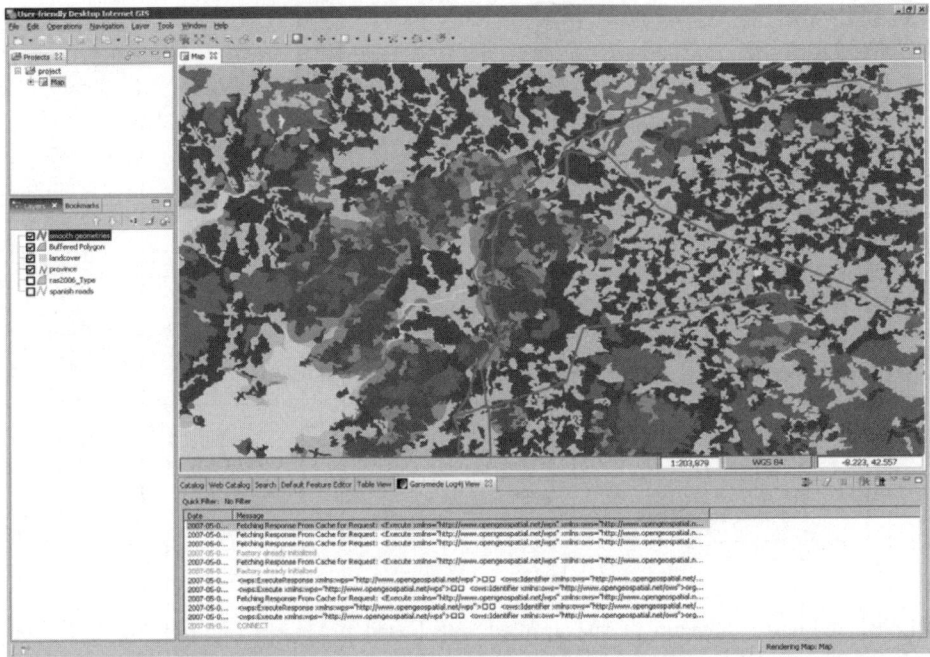

Fig. 7. Screenshot of the final result as described in the scenario

these services (via GetCapabilties & DescribeProcess) automatically. Afterwards the client is able (based on additional parameters for the processes) to perform a sequence of processes as proposed in this scenario and depicted in Figure 8.

5 Outlook and Conclusion

During the implementation and the testing we got a lot of insights into the WPS specification and the architecture of uDig. While uDig is in principal a real powerful tool, the concept of inversion of control can sometimes limit its capabilities. Comparing finally our experiences with JUMP [6] with our recent experiences gained from uDig, it becomes clear that uDig's capabilities to access data services eased the development significantly. In the case of JUMP we would have had to implement the connection to the data services from scratch. So this was a big advantage of uDig. Also the visualization part is more pleasing than in JUMP. However the developing effort of the wizard pages in uDig to configure the connection to the WPS in a user-friendly and flexible way was significantly higher, than it would be in JUMP, due to the limitations caused by the concept of inversion of control (details see in Section 3).

The WPS has also demonstrated the powerful feature of referencing distributed services. However some aspects of the specification are still not really sufficient. For instance, it is not possible to handle typed references. Such

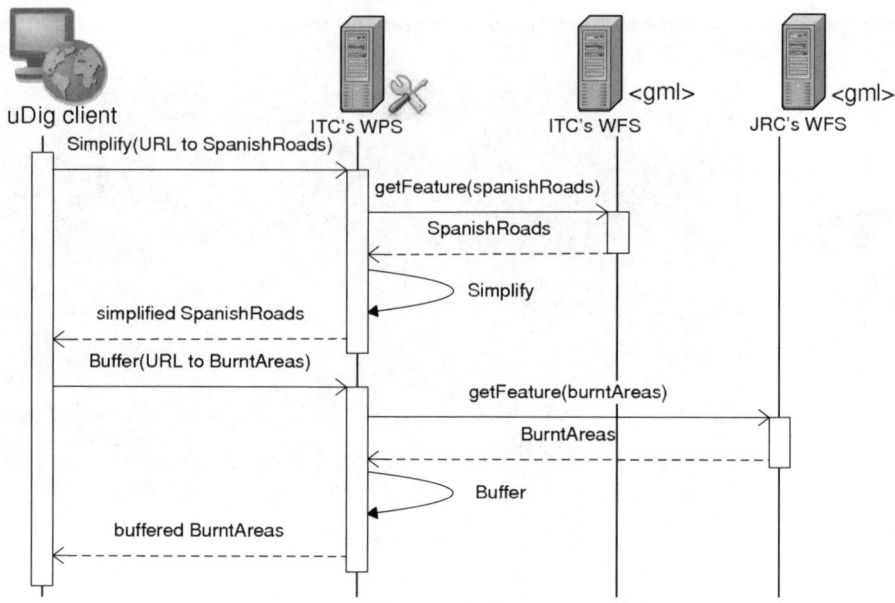

Fig. 8. Sequence diagram of the applied web-based processes (simplifying & buffering) incorporating the effect of automatically referencing the distributed data sources (spanish roads & burnt areas)

references would indicate the type of service is referenced in the ComplexValueReference and would thereby enable the WPS to apply different strategies of retrieving and caching the data.

In a further stage of the research we want to incorporate more service chaining capabilities in the client to model, deploy and trigger complex chains of processing and data services based upon WPS and BPEL.

Overall this study presents a client which can be applied for information integration in a real-time risk management scenario based upon a distributed service architecture as proposed in the Orchestra Project [11] or in the AFIS project [10]. The scenario demonstrates that processing is required to extract the required information (e.g. buffered area, simplified road geometries). Udig has proven to be a suitable platform to integrate data and processing services automatically in a user-friendly way.

The presented scenario in this paper is described in a hands-on tutorial and demonstrated in a screencast. Both plus the ready-to-use uDig plug-in are available through the 52° North website.

Acknowledgements

We highly appreciate the contribution of Dr. Rob Lemmens and Barend Köbben (both colleagues at ITC), who supported us presenting our work at the AGILE workshop 2007 about a Test-bed for geospatial web service interoperability. We also want to thank the Joint Research Center (JRC) for providing the data for the scenario. The work of Theodor Foerster is covered by the RGI 002 project[9].

Finally we want to thank the anonymous W2GIS reviewers for their constructive remarks.

References

1. Gottschalk, K., Graham, S., Kreger, H., Snell, J.: Introduction to web service architecture. IBM Journal 41, 170–177 (2002)
2. McLaughlin, J., Groot, R.: Geospatial data infrastructure: concepts, cases and good practice. In: Spatial Information Systems and Geostatistics Series, Oxford University Press (2000)
3. O.G.C: OpenGIS Web Processing Service. OGC discussion paper OGC 05-007r4, Open Geospatial Consortium (2005)
4. Foerster, T.: An open software framework for web service-based geo-processes. In: Free and Open Source Software for Geoinformatics, Lausanne, Switzerland (2006)
5. Cepicky, J.: Grass goes web: PyWPS. In: Free and Open Source Software for Geoinformatics, Lausanne, Switzerland (2006)
6. Foerster, T., Stoter, J.: Establishing an OGC web processing service for generalization processes. In: ICA workshop on Generalization and Multiple Representation (2006)
7. Kiehle, C.: Business logic for geoprocessing of distributed data. Computers & Geosciences (2006)
8. Johansson, J.: Standardized access to geospatial information and services. Master thesis, Department of Computing Science, Umea University, Sweden (Spring 2006)
9. Ramsey, P.: udig desktop application framework. In: Free and Open Source Software for Geoinformatics, Lausanne, Switzerland (September 11-15, 2006)
10. McFerren, G., Roos, S., Terhorst, A.: Fire Alerts for the Geospatial Web. In: The Geospatial Web, Springer, Heidelberg (2006)
11. Annoni, A., Bernard, L., Douglas, J., Greenwood, J., Laiz, I., Lloyd, M., Sabeur, Z., Sassen, A.M., Serrano, J.J., Uslaender, T.: Orchestra: Developing a unified open architecture for risk management applications. In: van Oosterom, P., Zlatanova, S., Fendel, E.M. (eds.) Geo-information for Disaster Management (2005)

---

[9] RGI 002 project website: www.durpondergronden.nl

Heuristic Evaluation of Usability of GeoWeb Sites

Jitka Komarkova, Ondrej Visek, and Martin Novak

Faculty of Economics and Administration, University of Pardubice, Studentska 95,
532 10 Pardubice, Czech Republic
{Jitka.Komarkova, Martin.Novak}@upce.cz, ondrej.visek@seznam.cz

Abstract. Usability of software belongs today to the top priorities of both managers and users of information technologies because the dependency of the majority of mankind activities on information and communication technologies in general is increasing rapidly. This requirement concerns geographic information systems in general but it principally concerns GeoWeb sites because spatial data and tools for their utilization are nowadays in great demand by end-users. GeoWeb sites can provide them an easy-to-use solution but they must be properly designed. The article describes proposal of heuristic evaluation of usability of GeoWeb sites and a case study: the proposed heuristic evaluation is applied to GeoWeb sites of all Czech regional authorities. ISO/IEC 9126 quality model is used as a framework and heuristic evaluation in according to Nielsen is used as a method for usability inspection. In the end some recommendations how user interface should be designed are proposed.

Keywords: GeoWeb, Internet GIS, Web-based GIS, Usability, Heuristic Evaluation, Software Quality.

1 Introduction

An increasing frequency of solving spatially oriented problems in both public and private sector is one of the reasons why accessibility of geographic information is in great demand today. Along with development of information society and increasing utilization of information for supporting decision-making processes, geographic information is more often required by the end users, like citizens, as well. Because of the special nature of geographic information special software tools are required to its treatment and analyses – geographic information systems (GIS) [17], [20], [27]. Nature of geographic data and spatial analyses allows utilization of various techniques for their treatment. Methods of artificial intelligence can be used as an example. Artificial neural network can be used to support GPS navigation by means of GPS data processing and provide map matching functionality [39]. Genetic algorithms can be used for solving the map-generalization problems [38]. The next significant problem is how to handle imprecision and uncertainty in geographic data. This is why possibility of utilization of the fuzzy set techniques for spatial data treatment and querying has been studied [15], [28]. In some cases, there can be a significant problem with some languages and their special letters and symbols too. Czech language can be given as an example [13].

J.M. Ware and G.E. Taylor (Eds.): W2GIS 2007, LNCS 4857, pp. 264–278, 2007.

But regardless of the previously shown problems, GIS have become an indispensable part of a variety of information systems used for supporting decision-making processes in business, public administration, and personal matters so an easy access of all potential users to geographic information and services is strongly required [6], [20], [26], [27]. Today, usability of information systems in general belongs to the top priorities of users [3] and designers [30]. So, applications are required to be intuitive because the digital literacy of many users is not very high. This requirement is even stricter in the case of Web applications. Their designers should remember that users do not read precisely the entire Web page and they are very often in a hurry [18].

Unfortunately, classic GIS software packages (desktop or professional GIS) are too sophisticated and they limit users at least for the following reasons [20], [26], [27]:

- Complicated user interface requires training so it usually disallows an immediate and fast solving of simple tasks.
- Every user has to buy a full license even if only a small part of available functions will be used.
- Desktop software is accessible only from the computer on which it is installed.
- Desktop GIS is still a proprietary technology which uses proprietary data formats.

The above stated problems along with spreading of the Internet and increasing demand for geographic information have driven a rapid process of geo-enabling the Web and a rapid development of various Internet GIS applications and technologies.

During last few years various geo-enabling technologies have been developed. The aim of all these technologies is to allow remote and easy access of end-users without any special education in the field of geoinformatics to geographic information by means of variety devices, e.g. computers, notebooks, pocket PCs, mobile phones, etc. Many various terms can be found which are used while talking about GIS applications on the Internet/intranet. The most common ones are: GIS on-line, distributed geographic information [29], Web-based GIS [26], Internet GIS, mobile GIS [20], [26] interactive mapping [6], [10], distributed GIservice [26], geo-enabled Web [24] GeoWeb, Internet map servers, and many others. In the framework of this paper the terms Internet GIS and GeoWeb will be used equally although attention will be paid only to the Web-based solutions.

As it was mentioned above, quality of information systems, including Web applications, is a very important issue today. But attempts to evaluate quality of information systems are quite old. The first widely recognized quality model was proposed in 1977 by McCall and the Boehm model followed in the next year. Because quality is a complex of many characteristics, these models use decompositional approach [8], [25]. ISO 9126 quality model again defined product quality as a set of its characteristics. After a revision ISO/IEC 9126 is now a four-part standard which defines quality model together with a set of internal metrics, external metrics, and quality-in-use metrics. This model is well-known and accepted by information technology community and it is widely used for both measuring architectures and intranet applications [4], [19], [21], [25]. ISO/IEC 9126 model defines quality characteristics and their sub-characteristics [4], [19], [25]:

- *Functionality*: suitability, accuracy, interoperability, security, functionality compliance.
- *Reliability*: maturity, fault tolerance, recoverability, reliability compliance.
- *Usability*: understandability, learnability, operability, attractiveness, usability compliance.
- *Efficiency*: time behavior, resource utilization, efficiency compliance.
- *Maintainability*: analyzability, changeability, stability, testability, maintainability compliance.
- *Portability*: adaptability, installability, coexistence, replaceability, portability compliance.

In fact, every specialist can propose his/her own quality model which will meet the needs of the given situation and the quality model do not need to follow hierarchical structure [2], [25].

2   GeoWeb

All Internet GIS solutions belong to the six main types of GIS software which are usually recognized: professional GIS, desktop GIS, mobile GIS, component GIS, viewers, and Internet GIS. Internet GIS solutions represent a quite new branch – they have risen in the end of the 20th century and they still undergo a rapid development. These solutions are mostly focused on casual end-users so in comparison with the other kinds of GIS software they are today used by the largest number of users for the lowest costs per user. They, of course, provide only limited number of functions but in fact they are expected to provide only this limited but required functionality [20], [26].

User interface of GeoWeb applications should be designed with a strong focus on the target group of users so it should be at least: visually balanced, enough contrast, typographically correct, readable, and using familiar presentations. Available tools should be clearly presented and the interface should support and guide users [34].

Another significant problem is that the users at the same level of digital literacy still individually differ so usability cannot be ensured by training or education. Users' diversity must be respected by design of the application as well [5].

2.1   System Architecture

The system architecture of Internet GIS follows architecture of the other information systems and it is usually based on the n-tier client/server architecture. At least the following parts can be usually recognized [1], [26]:

- *Data layer* – data management system which is able to store and provide both spatial and non-spatial data.
- *Application layer* (business logic) – processing and analytical functionality (at minimum map server and Web server must be available).
- *Presentation layer* – users interface.

Today, various Internet GIS applications are used as a tool for fast providing terabytes of data from various sources including data from remote sensing and various kinds of maps from various sources, e.g. in libraries [32]. Due to this demand and thanks to the existence of interoperability standards which are accepted by a wide community, Internet GIS were found as a suitable domain for application of ideas of parallel and distributed/grid computing too [11], [26], [36].

2.2 User Groups

Internet GIS now have the highest number of users in comparison with the other types of GIS applications (including desktop GIS). User classification varies from author to author and it depends on the purpose of classification too. Anyway, at least the following basic types of users are usually distinguished [20], [26], [31]:

- *Casual end-users*, e.g. citizens, business partners, tourists, managers, etc. They use GeoWeb application irregularly and casually. Their digital literacy may be very low. On the other side, only a few functions are required by these users. They usually need to set region of interest, select appropriate data layers, display geographic information, change scale, run very simple queries, and print outputs or save result maps. It is supposed that they can use various platforms including various Web browsers. They may not be able to install any software, and their Internet connection can be slow.
- *Regular end-users*, e.g. employees (like civil servants, managers, controllers), regular customers, cooperating partners, etc. Regular, everyday use of GeoWeb is typical for this group of the users. Again, they usually need only several functions. All needed functions are known in advance and they are used repeatedly. The users can be trained in advance if it is necessary. They use appointed Web browser or other defined client software. It means their working environment is known in advance and can be influenced in the case of necessity.
- *High-end users*, usually GIS specialists who treat data, run spatial analyses, and provide the results of their work to the other users by means of Internet GIS.
- *Mobile users*, i.e. people who use mobile devices. The users can vary from casual low-end users to high-end users. The set of demanded functions is limited but some special function, like disconnected editing of data, may be required.

3 Usability and Its Heuristic Evaluation

Today, it is preferred to "make the design to fit the users" to the previous attempt: "make the user to fit the design" [30]. The "fitness for use" definition takes customer's requirements and expectations into account, which involve whether the products or services fit their needs. Quality of design and quality of conformance are two most important parameters [14].

Many quality models for measuring quality and customer satisfaction have been proposed as it was discussed above. In many of the quality models usability is one of the characteristics of software quality. For this study ISO/IEC 9126 model will be used as a framework.

Usability is the extent to which the system is convenient and practical to use. Sometimes it is understood as a user-friendliness. Formally it can be defined as the probability that a user of a system will not experience problems with a user interface during a given period and under a given operational profile [8]. The aim of the usability evaluation is to collect empirical data about the tested application. In the framework of human-computer interaction research usability is studied by changing interface variables and measuring user performance. It means representative end-users are observed when they use the application to perform selected typical tasks. Usually, efficiency, effectiveness, and satisfaction are used as indicators of interface usability and user satisfaction [2], [5], [9], [29].

Many usability evaluation methods and tools have been developed. They can be divided into three types [9]:

- *Usability testing* – representative end-users use the evaluated system to accomplish typical tasks. They are usually observed and their results are evaluated by evaluators. Question-asking protocol (Dumas and Redish), retrospective testing (Nielsen), and thinking aloud protocol (Nielsen) can be used as examples.
- *Usability inspection* – evaluators evaluate whether each element of user interface follow given usability principles which are given in advance. Heuristic evaluation (Nielsen), cognitive walkthrough (Wharton/Rowley), and perspective-based inspection (Zhang) can be used as examples.
- *Usability inquiry* – evaluators talk to the users and observe them in a real work. Proactive field study (Nielsen) can be given as an example.

Various methods of usability evaluation have been used to improve many systems including Web-based systems. Usability testing was used as a one of techniques used to improve Web-based learning environment for the high school students. The aim of the project was to support and increase students' motivation to learn science [35], [37]. Usability testing was successfully used to evaluate and improve two interactive systems for monitoring and improving the home self-care of patients suffering with asthma and dyslipidemia [7]. The United States Computer Emergency Readiness Team (part of the Department of Homeland Security of the USA) changed significantly its Web site after the first usability test. Then, the second usability test of the new Web site showed that technical users' success was improved by 24%, their satisfaction was improved by 16%. The non-technical (home) users' success was improved by 20%, and their satisfaction was improved by 93% [33].

Heuristic evaluation can be done by means of numerous sets of heuristics. Most of them are derived from guidelines and only a few of them are based on cognitive theory [9], [12]. In according to [9] heuristic (or checklist) based approach can significantly improve design of information system, i.e. it can be significantly helpful before the system is programmed and available for usability testing by end-users by means of e.g. thinking aloud protocol. Heuristic evaluation can be done in the two different ways of reporting, on paper or with support of some software tool. Utilization of Web-based tool is recommended by study [12] because it allows faster data collection and evaluation although it needs users to switch between applications and it allows more noise to be reported. But only a few suitable software tools have been developed.

Heuristic evaluation was found as the second most useful way of usability evaluation of information systems. Usability user testing was found as the most useful one [22]. Very important advantage of heuristic evaluation is that only a few evaluators take part in the process of evaluation in comparison to the usability user testing where some users have to take part too together with evaluators [23].

4 Proposal of Heuristic Evaluation of GeoWeb Applications

Heuristic evaluation is proposed in the framework of ISO/IEC 9126 quality model, in according to the Nielsen's approach and with utilization of pen and paper, without any special software tool.

At first, aims of evaluation must be determined. Aims of usability evaluation of GeoWeb applications can be:

- System services enhancement
- System simplification, namely from the user's point of view
- Increase of a number of visitors of a GeoWeb application
- Increase of information dissemination

These aims are, of course, general examples and they must be refined for each particular study.

Then, a set of heuristics must be selected or proposed by usability and software development experts. In the next step heuristic evaluation is done by evaluators who obtain a set of heuristics and then they have to decide which heuristics are fulfilled, which are not fulfilled and what is not available. Later they have to assess not fulfilled heuristics and not available items to decide how important problem of usability they represent. So, they rate each heuristic by numbers from range 0 – 4 where 0 means that this is not a problem of usability at all and 4 means that this is the most serious problem of usability.

For heuristic evaluation of GeoWeb application set of 138 heuristics was proposed (see Appendix). The set can be divided into the following parts:

- *Environment and user interface* – application should provide all necessary basic navigation items and actual data in understandable form.
- *Utilization and support of various technologies* – users should be able to use different browsers and different user settings.
- *Error handling* - application should provide users information about errors and possibility of their solving.
- *Flexibility, design and its aesthetics* - all parts of application should occupy an appropriate part of viewport. All control elements should be clear and well-arranged. Everything has to be in color accordance and easy for users at first sight.
- *User-friendliness* - application should by easy to use, functional and intuitive for user.
- *User privacy* - unauthorized persons should not be able to access application.
- *Manuals (help) and other documentation* – high quality help and other documentation improve usability too although application should be controlled intuitively.
- *User control of the application, user freedom and skills* – user should be able to control application easily and by means of commonly used and well-known basic functions.
- *List of available services* - Provider should provide an easily accessible and well-arranged list of maps and have to provide sufficient number of maps.

5 Case Study: Heuristic Evaluation of GeoWeb Applications of the Czech Regional Authorities

In the framework of the study 14 GeoWeb applications were tested. All the applications are run by the Czech regional authorities (14 in total) so they can be understood as comparable ones. Target user group of these applications is the public, e.g. citizens, tourists, businessmen. In other words, the applications are aimed at casual users without any special GIS knowledge and skills and sometimes with a low level of digital literacy.

5.1 Experimental Conditions

Evaluator
University student, male, 23 years old. Study branch: informatics in public administration. High level of digital literacy and good practical skills including good knowledge of GIS.

Set of testing tasks
Each application was evaluated in according to the proposed set of 138 heuristics (see Appendix).

5.2 Results

Each GeoWeb application was evaluated by evaluator and in the case of a problem application obtained relevant points. Resulting order of GeoWeb applications was created in according to the sum of obtained points – application with the highest number of obtained points is the application with the highest number of usability problems. Results are given on the Fig. 1.

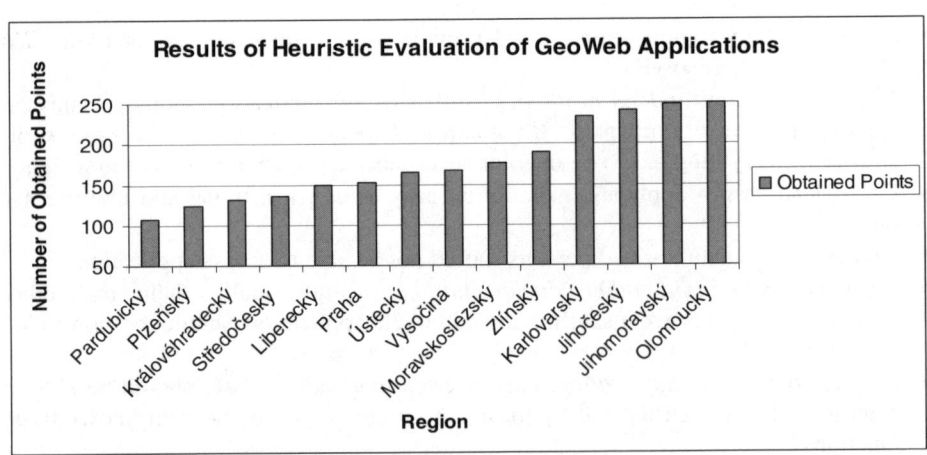

Fig. 1. Results of heuristic evaluation of GeoWeb applications of Czech regional authorities (source: authors)

Results of heuristic evaluation can be compared with results of our previous study - usability user testing [16], to verify appropriateness of the proposed set of heuristics. There are only small differences, mostly in the beginning of the list, e.g. Pardubicky region was the second one in usability testing and it is the best in heuristic evaluation. The four most problematic applications reached the same position in the usability testing and in heuristic evaluation.

5.2.1 General Remarks

Our study shown that many GeoWeb applications are not properly designed and do not provide user-friendly interface. Some of them are too complicated so users need to spend long time trying to find out how the system works or they would have to read help. Both are annoying for many people so they may prefer to leave the application. Tested applications provided by regional authorities were designed in many different ways and using different technologies and application map servers so users usually cannot use skills and experience gained from one system to work with the second one.

But the most important fact is that common users do not know principles of geographic information systems so they very often do not understand why they have to set active layers for the following treatment and then they have to select layers to be displayed. HTML form elements (checkbox and radio button) are usually used but they must be used carefully to do not confuse users.

Fast response time belongs to the crucial requirements on GeoWeb application. Response time may be influenced by many factors, e.g. by number of simultaneously working users, hardware, network, amount of data, panning and moving in the map. Response time must be considered during design because slow response time can discourage users from using application. AJAX technology (Asynchronous JavaScript and XML) can be used to shorten response time thanks to loading the neighborhood to the cache in advance so panning can be done faster.

A map should occupy the biggest area of viewport. A map and the basic navigation are the most important parts so they should be permanently available. Other items which users do not need so often should be hidden or minimized to maximize space for the map. If navigation or some other informational box needs a larger space, it can be solved for example by showing/hiding menus by means of Cascading Style Sheets (CSS) and JavaScript.

Graphical scale bar and north arrow should be clearly presented to help users understand the map. Each map should contain its factual, spatial and temporal description.

5.2.2 Recommendations

Each target group of users has different requirements, different skills and different equipment. All these facts have to be collected before the design of the application and they have to be taken into account during the system analysis and design process. The study was focused on casual users so following recommendations are targeted to this group of users:

- Initial scale should be provided for each data layer because utilization of data in an inappropriate scale can confuse users and overload server. For these reasons minimum and maximum scale for each layer should be set in the application.
- All used icons should be well-known and understandable.
- Application should work correctly with all Web browsers. Application which works only with the only selected browser can again discourage some users.
- Additional software and plug-ins should not be required.
- Map panning should be available by means of hand tool and key arrows. Panning with preloading the nearest neighborhood into the server cache can have a great positive impact on satisfaction of user.
- A special view or page should be provided for printing the output maps. Users do not want to print useless icons; they want to print only a map and desired information.
- Users should be able to get Uniform Resource Locator (URL) of displayed map.
- Utilization of new Web browser windows without menu and icons to provide the largest possible area for map is arguable because it makes user interaction with Web browser more difficult.

6   Conclusions and Future Work

Utilization of geographic information is highly required by end-users today so technologies providing easy access to geographic data are in great demand. Along with increasing number of users requirement on GeoWeb application quality is increasing too. User-centric design is now preferred by many software developers to provide high level quality software. Usability is today very often used to measure how software satisfies users. Many methods of usability evaluation have been developed. Usability user testing was found as the most useful one but it is demanding and costly method so heuristic evaluation can be used for decrease costs of evaluation.

In the paper a heuristic evaluation of GeoWeb applications was proposed with focus on casual users as a target user group. For this purpose at first aims of evaluation were stated. Then, a set of heuristics with their rating was developed. In the end GeoWeb applications of all Czech regional authorities were evaluated by means of the proposed heuristic evaluation. The obtained results were found very similar to the results obtained by means of usability user testing done by means of thinking-aloud protocol [16] so the proposed set of heuristics was found as a suitable one for usability evaluation of GeoWeb applications.

The most problematic issues found during the heuristic evaluation are briefly discussed. Layered GIS approach may not be understandable for many users. Next, a necessity to install something on their computer can be frustrating for the users. Cartographic rules should be followed too. General principles for designing Web applications should be obeyed too.

For the future work, a deep multicriterial assessment of obtained results will be done and then a user-friendly user interface will be proposed and designed with utilization of available modern technologies and general principles of creating Web sites because GeoWeb sites are still web sites.

References

1. Alter, S.: Information Systems: Foundation of E-Business, 4th edn. Prentice-Hall, Upper Saddle River (2002)
2. Azuma, M.: Software Products Evaluation System: Quality Models, Metrics and Processes - International Standards and Japanese Practice. Information and Software Technology 38, 145–154 (1996)
3. Boehm, B.: The Future of Software Processes. In: Li, M., Boehm, B., Osterweil, L.J. (eds.) SPW 2005. LNCS, vol. 3840, pp. 10–24. Springer, Heidelberg (2006)
4. Côté, M.A., Suryn, W., Martin, R., Laporte, C.Y.: Evolving a Corporate Software Quality Assessment Exercise: A Migration Path to ISO/IEC 9126. Software Quality Professional 6, 4–17 (2004)
5. Dillon, A.: Spatial-Semantics: How Users Derive Shape from Information Space. J. of Am. Soc. for Information Science 51, 521–528 (2000)
6. Doyle, S., Dodge, M., Smith, A.: The potential of Web-based mapping and virtual reality technologies for modelling urban environments. Computers, Environment and Urban Systems 22, 137–155 (1998)
7. Farzanfar, R., Finkelstein, J., Friedman, R.H.: Testing the Usability of Two Automated Home-Based Patient-Management Systems. Journal of Medical Systems 28, 143–153 (2004)
8. Fenton, N.E., Pfleeger, S.L.: Software metrics: A Rigorous and Practical Approach, 2nd edn. PWS Publishing Company, Boston (1997)
9. Folmer, E., Bosch, J.: Architeting for usability: a survey. Journal of Systems and Software 70, 61–78 (2004)
10. Friedl, M.A., McGwire, K.C., Star, J.L.: MAPWD: An interactive mapping tool for accessing geo-referenced data sets. Computers & Geosciences 15, 1203–1219 (1989)
11. Hawick, A.K., Coddington, P.D., James, H.A.: Distributed frameworks and parallel algorithms for processing large-scale geographic data. Parallel Computing 29, 1297–1333 (2003)
12. Hvannberg, E.T., Law, E.L.Ch., Lárusdóttir, M.K.: Heuristic evaluation: Comparing ways of finding and reporting usability problems. Interacting with Computers 19, 225–240 (2007)
13. Janakova, H.: Text categorization with feature dictionary problem of Czech language. WSEAS Transactions on Information Science and Applications 1, 368–372 (2004)
14. Kan, S.H.: Metrics and Models in Software Quality Engineering, 2nd edn. Pearson Education, Upper Saddle River (2003)
15. Kollias, V.J., Voliotis, A.: Fuzzy reasoning in the development of geographical information systems. FRSIS: a prototype soil information system with fuzzy retrieval capabilities. Int. J. of Geographical Information Systems 5, 209–223 (1991)
16. Komarkova, J., et al.: Usability of GeoWeb Sites: Case Study of Czech Regional Authorities Web Sites. In: Abramowicz, W. (ed.) BIS 2007. LNCS, vol. 4439, pp. 411–423. Springer, Heidelberg (2007)
17. Konecny, G.: Remote Sensing, Photogrammetry and Geographic Information Systems. 1st publ. Taylor & Francis, London (2003)
18. Krug, S.: Don' t Make Me Think: A Common Sense Approach to Web Usability, 2nd edn. New Riders Press, Indianapolis (2005)
19. Leung, H.K.N.: Quality Metrics for Intranet Applications. Information & Management 38, 137–152 (2001)
20. Longley, P.A.: Geographic Information Systems and Science, 1st edn. John Wiley & Sons, Chichester (2001)

21. Losavio, F., Chirinos, L., Matteo, A., Lévy, N., Ramdane-Cherif, A.: ISO Quality Standards for Measuring Architectures. Journal of Systems and Software 72, 209–223 (2004)
22. Nielsen, J.: Useit.com: Technology Transfer of Heuristic Evaluation and Usability Inspection [online]. [cit 2007-03-05] (1995), Available from http://www.useit.com/papers/heuristic/learning_inspection.html
23. Nielsen, J., Mack, R.L.: Usability inspection methods. John Wiley & Sons, New York (1994)
24. Open Geospatial Consortium [online]. [cit, 2007-06-07], Available from http://www.opengeospatial.org/about/
25. Ortega, M., Pérez, M., Rojas, T.: Construction of a Systemic Quality Model for Evaluating a Software Product. Software Quality Journal 11, 219–242 (2003)
26. Peng, Z.-R., Tsou, M.-H.: Internet GIS: Distributed Geographic Information Services for the Internet and Wireless Networks. John Wiley & Sons, Hoboken (2003)
27. Peng, Z.-R., Zhang, C.: The roles of geography markup language (GML), scalable vector graphics (SVG), and Web feature service (WFS) specifications in the development of Internet geographic information systems (GIS). J. Geograph. Syst. 6, 95–116 (2004)
28. Petry, F.E., Cobb, M.A., Wen, L., Yang, H.: Design of system for managing fuzzy relationships for integration of spatial data in querying. Fuzzy Sets and Systems 140, 51–73 (2003)
29. Plewe, B.S.: GIS Online: Information Retrieval, Mapping, and the Internet. OnWord Press, Santa Fe (1997)
30. Rubin, J.: Handbook of Usability Testing: How to Plan, Design, and Conduct Effective Tests. John Wiley & Sons, New York (1994)
31. Schaller, J.: GIS on the Internet and environmental information and planning. In: 13th ESRI European User Conference, 7. – 9. 10. 1998, Firenze, Italy, [online]. [cit. 2007-06-01]. ESRI (2002), Available from http://gis.esri.com/library/userconf/europroc98/proc/idp27.html
32. Shawa, T.W.: Building a System to Disseminate Digital Map and Geospatial Data Online. Library Trends 55, 254–263 (2006)
33. US-CERT Usability Lessons Learned [online]. [cit. 2007-02-10], Available from http://www.us-cert.gov/usability/
34. Voženílek, V.: Cartography for GIS: Geovisualization and Map Communication. Univerzita Palackeho, Olomouc (2005)
35. Wang, S.-K., Reeves, T.C.: The Effects of a Web-Based Learning Environment on Student Motivation in a High School Earth Science Course. Educational Technology, Research and Development 54, 597–621 (2006)
36. Wang, J., et al.: Design of GridGIS Architecture. In: Min, G., Di Martino, B., Yang, L.T., Guo, M., Ruenger, G. (eds.) Frontiers of High Performance Computing and Networking – ISPA 2006 Workshops. LNCS, vol. 4331, pp. 628–636. Springer, Heidelberg (2006)
37. Wang, S.-K., Yang, Ch.: The Interface Design and the Usability Testing of a Fossilization Web-Based Learning Environment. Journal of Science Education and Technology 14, 305–313 (2005)
38. Wilson, I.D., Ware, J.M., Ware, J.A.: A Genetic Algorithm approach to cartographic map generalisation. Computers in Industry 52, 291–304 (2003)
39. Winter, M., Taylor, G.: A Modular Neural Network Approach to Improve Map-Matched GPS Positioning. In: Carswell, J.D., Tezuka, T. (eds.) Web and Wireless Geographical Information Systems. LNCS, vol. 4295, pp. 76–89. Springer, Heidelberg (2006)

Appendix: Proposed Set of Heuristics

	Heuristics	Rating points
	Environment and user interface	
1.	Does history of user activity exist?	2
2.	Does button "back" exist?	1
3.	Does button "forward" exist?	1
4.	Can be selected icon easily distinguished from the other icons?	4
5.	Has every Web page a title or headline which describes showed content?	4
6.	Is cursor placed in the field which is the most often needed when user enters the Web page or dialog window?	3
7.	Is speed of loading of a map adequate (i.e. fast enough)?	4
8.	Is size of used font adequate so all texts are readable?	4
9.	Are all data up-to-date?	4
10.	Do all maps use the same user interface?	4
11.	Are the used terms commonly known?	2
12.	Is there any item which informs about loading the page?	4
13.	Are dates of pages updates correct?	1
	Utilization and support of various technologies	
14.	No plug-in is necessary.	4
15.	No additional program (e.g. Java) is necessary.	4
16.	If a plug-in or an additional program necessary, is its download link available from the page?	4
17.	Does application work in MS Internet Explorer correctly?	4
18.	Does application work in Firefox correctly?	4
19.	Does application work in Safari correctly?	2
20.	Does application work in Opera correctly?	3
21.	Is application viewable when display resolution is 800 x 600?	3
22.	Is application viewable when display resolution is 1024 x 768?	3
23.	Is application viewable when display resolution is 1280 x 1024?	3
24.	HTML frames are not used.	3
	Error handling	
25.	Is a sound used to signal an error?	1
26.	Are error messages stated in so that the system, not the user, takes the blame?	1
27.	Are error messages grammatically correct?	3
28.	Do error messages avoid the use of exclamation mark?	1
29.	Do error messages avoid the use of violent or hostile words?	1
30.	Do messages place users in control of the system?	1
31.	If an error is detected in a data entry field, does the system place the cursor in that field or highlight the error?	2

32.	Do error messages inform the user of the error's severity?	1
33.	Do error messages suggest a possible cause of the problem?	4
Flexibility, design and its aesthetics		
34.	Are icons separated into thematic groups?	4
35.	Are icons representative enough in relation to their functions?	3
36.	Are not icons too much detailed?	2
37.	Are constituent elements of the page placed on the page suitably?	4
38.	Are suitable colors used for constituent elements of the page?	4
39.	The size of not unused space is not higher than 5%?	2
40.	Does the real map occupy at minimum 40%?	2
41.	Does not description and symbology of layers occupy more than 20%?	2
42.	Does not overview map occupy more than 20%?	2
43.	Does overview map occupy at least 3%?	2
44.	Do layers and their description and symbology occupy at minimum 10%?	2
45.	Is a place appointed to the constituent elements of a page relevant to their importance?	4
46.	Is a name (heading) of a page placed in the upper edge of page?	4
47.	Is enough free space around text areas?	3
48.	Is an overview map in accordance to the distribution?	3
49.	Is there a good contrast, brightness and color harmony between pictures page elements and background?	3
50.	Does the whole application use one color scheme?	3
51.	Are descriptions of all fields familiar, brief, polite and descriptive?	4
52.	Pop-up windows are not opened for layers.	4
User-friendliness		
53.	Are icons ordered by their importance?	2
54.	Are all important items on the page highlighted?	4
55.	Are layers divided into thematic groups?	3
56.	Does layer list contain legend as well?	3
57.	Is there a graphic scale bar?	4
58.	Is there a numeric scale bar?	1
59.	Is there any scale factor also outside of status bar of Web browser?	4
60.	Are all items of newly opened window linked back to the origin window with map list?	4
61.	Is there possibility to choose another map from windows with current map list?	4
62.	Is there link to help section?	4
63.	Is there link to information of used technology?	2
64.	Is there link or information about author of application?	1

65.	Are items inserted into status bar also somewhere else on the page?	4
66.	Are results of searching linked back to the map list?	4
67.	Is searching through all layers automatic?	3
68.	Are there details about object (address etc.)?	4
69.	Is there any print preview?	4
70.	Is map on screen same as print version?	4
71.	Does the print version contain additional information?	2
72.	Does search automatically complete search query?	3
73.	Are hotkeys reserved?	2
74.	Can user more often initiate actions then only respond to the system?	3
75.	Are labels in Czech language?	4
76.	Are labels grammatically correct?	3
77.	Are used generally known and clear abbreviations for all terms?	3
78.	Does list of maps in map field correspond with list of maps in web pages?	4
79.	Entering search queries is easy.	4
80.	Is search not case-sensitive?	4
81.	Can user save the URL of the actually shown map?	3
82.	Is it possible to select size of map viewport?	2
83.	Are appropriate scales recommended for each data layer?	4
84.	Are limit scales set for each data layer?	4
85.	Does map contain factual, spatial and temporal description?	3
User privacy		
86.	Are private maps available through web interface?	4
87.	Are private maps completely protected?	4
88.	Is there any possibility to get access into private area after registration?	1
Manuals (help) and other documentation		
89.	Is it easy to go into help and back to the application?	4
90.	Are information relevant?	3
91.	Are information goal oriented (what can I do with this program)?	2
92.	Are information descriptive (what is this thing for)?	2
93.	Are information procedural (how do I do this task)?	2
94.	Are information interpretive (why did that happen)?	2
95.	Are information navigational (where am I)?	2
96.	Are icons followed with context help?	4
97.	Is it easy to switch between task and help?	4
User control of the application, user freedom and skills		
98.	Is it possible to set restrictive scale to show layers?	3
99.	Is it possible to choose layers?	4
100.	Is it possible to set automatic actualization of map field?	3

101.	Is it possible to switch off automatic actualization of map field?	3
102.	Is it possible to move by mouse in map?	4
103.	Is it possible to move by keyboard in map?	2
104.	Is it possible to move by arrows in map field?	4
105.	Is it possible to choose quality of map?	3
106.	Is it possible to zoom in by double-click?	4
107.	Is it possible to zoom in by selection of area?	4
108.	Is it possible to measure distance in a beeline?	4
109.	Is it possible to measure distance of lines?	3
110.	Is it easy to cancel selection?	3
111.	Is it possible to switch to full screen?	3
112.	Is it possible to center map to selected item?	3
113.	Is it possible to show whole map?	3
114.	Is it possible to use tool which show information about object?	4
115.	Is there "hotlink"?	4
116.	Are there more databases from which "hotlink" getting information?	3
117.	Is it possible to set exact user-defined scale?	3
118.	Is a searching easy?	4
119.	Is it possible to save map as an image?	4
120.	Is there possibility of additional settings for print formatting?	4
121.	Does search provide possibilities?	3
122.	Is it possible to print map?	4
123.	Is it possible to make more complex query?	4
124.	Is it possible to search in roll-up menus by first letter?	2
125.	Is it possible to set units?	3
126.	Is it possible to measure area of polygons?	3
List of available services		
127.	Is there any link to GIS maps from main page?	4
128.	Is the map list accessible quick and easy?	4
129.	Isn't necessary to use searching?	3
130.	Are the maps divided by topics?	3
131.	Are the maps divided into secured and unsecured ones?	3
132.	Is there a possibility to download any geographic data?	1
133.	Is web mapping service (WMS) provided?	3
134.	Is there online reference?	3
135.	Is there description of available data layers?	3
136.	Are all expected topics covered by provided services?	3
137.	Is it possible to send a message to administrator?	1
138.	Is there possibility to send feedback from user for improving application?	1

GeminiMap – Geographical Enhanced Map Interface for Navigation on the Internet

Masayoshi Hirose, Ryoko Hiramoto, and Kazutoshi Sumiya

School of Human Science and Environment
University of Hyogo
1–1–12 Shinzaike-honcho, Himeji, Hyogo 670-0092, Japan
{na02x165@stshse, nd06c019@stshse, sumiya@shse}.u-hyogo.ac.jp

Abstract. In this paper, we propose a method of enhancing digital map interface to reflect users' intentions by automatically displaying ancillary maps next to the original one. The users' intentions are extracted from the sequences of the users' map operations (zooming, centering, and panning) and determine the kinds of objectives based on the sequence patterns. We introduce an operation chunk mechanism to express the meaning of the operation sequences and a complex chunk mechanism to describe the users' intentions by combining the operation chunks. We have developed a prototype system called *GeminiMap* based on the method using *two* engaging maps on the browser and evaluated the appropriateness of the definition of the operation chunks and the complex chunks. The experimental results have shown that the chunk mechanism was defined appropriately, and that the automatically displayed ancillary map was sufficiently effective.

1 Introduction

Digital maps on the Web are becoming increasingly popular in part because users can easily change the area of the map displayed and adjust the scale as they search for target geographic locations[1,2,3]. Various sites provide accurate detailed map information, links to Web pages and can store retrieved data[1]. Although users can easily manipulate digital maps, their operations are directly reflected only in the currently displayed map. If the users' intentions could be inferred, a map could also be displayed that was appropriate for the intentions. For example (see Fig. 1), if a user looks for a route from a station (A in Fig. 1) to a destination (B), the current map screen may show only the destination. However, a map screen that shows both the start point and the destination point on the sane screen would be more useful.

To enable users to browse both the original map screen and another one reflecting the users' intentions, we have developed a method for processing a displayed map and then by inferring the users' intention, a map which focuses

---

[1] GoogleMaps: http://maps.google.com/, YahooMaps: http://maps.yahoo.com/, LiveSearch: http://maps.live.com/

J.M. Ware and G.E. Taylor (Eds.): W2GIS 2007, LNCS 4857, pp. 279–292, 2007.

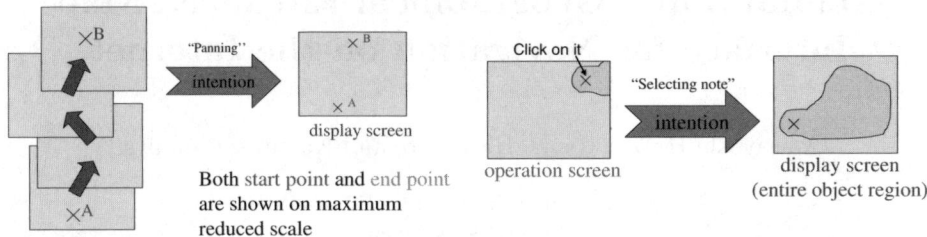

Fig. 1. Motivating Example   **Fig. 2.** Selecting-Region

on that intention in displayed. This method is based on the concept of *operation chunks*, which are defined by the pauses in the users' operations. Operation chunks are the shortest sequences of operation from which some meaning can be reliably inferred. The ancillary map is more effective than the original one if it indeed matches the users' intention. Using this method, we developed a prototype system called *GeminiMap* (Geographical Enhanced Map Interface for Navigation on the Internet) and used it for evaluating our proposed method.

This paper is structured as follows. Section 2 discusses our approach and related work. Section 3 describes how users' intentions are inferred. Section 4 describes how the information is processed and displayed and our prototype system. Section 5 describes the evaluation and results. Section 6 concludes with a brief summary and a look at future work.

2   Our Approach

2.1   Motivation

In our approach, users manipulate a map on a screen (the *original map screen*), and the system automatically displays another map based on the inferred users' intention (the *desired map screen*). As shown in Fig. 3, the system analyzes the operations used to manipulate the map automatically and infers the users' intentions. Finally, it displays a map appropriate for the inferred intention. The users can thus view two maps; the original map and a map created by the system that focuses on their intention.

The users' intention is inferred on the basis of five operations that the users typically use when manipulating digital maps: zooming-in and out (which changes the scale), panning (which moves the screen), centering (which places an arbitrary point at the center of the map), and selecting-note[2] (which selects a note on the map and places it at the center). While a selecting-note function is not provided in conventional digital map systems, it can be used by clicking on the selecting-note operation button in our proposed method.

Maps are generally manipulated by users to obtain specific information, and these manipulations provide clues to their intentions. However, the intention

---
[2] Notes are the names of places and facilities on the map.

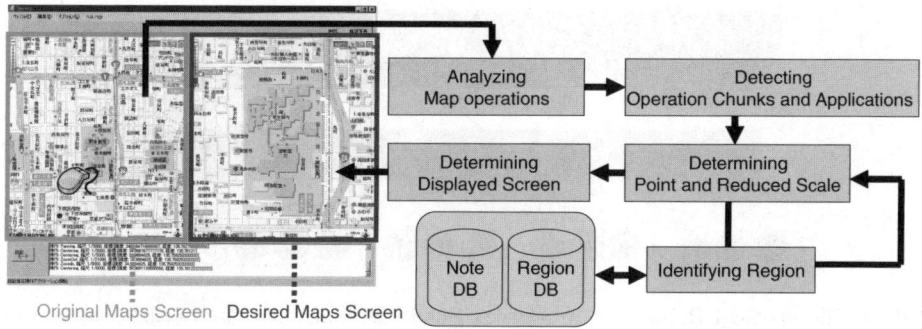

Original Maps Screen   Desired Maps Screen

Fig. 3. Our Approach

cannot be reliably inferred from a single operation. For example, if a user zooms in on a detail, the displayed range is narrowed. The user's intention might be to see detailed information, or it might be to narrow the range displayed. We therefore defined the shortest sequence of operations from which some meaning can be reliably inferred as an *operation chunk*. We defined four types of operation chunks and generated *complex chunks* based on them. We implemented several applications using this complex chunk mechanism. It is possible to generate two or more complex chunks by using these operation chunks. The hierarchical relationship between users' operations, operation chunks, and complex chunks is illustrated in Fig. 4 and 5. Each operation chunk comprises a single sequence of operations, and a complex chunk comprises a single sequence of operations and at least one operation chunk.

There are two screens in this system: one is inputted by the user; the other is processed from their input and desired maps are displayed by analyzing their intentions. Using the generated complex chunks, the system generates various outputs. In this paper, we describe seven complex chunks and the corresponding outputs. And we introduce a basic mechanism to suit and fit the displayed screen.

2.2   Related Work

Our former research[4,5] proposed a Web information retrieval method that detected users' intentions from their manipulations of digital maps and automatically generated queries to retrieve Web pages. In the research, we confirmed that the users' intentions can be inferred from a sequence of map operations if the sequence has some meaning. The intentions can then be used to formulate queries for retrieving appropriate Web pages. We have now applied this method of inferring the users' intentions to display maps that focus on the in intentions.

There has been other work on intentions inferred from map manipulations. Weakliam et al.[6,7] proposed a system that generated a personalized map based on a users' map manipulations and defined map operations. Our work resembles theirs in that map operations are reflected in the focused map. However, ours differs greatly from theirs in that they infer the intention from a single map

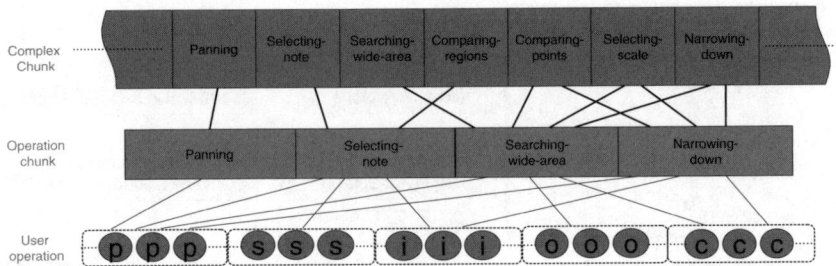

Fig. 4. Hierarchical Relationship between Single Operations, Operation Chunks, and Complex Chunks

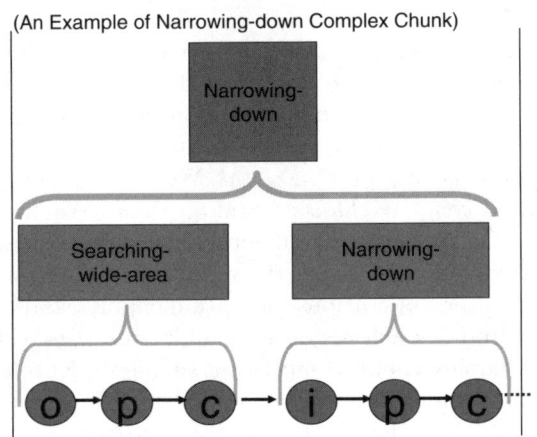

Fig. 5. An Example of Complex Chunk

operation. The defined meaning of each operation is simple; for example, if it is zooming-in, the intention must mean the users' interest in something. The output tends to become an emphasis of this interested in object. In contrast, our system considers the sequence of the users' operations, so it can infer more complex intentions by relating present manipulations to past manipulations. It also has a wider variety of outputs based on the intentions.

Tezuka et al.[8] proposed a system that detects Web pages containing appropriate details based on the users' map manipulations by relating the details on the map to those in the content of the Web pages. Our work resembles theirs in that the target information is inferred from the users' map manipulations. However, while they focused on retrieving Web pages, we focus on providing more focused maps.

Several systems enable users to obtain new information without having to explicitly input anything. They resemble our system in that they automatically offer new information. However, their input and output differs from those of ours. Ours provides maps that match the users' intentions to their manipulations of digital maps.

Yang et al.[9,10] proposed a framework that personalizes Web navigational experiences by using spatial entities embedded in Web documents. It combines semantic similarity, spatial proximity, and k-order Markov chains to predict the next spatial entity which is likely to be in an interaction with a given user. The semantic similarity reflects to which degree a spatial entity is close to another in the semantic domain. The spatial proximity gives a contextual form of inverse distance between two spatial entities. Markov chains implicitly monitors and records the users' trails on the Web, and derives navigational patterns and knowledge in order to predict the users' interactions on the Web.

The SUITOR system[11] analyzes the users' behavior and displays timely and suitable information. It builds a model based on the users' behavior such as what applications they are running and what text they are currently writing or reading. It then determines their interest. The Bilingual Comparative Web Browser[12] retrieves information from news sites in addition to retrieving information from the site currently being viewed. It does this by analyzing the users' operations and then accessing a different news site with the same content.

Henzinger et al.[13] and Ma et al.[14] proposed systems that retrieve related Web information by using the news streams currently being viewed. Ma et al.'s system automatically retrieves Web news on the basis of the TV-news-streams' closed captions, and Henzinger et al.'s uses the metadata delivered with digital TV broadcasts.

Lieberman et al.[15] proposed a system that retrieves Web pages related to the page currently being browsed. This system analyzes the pages' hyperlinks and retrieves other pages in which the user may be interested. Kitayama et al.[16] proposed a system that retrieves blogs with the same evaluations as the Web news streams being browsed. The system calculates the evaluation of a news stream by using its closed captions and retrieves blogs that are related to the news and have the same evaluations.

3   Users' Intentions and Chunk Mechanism

3.1   Definition of Operation Chunks

We define certain sequences of the users' operations as operation chunks. As described above, there are five kinds of operations used to manipulate digital maps: *zooming-in* (i) and *zooming-out* (o) to change the scale, *panning* (p) to shift the screen, *centering* (c) to place an arbitrary point at the center on the screen, and *selecting-note* (s) to place a particular note at the center of the map.

The users' intentions in obtaining information from the map are considered to be indicated by the sequences of map operations. Our proposed method infers the intentions by detecting and analyzing these sequences. Operation sequences that carry meaning are defined as operation chunks.

There are four types of operation chunks: *searching-wide-area*, *narrowing-down*, *panning*, and *selecting-note*. As shown in Table 1, each is expressed as a sequence of operations. The operation sequences are described by regular

Table 1. Operation Chunk

Operation Chunk	Abbr.	User Operations
Searching-wide-area	W	$o^+p^*c^+$
Narrowing-down	N	$i^+p^*c^+$
Panning	P	p^+
Selecting-note	S	i^+p^*s

Table 2. Complex Chunk

Complex Chunk	User Operations	Combination of Operation Chunk
Searching-wide-area	$o^+p^*c^+$	W
Narrowing-down	$o^+p^*c^+\ i^+p^*c^+$	W and N
Panning	p^+	P
Comparing-points	$c^+o^+[pc]^*\ (i^+p^*c^+)\ o^+$	W and N
Comparing-regions	$i^+p^*s\ [pco]^*\ i^+p^*s\ o^+$	S
Selecting-scale	$i^+p^*c^+\ o^+p^*c^+\ i^+$	N and W
Selecting-note	i^+p^*s	S

expression. An asterisk means the operation may or may not be repeated, and a plus sign means the operation is repeated at least once.

3.2   Dynamic Features of Complex Chunks

The complex chunks, which are based on operation chunks, reflect the users' intentions. Any kind of complex chunk can be easily defined in several specific domains. They are built using the following procedure and have several dynamic features.

1. Define operation sequences as components of a complex chunk.
2. Determine output suitable for the users' intention.
3. Determine timing of trigger[3].

A complex chunk must satisfy these three conditions:

- Contain at least one operation chunk
- Be set to trigger at an arbitrary locus
- Permit the inclusion of other complex chunks in the complex chunk

Various formulas are used for defining the various complex chunks in accordance with the need. For example, Formula (1) is used to obtain the center point, (X, Y), of the n points. It is used when the system determines the output

---

[3] When two or more complex chunk triggers are placed in the same position, one common output should be defined.

of complex chunks. Here, x_i is the latitude of each point i, and y_i is the longitude of each point i (see Fig. 1).

$$(X, Y) = \left(\frac{\sum_{i=1}^{n} x_i}{n}, \frac{\sum_{i=1}^{n} y_i}{n} \right) \tag{1}$$

Formula (2) is used to determine the range of the display screen, i.e., the region for the entire selected object. When the user clicks on part of a certain object, it might be more useful to display the entire region of the object than to display only the region for the point clicked (see Fig. 2). Here, M means the output screen. D means the range of the display screen, sc means reduced scale, and R means the region of the object.

$$M = \{D_{sc} \mid max\left(\{sc \mid D_{sc} \supset R\}\right)\} \tag{2}$$

The regular expression for the complex chunks are listed in Table 2. The $[\cdots]$ means selecting any elements, and $(\cdots)$ means maintaining the order of the elements.

4   Inferring Users' Intentions

4.1   Kinds of Operation Chunk

Search-wide-area (W)
This comprises a sequence of operations that reflects the users' interest in a particular area of the map, for example, zooming-out to get an overview and then centering to set a certain range.

Narrowing-down (N)
This comprises a sequence of operations that reflects the users' interest in more detail about a particular map area, for example, zooming-in and centering on a particular point of interest to narrow down the range.

Panning (P)
This comprises a sequence of operations that reflects the users' interest in the overall map, for example, panning around the map to determine its contents, zooming-in on various areas, and panning to different positions.

Selecting-note (S)
This comprises a sequence of operations that reflects the users' interest in place or facility names. Although notes that are clicked on are usually displayed at the center of the screen, the user may not be interested in where they are located but rather in what they say.

4.2 Kinds of Complex Chunk

Searching-wide area: $o^+p^*c^+$

This can use its corresponding operation chunk. After zooming-out from a certain point, the users generally center on a new point of interest. The system then presents detailed information by automatically zooming-in on the new point and increasing the scale. For example, if the user is looking for the White House in Washington, D.C., and centers on a point near it after displaying the map for all of Washington by zooming-out, the system infer that the user is interested in the area near the White House. The system automatically zooms in on that area. The trigger is set to the centering operation.

Narrowing-down: $o^+p^*c^+$ $i^+p^*c^+$

The automatically zooms in on the point that was clicked on in the last centering operation during the narrowing-down operation chunk, and when the narrowing-down operation chunk was detected immediately after a wide area had been searched. When users want to search for a single point, they usually zoom out to view a wide area, then pan and/or center to look for the point, and finally center on the point of interest because they are most interested in the point that was last clicked on in the centering operation; therefore the system displays the point in detail by automatically zooming-in on it. The trigger is set to the centering operation.

Panning: p^+

This may be a panning operation chunk. Users search for information on the map by panning around. The system displays the start and end points of panning on the map to show their relative positions by zooming-out to a reduced scale. The center point is calculated using (1). The start point is connected with the end point by a straight line, and a quadrangle with the straight line as a diagonal is used as region R in (2). The trigger is set to the panning operation.

Comparing-points: $c^+o^+[pc]^*$ $(i^+p^*c^+)$ o^+

This shows the relative positions of two or more points in which users are interested in. After users have found one point and want to display another point in which they are interested in but cannot find in the current view of the map, they generally zoom out to find it; therefore they determine the point of interest by using the narrowing-down operation chunk.

When users zoom out after having narrowed down their search, the operation sequence indicates that they want to look around the map or see the relative positions between displayed points and previously displayed points. The system displays a map showing the current position and two or more previously displayed points on the maximum scale. The center point on the map is calculated using (1). If there are two points, they are connected by a straight line, and a quadrangle with the straight line as diagonal is used as region R in (2). If there are three or more points, the ones that maximize the area when they are connected by a straight line are used to form the diagonal line. The display screen is thereby determined. The trigger is set to the zooming-out operation immediately after the narrowing-down operation chunk.

Comparing-regions: i^+p^*s [pco]* i^+p^*s o^+

This shows the relative positions of two or more regions in which the user is interested in. The users' interest is considered to be high for a region from which notes are selected. The relative positions of regions that have been previously displayed are considered to contain additional important information.

When users zoom out after having selected a note, the operation sequence indicates that they want to look around the map or see the relative positions of previously displayed regions. The system displays a map showing two or more previously displayed regions on the maximum scale. The display screen is then determined using (1) and (2). The trigger is set to the last zooming-out operation in the complex chunk.

Selecting-scale: $i^+p^*c^+$ $o^+p^*c^+$ i^+

This determines an appropriate reduced scale and automatically zooms in when users zoom in after a searching-wide-area operation chunk is detected immediately after a narrowing-down operation chunk has ended. When a searching-wide-area operation chunk is detected immediately after a narrowing-down operation chunk to obtain more information about a point has ended, users generally begin searching for the next point of interest by searching a wide area. A reduced scale when the narrowing-down operation chunk has ended is therefore considered to be better for obtaining information about that point.

Users may be interested in a point that was clicked on when using centering in a searching-wide-area operation chunk. When users zoom in to search in detail, the appropriate reduced scale at that time and the final reduced scale in the last narrowing-down operation chunk are considered to be the same. When they zoom in after a searching-wide-area operation chunk is detected, the system automatically reduces the scale to that of when the last narrowing-down operation chunk ended. The trigger is set to the zooming-in operation immediately after a searching-wide-area operation chunk.

Selecting-note: i^+p^*s

This can use its selecting-notes operation chunk. Maps include much note information, i.e., place and facility names. Notes must include the region as well as the name. For example, the note "Great Lakes" includes an entire region, and the note "Lake Ontario" also includes an entire region. When users click on a note on a map, the point they have clicked on is usually displayed at the center of the screen (like centering) without relating to the note. The user is considered to be interested in the object to which the note refers. We therefore defined regions to which notes refer as corresponding to user intentions.

The trigger in this complex chunk is set to selecting-note operation. For example, when users look at a map of a lake, they are not interested in the lake's name, but in what the name represents. The system shows them two map screens. The

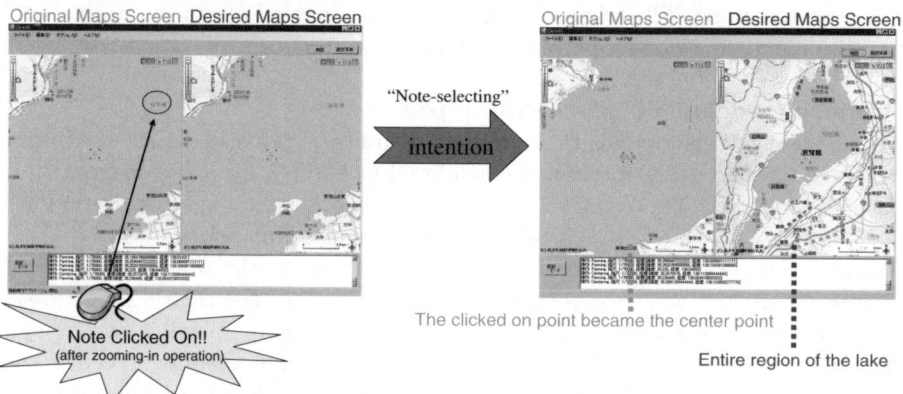

Fig. 6. Selecting-note Complex Chunk

left screen shows the map for which the clicked on point became the center point, and the right screen shows a map for the entire lake (see Fig. 6). The display screen is then determined using (1) and (2).

4.3   Prototype System

We have developed a prototype system called GeminiMap using Visual C♯.Net (see Fig. 7). The interface has two screens: an input screen and an output screen that reflects the users' intentions. These screens were taken from Yahoo! Maps. The operations are identified by analyzing the Java script. There is a button for the user to utilize the select-notes mode.

We have also developed note and region databases. The note database stores notes and their locations, and the region database stores the region of each note. The region database, which contains the latitude and longitude, was prepared beforehand.

The system infers users' intentions by matching the operations used to manipulate the map with the defined operation sequences. Next, it determines the map screen to be displayed on the basis of the inferred intentions. If necessary, it calculates the center point or reduced scale. When the region is needed, it retrieves the region from the region database. Finally, it determines the map to display to the users (see Fig. 3).

5   Evaluation

5.1   Evaluation Experiments

We evaluated using a number of examinees in two experimentshow well our prototype system performed. They manipulated the maps and subjectively judged how well the maps presented to them matched their intentions. We then evaluated the appropriateness of the inferred intentions and the effectiveness of the presented maps.

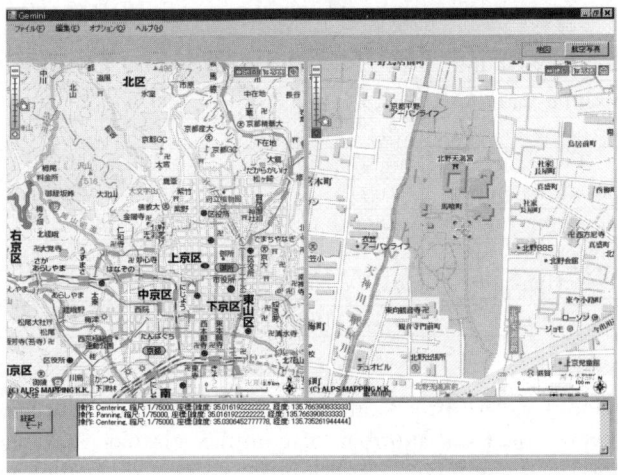

Fig. 7. Screen Image of GeminiMap

5.2 Precision Ratio for Defined Complex Chunks

First, we evaluated the appropriateness of the intentions detected by each type of complex chunk. We focused our evaluation on operation chunks because complex chunks are changed by the combinations of operation chunks. However, operation chunks do not reflect absolute intentions. Therefore, we examined whether the operation sequences of operation chunks reliably reflect intentions by evaluating complex chunks.

We asked the eight examinees to manipulate the maps while thinking about various situations in the same environment during the seven complex chunks:

- First, search for detail about a certain point. Next, search for another point some distance from the first in detail.
- First, search for detail about a certain point. Next, perform operations to select another point some distance from the first.
- First, search for detail about a certain point. Next, perform operations to search for another point some distance from the first.
- First, select one point. Next, select another point some distance from the first. Finally, perform operations to see their relative positions.
- Perform operations to pan from point A to point B.
- First, select one note. Next, select another note some distance from the first. Finally, perform operations to see their relative positions.
- Perform operations to select notes.

We asked the examinees whether the intentions inferred by the system were appropriate. They evaluated them by awarding points. If the intention was appropriate, one point was awarded; if not, no point was awarded. We explained the intentions inferred by the system, and the examinees judged whether they matched their intentions or not.

Table 3. Precision Ratio of Inferred Intentions and Effectiveness of Output

Complex Chunk	Precision Ratio	Degree of Effectiveness
Searching-wide-area	88.9% (32/36)	96.7% (29/30)
Narrowing-down	79.0% (15/19)	70.0% (21/30)
Panning	68.0% (53/78)	100.0% (30/30)
Comparing-points	100.0% (1/1)	96.7% (29/30)
Comparing-regions	20.0% (1/5)	93.3% (28/30)
Selecting-scale	63.6% (7/11)	73.3% (22/30)
Selecting-note	81.1% (60/74)	73.3% (22/30)
Total	75.4% (169/224)	86.2% (181/210)

We then calculated the precision ratio for the seven complex chunks. The denominators were the total number of complex chunks that the system had detected during the examinees' manipulations. The molecules were the total points that the examinees had awarded. The results are listed in Table 3.

The average precision ratio for the seven complex chunks was 169 appropriate operation sequences to 224 detected complex chunks, or 75.4%, which is sufficiently accurate. The panning complex chunk was detected the most frequently because it is a basic operation, and its precision ratio was high. Few comparing-points complex chunks were detected. This might be because the participants rarely searched for two or more positions, and/or the operation sequence for this complex chunk might have been improperly defined.

5.3 Effectiveness of Output

We then evaluated the effectiveness of the output for each complex chunk. Thirty examinees manipulated a map on the basis of the defined operation sequences while thinking about situations in which each complex chunk might be used. For example, we assumed they had found Kiyomizudera Temple while searching around Kyoto Station and had been presented with the relative positions of points for the station and temple. We asked them whether the map displayed on the output screen was effective or not. If it was, they awarded one point; if not, they awarded no points. The effectiveness was given by the total points awarded divided by the number of outputs. The results are listed in Table 3.

The total degree of effectiveness was 86.2%, which is quite high. The examinees stated that the panning, searching-wide-area, comparing-regions and comparing-points complex chunks were useful. Some could not understand why the selecting-note complex chunks' outputs were displayed because the display style differed significantly from the conventional one. Nevertheless, the overall evaluations were sufficiently effective.

6   Conclusion and Future Work

We have described a method for reflecting the intentions of users manipulating digital maps that is based on the operations used for the manipulations.

Operation chunks are defined as a sequence of operations from which some meaning can be reliably inferred, and complex chunks are defined using these operation chunks. Operation chunks express the meanings of the operation sequences, and complex chunks describe the users' intentions.

We implemented this method into the GeminiMap and evaluated it. In the evaluation, the examinees were generally presented with suitable maps. The precision ratio for inferred intentions was about 75%, and the degree of effectiveness was about 86%. However, we need to improve the method for defining operation chunks to enable the number of complex chunks to be increased. We plan to conduct experiments to evaluate the use of this method for various purposes in a variety of application fields. We also plan to develop an interactive method of processing two map screens bi-directionally.

References

1. Masui, T., Minakuchi, M., Borden IV, G.R., Kashiwagi, K.: Multiple-View Approach for Smooth Information Retrieval. In: ACM UIST 1995. Proc. of ACM Symposium on User Interface Software and Technology, pp. 199–206 (1995)
2. Burigat, S., Chittaro, L.: Visualizing the results of interactive queries for geographic data on mobile devices. In: ACM GIS 2005. Proc. of the 13th International Symposium of Advances in Geographic Information Systems, pp. 277–284 (2005)
3. Burigat, S., Chittaro, L., Marco, L.D.: Bringing dynamic queries to mobile devices: a visual preference-based search tool for tourist decision support. In: Costabile, M.F., Paternó, F. (eds.) INTERACT 2005. LNCS, vol. 3585, pp. 213–226. Springer, Heidelberg (2005)
4. Hiramoto, R., Sumiya, K.: A Web Search Method using Digital Maps. In: FMUIT 2006. Proc. of the 1stInternational Workshop on Future Mobile and Ubiquitous Information Technologies, pp. 105–109 (2006)
5. Hiramoto, R., Sumiya, K.: Web Information Retrieval Based on User Operation on Digital Maps. In: ACM GIS 2006. Proc. of the 14th International Symposium of Advances in Geographic Information Systems, pp. 99–106 (2006)
6. Weakliam, J., Bertolotto, M., Wilson, D.: Implicit interaction profiling for recommending spatial content. In: ACM GIS 2005. Proc. of the 13th International Symposium of Advances in Geographic Information Systems, pp. 285–294 (2005)
7. Weakliam, J., Lynch, D., Doyle, J., Bertolotto, M., Wilson, D.: Delivering Personalized Context-Aware Spatial Information to Mobile Devices. In: Li, K.-J., Vangenot, C. (eds.) W2GIS 2005. LNCS, vol. 3833, pp. 194–205. Springer, Heidelberg (2005)
8. Tezuka, T., Yokota, Y., Iwaihara, M., Tanaka, K.: Extraction of Cognitively-Significant Place Names and Regions from Web-Based Physical Proximity Co-occurrences. In: Zhou, X., Su, S., Papazoglou, M.M.P., Orlowska, M.E., Jeffery, K.G. (eds.) WISE 2004. LNCS, vol. 3306, pp. 113–124. Springer, Heidelberg (2004)
9. Yang, Y., Claramunt, C.: A Flexible Competitive Neural Network for Eliciting User's Preferences in Web Urban Spaces. In: Proc. of the 11th International Spatial Data Handling Conference, pp. 41–57. Springer, Heidelberg (2004)
10. Yang, Y., Claramunt, C.: A Hybrid Approach for Spatial Web Personalization. In: Li, K.-J., Vangenot, C. (eds.) W2GIS 2005. LNCS, vol. 3833, pp. 206–221. Springer, Heidelberg (2005)

11. Maglio, P.P., Barrett, R., Campbell, C.S., Selker, T.: Suitor: An attentive information system. In: IUI 2000. Proc. of the 5th International Conference on Intelligent User Interfaces, pp. 169–176 (2000)
12. Nadamoto, A., Tanaka, K.: A Comparative Web Browser (CWB) for Browsing and Comparing Web Pages. In: WWW 2003. Proc. of The 12th International World Wide Web Conference, pp. 727–735 (2003)
13. Henzinger, M., Chang, B.W., Milch, B., Brin, S.: Query-Free News Search. In: WWW 2003. Proc. of the 12th International World Wide Web Conference, pp. 1–10 (2003)
14. Ma, Q., Sumiya, K., Tanaka, K.: WebTelop: A Dynamic TV-content Augmentation by Using Web Pages. In: ICME 2003. Proc. of IEEE International Conference on Multimedia and Expo, vol. 2, pp. 173–176 (2003)
15. Lieberman, H., Fry, C., Weitzman, L.: An agent that assists web browsing. In: IJCAI 1995. Proc. of the Fourteenth International Joint Conference on Artificial Intelligence, pp. 924–929 (1995)
16. Kitayama, D., Sumiya, K.: An Evaluation System for News Video Streams and Blogs. In: SAC 2006. Proc. of the 21st ACM Symposium on Applied Computing, vol. 2, pp. 1361–1368 (2006)

Author Index

Printing: Mercedes-Druck, Berlin
Binding: Stein+Lehmann, Berlin

Lecture Notes in Computer Science

Sublibrary 3: Information Systems and Application, incl. Internet/Web and HCI

For information about Vols. 1– 4430
please contact your bookseller or Springer

Vol. 4654: I.-Y. Song, J. Eder, T.M. Nguyen (Eds.), Data Warehousing and Knowledge Discovery. XVI, 482 pages. 2007.

Vol. 4653: R. Wagner, N. Revell, G. Pernul (Eds.), Database and Expert Systems Applications. XXII, 907 pages. 2007.

Vol. 4636: G. Antoniou, U. Aßmann, C. Baroglio, S. Decker, N. Henze, P.-L. Patranjan, R. Tolksdorf (Eds.), Reasoning Web. IX, 345 pages. 2007.

Vol. 4611: J. Indulska, J. Ma, L.T. Yang, T. Ungerer, J. Cao (Eds.), Ubiquitous Intelligence and Computing. XXIII, 1257 pages. 2007.

Vol. 4607: L. Baresi, P. Fraternali, G.-J. Houben (Eds.), Web Engineering. XVI, 576 pages. 2007.

Vol. 4606: A. Pras, M. van Sinderen (Eds.), Dependable and Adaptable Networks and Services. XIV, 149 pages. 2007.

Vol. 4605: D. Papadias, D. Zhang, G. Kollios (Eds.), Advances in Spatial and Temporal Databases. X, 479 pages. 2007.

Vol. 4602: S. Barker, G.-J. Ahn (Eds.), Data and Applications Security XXI. X, 291 pages. 2007.

Vol. 4601: S. Spaccapietra, P. Atzeni, F. Fages, M.-S. Hacid, M. Kifer, J. Mylopoulos, B. Pernici, P. Shvaiko, J. Trujillo, I. Zaihrayeu (Eds.), Journal on Data Semantics IX. XV, 197 pages. 2007.

Vol. 4592: Z. Kedad, N. Lammari, E. Métais, F. Meziane, Y. Rezgui (Eds.), Natural Language Processing and Information Systems. XIV, 442 pages. 2007.

Vol. 4587: R. Cooper, J. Kennedy (Eds.), Data Management. XIII, 259 pages. 2007.

Vol. 4577: N. Sebe, Y. Liu, Y.-t. Zhuang, T.S. Huang (Eds.), Multimedia Content Analysis and Mining. XIII, 513 pages. 2007.

Vol. 4568: T. Ishida, S. R. Fussell, P. T. J. M. Vossen (Eds.), Intercultural Collaboration. XIII, 395 pages. 2007.

Vol. 4566: M.J. Dainoff (Ed.), Ergonomics and Health Aspects of Work with Computers. XVIII, 390 pages. 2007.

Vol. 4564: D. Schuler (Ed.), Online Communities and Social Computing. XVII, 520 pages. 2007.

Vol. 4563: R. Shumaker (Ed.), Virtual Reality. XXII, 762 pages. 2007.

Vol. 4561: V.G. Duffy (Ed.), Digital Human Modeling. XXIII, 1068 pages. 2007.

Vol. 4560: N. Aykin (Ed.), Usability and Internationalization, Part II. XVIII, 576 pages. 2007.

Vol. 4559: N. Aykin (Ed.), Usability and Internationalization, Part I. XVIII, 661 pages. 2007.

Vol. 4558: M.J. Smith, G. Salvendy (Eds.), Human Interface and the Management of Information, Part II. XXIII, 1162 pages. 2007.

Vol. 4557: M.J. Smith, G. Salvendy (Eds.), Human Interface and the Management of Information, Part I. XXII, 1030 pages. 2007.

Vol. 4541: T. Okadome, T. Yamazaki, M. Makhtari (Eds.), Pervasive Computing for Quality of Life Enhancement. IX, 248 pages. 2007.

Vol. 4537: K.C.-C. Chang, W. Wang, L. Chen, C.A. Ellis, C.-H. Hsu, A.C. Tsoi, H. Wang (Eds.), Advances in Web and Network Technologies, and Information Management. XXIII, 707 pages. 2007.

Vol. 4531: J. Indulska, K. Raymond (Eds.), Distributed Applications and Interoperable Systems. XI, 337 pages. 2007.

Vol. 4526: M. Malek, M. Reitenspieß, A. van Moorsel (Eds.), Service Availability. X, 155 pages. 2007.

Vol. 4524: M. Marchiori, J.Z. Pan, C.d.S. Marie (Eds.), Web Reasoning and Rule Systems. XI, 382 pages. 2007.

Vol. 4519: E. Franconi, M. Kifer, W. May (Eds.), The Semantic Web: Research and Applications. XVIII, 830 pages. 2007.

Vol. 4518: N. Fuhr, M. Lalmas, A. Trotman (Eds.), Comparative Evaluation of XML Information Retrieval Systems. XII, 554 pages. 2007.

Vol. 4508: M.-Y. Kao, X.-Y. Li (Eds.), Algorithmic Aspects in Information and Management. VIII, 428 pages. 2007.

Vol. 4506: D. Zeng, I. Gotham, K. Komatsu, C. Lynch, M. Thurmond, D. Madigan, B. Lober, J. Kvach, H. Chen (Eds.), Intelligence and Security Informatics: Biosurveillance. XI, 234 pages. 2007.

Vol. 4505: G. Dong, X. Lin, W. Wang, Y. Yang, J.X. Yu (Eds.), Advances in Data and Web Management. XXII, 896 pages. 2007.

Vol. 4504: J. Huang, R. Kowalczyk, Z. Maamar, D. Martin, I. Müller, S. Stoutenburg, K.P. Sycara (Eds.), Service-Oriented Computing: Agents, Semantics, and Engineering. X, 175 pages. 2007.

Vol. 4500: N.A. Streitz, A.D. Kameas, I. Mavrommati (Eds.), The Disappearing Computer. XVIII, 304 pages. 2007.

Vol. 4495: J. Krogstie, A. Opdahl, G. Sindre (Eds.), Advanced Information Systems Engineering. XVI, 606 pages. 2007.

Vol. 4480: A. LaMarca, M. Langheinrich, K.N. Truong (Eds.), Pervasive Computing. XIII, 369 pages. 2007.

Vol. 4473: D. Draheim, G. Weber (Eds.), Trends in Enterprise Application Architecture. X, 355 pages. 2007.

Vol. 4471: P. Cesar, K. Chorianopoulos, J.F. Jensen (Eds.), Interactive TV: A Shared Experience. XIII, 236 pages. 2007.

Vol. 4469: K.-c. Hui, Z. Pan, R.C.-k. Chung, C.C.L. Wang, X. Jin, S. Göbel, E.C.-L. Li (Eds.), Technologies for E-Learning and Digital Entertainment. XVIII, 974 pages. 2007.

Vol. 4443: R. Kotagiri, P. Radha Krishna, M. Mohania, E. Nantajeewarawat (Eds.), Advances in Databases: Concepts, Systems and Applications. XXI, 1126 pages. 2007.

Vol. 4439: W. Abramowicz (Ed.), Business Information Systems. XV, 654 pages. 2007.